Information Technologies Applied on Healthcare

Information Technologies Applied on Healthcare

Editors

Giner Alor-Hernández
Jezreel Mejía-Miranda
José Luis Sánchez-Cervantes
Alejandro Rodríguez-González

Basel • Beijing • Wuhan • Barcelona • Belgrade • Novi Sad • Cluj • Manchester

Editors

Giner Alor-Hernández
Instituto Tecnológico de Orizaba
Orizaba
Mexico

Jezreel Mejía-Miranda
Research Center in
Mathematics A.C
Zacatecas
Mexico

José Luis Sánchez-Cervantes
Instituto Tecnológico de Orizaba
Orizaba
Mexico

Alejandro
Rodríguez-González
Technical University of Madrid
Madrid
Spain

Editorial Office
MDPI
St. Alban-Anlage 66
4052 Basel, Switzerland

This is a reprint of articles from the Special Issue published online in the open access journal *Healthcare* (ISSN 2227-9032) (available at: https://www.mdpi.com/journal/healthcare/special_issues/Information_Technologies_Healthcare).

For citation purposes, cite each article independently as indicated on the article page online and as indicated below:

Lastname, A.A.; Lastname, B.B. Article Title. *Journal Name* **Year**, *Volume Number*, Page Range.

ISBN 978-3-7258-1127-4 (Hbk)
ISBN 978-3-7258-1128-1 (PDF)
doi.org/10.3390/books978-3-7258-1128-1

© 2024 by the authors. Articles in this book are Open Access and distributed under the Creative Commons Attribution (CC BY) license. The book as a whole is distributed by MDPI under the terms and conditions of the Creative Commons Attribution-NonCommercial-NoDerivs (CC BY-NC-ND) license.

Contents

About the Editors vii

Preface ix

Aman Upadhyay, Niha Kamal Basha and Balasundaram Ananthakrishnan
Deep Learning-Based Yoga Posture Recognition Using the Y_PN-MSSD Model for Yoga Practitioners
Reprinted from: *Healthcare* **2023**, *11*, 609, doi:10.3390/11040609 1

Mohamed E. Issa, Ahmed M. Helm, Mohammed A. A. Al-qaness, Abdelghani Dahou, Mohamed Abd Elaziz and Robertas Damaševičius
Human Activity Recognition Based on Embedded Sensor Data Fusion for the Internet of Healthcare Things
Reprinted from: *Healthcare* **2022**, *10*, 1084, doi:10.3390/healthcare10061084 20

Agne Paulauskaite-Taraseviciene, Julius Siaulys, Kristina Sutiene, Titas Petravicius, Skirmantas Navickas, Marius Oliandra, et al.
Geriatric Care Management System Powered by the IoT and Computer Vision Techniques
Reprinted from: *Healthcare* **2023**, *11*, 1152, doi:10.3390/healthcare11081152 36

Leonor Teixeira, Irene Cardoso, Jorge Oliveira e Sá and Filipe Madeira
Are Health Information Systems Ready for the Digital Transformation in Portugal? Challenges and Future Perspectives
Reprinted from: *Healthcare* **2023**, *11*, 712, doi:10.3390/healthcare11050712 59

Jorge Pool-Cen, Hugo Carlos-Martínez, Gandhi Hernández-Chan and Oscar Sánchez-Siordia
Detection of Depression-Related Tweets in Mexico Using Crosslingual Schemes and Knowledge Distillation
Reprinted from: *Healthcare* **2023**, *11*, 1057, doi:10.3390/healthcare11071057 79

Chien-Ta Ho and Cheng-Yi Wang
A Robust Design-Based Expert System for Feature Selection and COVID-19 Pandemic Prediction in Japan
Reprinted from: *Healthcare* **2022**, *10*, 1759, doi:10.3390/healthcare10091759 97

Mouz Ramzan, Muhammad Hamid, Amel Ali Alhussan, Hussah Nasser AlEisa and Hanaa A. Abdallah
Accurate Prediction of Anxiety Levels in Asian Countries Using a Fuzzy Expert System
Reprinted from: *Healthcare* **2023**, *10*, 1594, doi:10.3390/healthcare11111594 115

Ibrahim Mutambik, John Lee, Abdullah Almuqrin and Zahyah H. Alharbi
Identifying the Barriers to Acceptance of Blockchain-Based Patient-Centric Data Management Systems in Healthcare
Reprinted from: *Healthcare* **2024**, *12*, 345, doi:10.3390/healthcare12030345 131

Ayogeboh Epizitone, Smangele Pretty Moyane and Israel Edem Agbehadji
A Systematic Literature Review of Health Information Systems for Healthcare
Reprinted from: *Healthcare* **2023**, *11*, 959, doi:10.3390/healthcare11070959 155

Xiaoyu Wang, Yi Xie, Xuejie Yang and Dongxiao Gu
Internet-Based Healthcare Knowledge Service for Improvement of Chinese Medicine Healthcare Service Quality
Reprinted from: *Healthcare* **2023**, *11*, 2170, doi:10.3390/healthcare11152170 168

About the Editors

Giner Alor-Hernández

Giner Alor-Hernández is a full-time researcher at the Division of Research and Postgraduate Studies in Orizaba's technological institute: Tecnológico de Orizaba. He holds an MSc and a Ph.D. in Computer Science from the Center for Research and Advanced Studies of the National Polytechnic Institute (CINVESTAV), Mexico. He has led 10 Mexican research projects granted by CONAHCYT, DGEST, and PRODEP. He is the author/co-author of around 250 journal and conference papers on computer science. He has also been a committee program member of around 60 international conferences sponsored by IEEE, ACM, and Springer. He also holds the position of editorial board member of eight indexed journals; he has been guest editor of several JCR-indexed journals. He is the main author of the book entitled Frameworks, Methodologies, and Tools for Developing Rich Internet Applications, published by IGI Global Publishing. He has been editor of nine books published by Springer Verlag, and one book published by IGI Global Publishing. He has 29 copyrights and 4 international patents. He has supervised 25 bachelor theses, 35 master theses, and 8 Ph.D. theses. His research interests include the Semantic Web, Intelligent Systems, Big Data, the Internet of Things, and Software Engineering. He is an IEEE Senior Member and an ACM Senior member. He is a regular member of the Artificial Intelligence Mexican Society, Computing Mexican Academy, and the Mexican Society of Computing Science. He is a National Researcher recognized at Level 3 by the National Council of Science & Technology of Mexico (CONACYT). His H-index is 19 in Scopus with more than 1,900 citations. ORCID: 0000-0003-3296-0981, Scopus Author ID: 17433252100, Web of Science Researcher ID: U-9203-2017.

Jezreel Mejía-Miranda

Jezreel Mejía-Miranda is a full-time researcher at the Research Center in Mathematics A.C., Zacatecas Unit, México. He holds a Ph.D. in Computer Science. His research field is Software Engineering. Dr. Mejía is also a member of the National Researcher System by CONACYT. His main research field is process improvement, focusing on the topics of multi-model environments, project management, acquisition and outsourcing processes, solicitation and supplier agreement development and agile methodologies, and Security Information Systems. He has published several technical papers on acquisition process improvement, project management, TSPi, CMMI, and multi-model environments. He has been a member of the team that has translated CMMI-DEV v1.2 and v1.3 to Spanish. He is the General Chair of the International Conference on Software Process Improvement (CIMPS). He has been a guest editor of JCR-indexed journals such as the International Journal of Software Engineering and Knowledge Engineering and IEEE LATIN AMERICA TRANSACTIONS. He has participated in several Program Committees at conferences such as conferences CERMA, CISTI, WORLCIST, ICMEAE, INFONOR, and CIMPS and international journals such as RISTI, IJSEKE, IEEE Latin American Transactions, and Revista de Ingeniería. ORCID: 0000-0003-0292-9318.

José Luis Sánchez-Cervantes

José Luis Sánchez-Cervantes is a full-time researcher at the Division of Research and Postgraduate Studies in Orizaba's technological institute: Tecnológico de Orizaba. He holds an MSc from Instituto Tecnológico de Orizaba, and he has a Ph.D. in Computer Science and Technology from the Carlos III University of Madrid, Spain. His degree was validated as a Ph.D.

in Artificial Intelligence by the Universidad Veracruzana (UV), Mexico. He is the author/co-author of several research papers in computer science published in journals indexed in the JCR, as well as in peer-reviewed journals and reports of national and international congresses. He has led three Mexican research projects granted by CONAHCYT and COVEICYDET. He is an IEEE Member and an ACM member. He is an adherent member of the Artificial Intelligence Mexican Society, Computing Mexican Academy, and Mexican Society of Computing Science. He is a National Researcher recognized at Level 1 by the National Council of Humanities, Science and Technology of Mexico (CONAHCYT). His research interests include the Semantic Web, Linked Open Data, Artificial Intelligence, Big Data, and the Internet of Things. His H-index is 16 in Scopus with more than 830 citations. ORCID: 0000-0001-5194-1263; Scopus Author ID: 36976388800; Web of Science Researcher ID: M-3453-2019.

Alejandro Rodríguez-González

Alejandro Rodríguez-González is a Full Professor at the Technical University of Madrid. He holds a Ph.D. in Computer Science, and his research experience includes the publication of more than 100 scientific papers. He has participated in more than 20 research and transfer projects at the national and international levels, principal investigator in several of them. He is currently the head of the Medical Data Analysis Laboratory (MEDAL) of the Biomedical Technology Center. His main line of research is the application of Artificial Intelligence (AI) in biomedicine, via the creation of AI models that can help in clinical practice and the application of data-driven approaches to better understand how diseases resemble and function via the concept of network medicine. In this last line, he focuses on applying this knowledge in drug repositioning, trying to find new uses for existing drugs. Since March 2023, he has been the president of the Spanish Society of Artificial Intelligence in Biomedicine (IABiomed).

Preface

The importance of healthcare has increased, and it will be central for recovery in several sectors at the global level, such as the economy, education, tourism, and e-commerce. According to analysis carried out by international organizations such as UNICEF and the European Union, by the year 2030, it is expected that there will be a huge increase in the use of e-health services by young people, where emergent technologies, including Artificial Intelligence, Big Data, the Web, IoT Technologies, and mobile devices, as well as governmental policies, will play a crucial role in successfully delivering relevant data to health professionals in order to obtain information and advice that benefit health consumers.

Therefore, the main objective of this Special Issue was to collect and consolidate innovative and high-quality research contributions regarding Information Technologies Applied on Healthcare to different disciplines and its challenges such as the systematization and standardization of healthcare information systems, the detection of diseases at early stages, open healthcare data, integrated health services, cybersecurity and data protection in healthcare, interoperability data health, information technologies in healthcare, human–computer interaction (HCI) in healthcare, intelligent medical devices and smart technologies, artificial intelligent techniques applied to healthcare, digital healthcare, telehealth (telemonitoring for diseases, remote consultation, and remote education and support), prognosis, diagnosis and treatment in healthcare, big data analytics for healthcare, computer games for healthcare, m-Health, smart technologies for healthcare, predictive modeling and analytics for healthcare, computer vision in healthcare, and healthcare decision support systems.

This Special Issue aims to provide insights into the recent advances in the aforementioned topics by soliciting original scientific contributions in the form of theoretical foundations, models, experimental research, surveys, and case studies of Information Technologies Applied on Healthcare. This Special Issue in *Healthcare* contains two types of contributions: research and review papers.

All accepted papers align with the scope of the Special Issue, and all of them provide interesting research techniques, models, and studies directly applied to the field of healthcare. Last but not least, we would also like to express our gratitude to the reviewers who kindly agreed to contribute to the evaluation of papers at all stages of the editing process. We equally and especially wish to thank MDPI Publishing for granting us the opportunity to edit this Special Issue and provide valuable comments to improve the selection of research works.

Giner Alor-Hernández, Jezreel Mejía-Miranda, José Luis Sánchez-Cervantes, and Alejandro Rodríguez-González
Editors

Article

Deep Learning-Based Yoga Posture Recognition Using the Y_PN-MSSD Model for Yoga Practitioners

Aman Upadhyay [1], Niha Kamal Basha [1,*] and Balasundaram Ananthakrishnan [2]

1 School of Computer Science and Engineering, Vellore Institute of Technology (VIT), Vellore 632014, India
2 School of Computer Science and Engineering, Center for Cyber Physical Systems, Vellore Institute of Technology (VIT), Chennai 600127, India
* Correspondence: niha.k@vit.ac.in; Tel.: +91-9842444786

Abstract: In today's digital world, and in light of the growing pandemic, many yoga instructors opt to teach online. However, even after learning or being trained by the best sources available, such as videos, blogs, journals, or essays, there is no live tracking available to the user to see if he or she is holding poses appropriately, which can lead to body posture issues and health issues later in life. Existing technology can assist in this regard; however, beginner-level yoga practitioners have no means of knowing whether their position is good or poor without the instructor's help. As a result, the automatic assessment of yoga postures is proposed for yoga posture recognition, which can alert practitioners by using the Y_PN-MSSD model, in which Pose-Net and Mobile-Net SSD (together named as TFlite Movenet) play a major role. The Pose-Net layer takes care of the feature point detection, while the mobile-net SSD layer performs human detection in each frame. The model is categorized into three stages. Initially, there is the data collection/preparation stage, where the yoga postures are captured from four users as well as an open-source dataset with seven yoga poses. Then, by using these collected data, the model undergoes training where the feature extraction takes place by connecting key points of the human body. Finally, the yoga posture is recognized and the model assists the user through yoga poses by live-tracking them, as well as correcting them on the fly with 99.88% accuracy. Comparatively, this model outperforms the performance of the Pose-Net CNN model. As a result, the model can be used as a starting point for creating a system that will help humans practice yoga with the help of a clever, inexpensive, and impressive virtual yoga trainer.

Keywords: convolutional neural network; deep learning; Mobile-Net SSD; Pose-Net; posture recognition; tensor flow lite

1. Introduction

Information technology and science [1] are progressing at breakneck speed, making human life more convenient than ever. Everyone currently recognizes the significance of it in daily life. The impact of computers and computer-powered technologies on healthcare and allied disciplines, as well as every other domain, is extensively established. Yoga, Zumba, martial arts, and other hobbies, in addition to standard medical procedures, are commonly recognized as strategies to improve one's health. Yoga is a set of practices related to a person's physical, mental, and spiritual well-being that originated in ancient India. Artificial intelligence technologies [2], such as Pose-Net and Mobile-Net SSD, as well as human posture detection, can be effective in incorporating tech yoga. Human body posture detection remains a difficult task despite extensive research and development in the fields of artificial intelligence and computer vision. Human posture detection has a wide range of applications, from monitoring health to enhancing the security of the public. Yoga has become a popular kind of exercise in modern society, and there is a desire for training on how to perform proper yoga poses. Since performing incorrect yoga postures can result in injury and tiredness, having a trainer on hand is essential, and since many people do not have the financial means for an instructor, this is where AI offers guidance. The instructor must correct and perform each pose for the individual, and only then can the practitioner perform each pose. However, in today's situation, more people have adapted to using online tools for their needs and are comfortable with this in their homes. This has led to demand for a tech-driven yoga trainer. As a result, the automatic assessment of yoga postures is proposed for yoga posture recognition, which can alert practitioners. Pose-Net and Mobile-Net SSD (together named as TFlite Movenet) play a major role in using the Y_PN-MSSD model. A Pose-Net layer takes care of the feature point detection, while the Mobile-Net SSD layer performs human detection in each frame.

The working of the model is categorized into three stages:

1. The data collection/preparation stage—where the yoga postures are captured with the consent of four users who are trained yoga professionals, as well as an open-source dataset with seven yoga poses.
2. The feature extraction stage—by using these collected data, the model undergoes training in which the feature extraction takes place by connecting key points of the human body using Pose-Net layer.
3. Posture recognition—using the extracted features, yoga postures are recognized using the Mobile-Net SSD layer, which assists the user through yoga poses by live-tracking them as well as correcting them on the fly. Finally, the performance of the Y_PN-MSSD model is analyzed using a confusion matrix and compared with the existing Pose-Net CNN model.

In the past, many researchers have proposed models for tech yoga practices, and these are discussed below under Section 2. The methodology to recognize yoga postures is given in Section 3. By incorporating the given methodology, the data were prepared, trained, and tested on the proposed model, which is explained in Section 4. Finally, as detailed in Section 5, the results were analyzed and compared with those of the existing model to justify their performance. Deep learning typically needs less ongoing human intervention. Deep learning can analyze images, videos, and unstructured data in ways that machine learning cannot easily do. One of the biggest advantages of using a deep learning approach is its ability to execute feature engineering by itself. In this approach, an algorithm scans the data to identify features that correlate, and then combines them to promote faster learning without being told to do so explicitly. This proposed work uses a unique fusion of Pose-Net and Mobile-Net for better pose estimation when compared with other contemporary works using conventional CNN networks. Additionally, the combination of Pose-Net and Mobile-Net provides better accuracy when compared with other works. These are discussed in detail in the Materials and Methods section and Experimental Results section, respectively.

2. Literature Review

Chen et al. [3], in their paper titled "PoseNet: A Convolutional Network for Real-Time 6-DOF Camera Re-localization", demonstrate a monocular, six-degrees-of-freedom delocalization system that is reliable and works in real-time. Graph optimization is not required, however, as additional engineering would be needed. Their system uses CNN to infer the six poses from a camera shot of a single RGB image (in an end-to-end way). The method can run both indoors and outdoors in real time and computes each frame in large-scale outside scenes. It achieved an accuracy of around 2 m and 6 degrees, and it can achieve an accuracy of up to 0.5 m and 10 degrees. This is accomplished with the use of a productive, 23-layer deep ConvNet, proving that ConvNets may be utilized to address challenging, out-of-image-plane regression issues. This was made possible by leveraging transfer learning from large-scale classification data. They demonstrate how the network can localize using high-level features, and its resistance to challenging lighting, motion blur, and various intrinsic camera points of view.

Islam et al. [4], in their paper "Adversarial PoseNet: A Structure-aware Convolutional Network for Human Pose Estimation", implemented a joint occlusion method for a human body, which overlapped frequently and led to incorrect pose predictions when used for human pose estimation in monocular images. These conditions may result in pose predictions that are biologically improbable. Human vision, on the other hand, may anticipate postures by taking advantage of the geometric limitations of joint interconnectivity. Alex et al. [5], in their paper on simple and lightweight human pose estimation, demonstrated using benchmark datasets that the majority of existing methods often aim for higher scores by utilizing complicated architecture or computationally expensive models, while neglecting the deployment costs in actual use. They examine the issue of straightforward and lightweight human posture estimation in this study. Chen et al. [6], in their paper on continuous trade-off optimization between fast and accurate deep face detectors, demonstrated that DNNs, i.e., deep neural networks, are more effective at detecting faces than shallow or hand-crafted models, but their intricate designs have more computational overheads and slower inference rates. They researched five simple methods in this context to find the best balance between speed and accuracy in face recognition.

Zhang et al. [7], in their paper "Yoga Pose Classification Using Deep Learning", proposed a persistent issue in machine vision that has presented numerous difficulties in the past. Many industries, including surveillance cameras, forensics, assisted living, at-home monitoring systems, etc., can benefit from human activity analysis. People typically enjoy exercising at home these days because of our fast-paced lives, but many also experience the need for an instructor to assess their workout form and guide them. Petru et al. [8], in their paper "Human Pose Estimation with Iterative Error Feedback" presented a deep neural network (ConvNets), a type of deep hierarchical extractor, offering outstanding performance on a range of classifications using only feed-forward neural processing. Although feed-forward architectures are capable of learning detailed descriptions of the input feature space, they do not explicitly describe interconnections in the output spaces, which are highly structured for tasks such as segmenting objects or estimating the pose of an articulated human. Here, they offer a framework that incorporates top-down feedback and broadens the expressive potential of hierarchical feature extractors to include both input and output regions.

Yoli et al. [9]. in their paper "Yoga Asana Identification: A Deep Learning approach", describe how yoga is a beneficial exercise that has its roots in India, can revitalize physical, mental, and spiritual wellbeing, and is applicable across all social domains. However, it is currently difficult to apply artificial intelligence and machine learning approaches to transdisciplinary domains such as yoga. Their work used deep-learning methods, such as CNN and transfer learning, to create a system that can identify a yoga position from an image or frame of a video. Andriluka et al. [10] used an approach for grading yoga poses presented using computerized visuals representing contrastive skeleton features. In order to assign a grade, the primary goal of the yoga pose classification was to evaluate the inputted yoga posture and match it with a reference pose. The research proposed a contrastive skeleton feature representation-based framework for analyzing yoga poses. In order to compare identical encoded pose features, the proposed method first identified skeleton key points of the human body using yoga position images, which act as an input, and their coordinates are encoded into a pose feature, which is used for training along with sample contrastive triplets.

Belagiannis et al.'s [11] work discusses the introduction of deep-learning-based camera pose estimation, and how deep learning models perform transfer learning based on knowledge gained using generic large datasets. As a result, researchers are able to create more specific tasks by fine-turning the model. They first described the issue, the primary metrics for evaluation, and the initial architecture (PoseNet). Then, they detected recent trends that resulted from various theories about the changes needed to enhance the initial solution. In order to promote their analysis, they explicitly suggested a hierarchical classification of deep learning algorithms that focused on addressing the pose estimation problem. They also explained the presumptions that drove the construction of representative architectures in each of the groupings that were identified. They also offered a thorough cross-comparison of more than 20 algorithms, which was comprised of findings (localization error) across different datasets and other noteworthy features (e.g., reported runtime). They evaluated the benefits and drawbacks of several systems in light of accuracy and other factors important for real-world applications.

Buehler et al. [12] in their paper have proposed a spatio-temporal solution that is essential to resolving depth and occlusion uncertainty in 3D pose estimation. Prior approaches have concentrated on either local-to-global structures that contained a pre-set spatio-temporal information or temporal contexts. However, effective suggestions to capture various concurrent and dynamic sequences and to perform successful real-time 3D pose estimation have not been implemented. By modelling local and global spatial information via attention mechanisms, the authors of this study enhanced the learning of kinematic constraints in the human skeleton, including posture, local kinematics linkages, and symmetry.

Chiddarwar et al. [13] concluded that the Pose-Net CNN model was good for yoga posture recognition and developed an android application for the same. The real time data had been captured and processed to detect yoga postures, which guided the practitioners to perform safe yoga by detecting the essential points. Apart from these models, the proposed Y_PN-MSSD model performed well on the captured postures performed by trained yoga practitioners.

Ajay et al. [14] proposed a system that provided consistent feedback to the practitioner, so that they can identify the correct and incorrect poses with relevant feedback. They had used a data set that consisted of five yoga poses (Natarajasana, Trikonasana, Vrikshasana, Virbhadrasana and Utkatasana), which acted as an input to the deep learning model, which utilized a convolutional neural network for yoga posture identification. This model identified mistakes in a pose and suggested solutions on how the posture can be corrected. Additionally, it classified the identified pose with an accuracy of 95%. This system prevents users from being injured as well as increases their knowledge of a particular yoga pose.

Qiao et al. [15] presented a real-time 2D human gesture grading system using monocular images based on OpenPose, which is a library for real-time multi-person keypoint detection. After capturing 2D positions of a person's joints and skeleton wireframe of the body, the system computed the equation of motion trajectory for every joint. The similarity metric was defined as the distance between motion trajectories of standard and real-time videos. A modifiable scoring formula was used for simulating gesture grading scenarios. The experimental results showed that the system worked efficiently with high real-time performance, low cost of equipment and strong robustness to the interference of noise.

A blockchain-based decentralized federated transfer learning [16] method was proposed for collaborative machinery fault diagnosis. A tailored committee consensus scheme was designed for optimization of the model aggregation process. Here, two decentralized fault diagnosis datasets were implemented for validations. This work was effective in data privacy-preserving collaborative fault diagnosis. This proposed work was suitable for real industry applications. A deep learning-based intelligent data-driven prediction method was incorporated in [17] to resolve sensor malfunctioning problem. Later on, a global feature extraction scheme and adversarial learning were introduced to fully extract information of sensors as well as extract sensor invariant features, respectively. This proposed work was suitable for real industry applications.

A two stream real-time yoga posture recognition system [18] was developed and a best accuracy of 96.31% was achieved. A yoga posture coaching system [19] using six transfer learning models (TL-VGG16-DA, TL-VGG19-DA, TL-MobileNet-DA, TL-MobileNetV2-DA, TL-InceptionV3-DA, and TL-DenseNet201-DA) was exploited for classification of poses, and the optimal model for the yoga coaching system was determined based on evaluation metrics. The TL-MobileNet-DA model was selected as the optimal model as it provided an overall accuracy of 98.43%.

Qian et al. [20] proposed a contactless perspective approach, which is a kind of vision-based contactless human discomfort pose estimation method. Initially, human pose data were captured from a vision-based sensor, and corresponding human skeleton information was extracted. Five thermal discomfort-related human poses were analyzed, and corresponding algorithms were constructed. To verify the effectiveness of the algorithms, 16 subjects were invited for physiological experiments. The pose estimation algorithm proposed extracted the coordinates of key points of the body based on OpenPose, and then used the correlation between the points to represent the features of each pose. The validation results showed that the proposed algorithms could recognize five human poses of thermal discomfort.

3. Materials and Methods

Deep learning is an integral and essential learning-based method (ANNs) that represents the working of neurons in the human brain with the help of weights. This helps in representing the connection between neurons. It is an e-architecture [21], this allows crucial information from images to be learned automatically.

CNN is one of the deep learning models that has been widely utilized for estimating yoga poses. However, they are judged only based on images rather than videos. Here, the proposed Y_PN-MSSD model uses TFlite Movenet for recognizing yoga poses. Initially, an input (yoga pose) has been captured and processed using trained TFlite Movenet, where the Pose-Net detects the feature point from the input and the Mobile-Net SSD detects the person in the captured frames. Finally, the model recognizes the correct/incorrect yoga pose. The working of the model is depicted in Figure 1. The detailed description of Pose-Net and Mobile-Net SSD are given below with their respective architectures depicted in Figures 2 and 3.

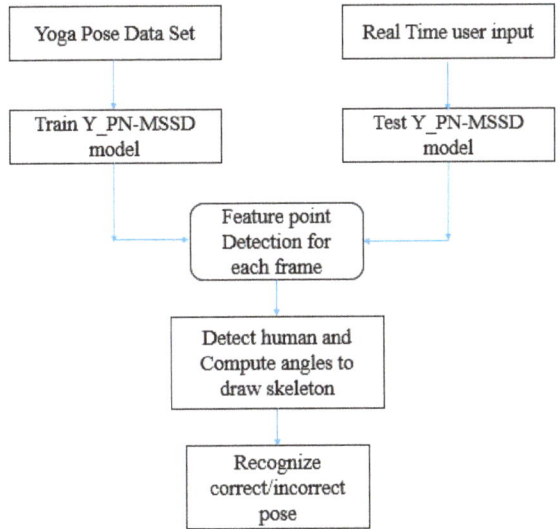

Figure 1. The working of the Y_PN-MSSD model.

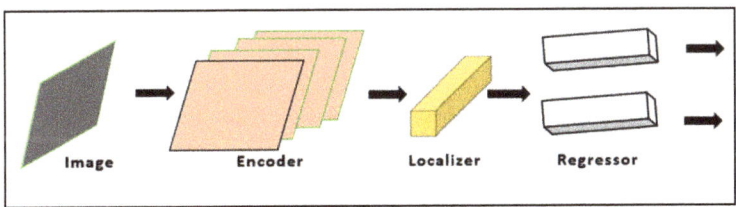

Figure 2. Pose-Net architecture.

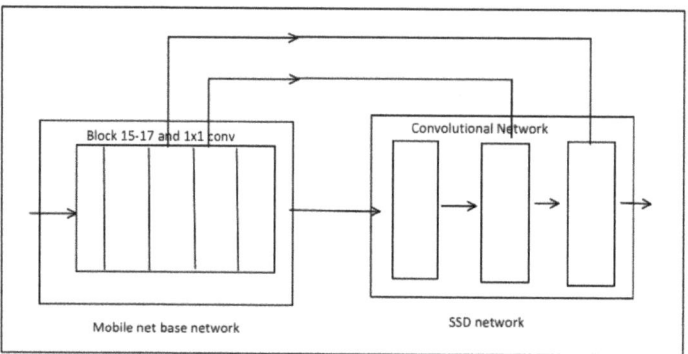

Figure 3. Mobile-Net SSD architecture.

3.1. Pose Net

It is a deep learning framework that identifies the positions of joints in images or video sequences involving human body movement. The joints are considered as key points, which are indexed by Part ID with a range from 0.0 to 1.0, with 1.0 as the highest score. Among different layers, soft-max activation function, which is at the last layer, has been replaced with a series of fully linked layers. Figure 1 depicts its high-level architecture. The performance of this model [22] depends and differs completely on the device and output stride. Because of this, the model works on posture positions in the entire scale of captured images, regardless of the size. Here, an encoder, localizer and regressor act as three components. The encoder generates the encoding vector, which contains the features of the input image represented as a 1024-dimensional vector. The localizer generates a vector with localized features. Finally, a regressor is used to perform regression on the final position using two linked layers in the model.

3.2. Mobile-Net SSD

It is a model that computes the input for making bounding box and categorize them from an input which is also called as object detection model. Single Shot Detector (SSD) in this model uses Mobile-Net [22] to achieve optimized object detection, which will work faster on mobile devices. This model contains offset values (cx, cy, w, h) where cx is the input, cy is the output and w is the weight and h is the score. The scores contain confidence values for 20 objects, and the background value is reserved as 0. Here, the extracted features are used as an arbitrary backbone and the resolution is calculated by reducing the extra layers. Then, this model computes the resolution by concatenating six levels of output. Finally, by utilizing non-maximum suppression (nms), the bounding boxes are filtered out. For a better understanding of the interface, users should first read the manual instruction (guide) of how the YogGuru-Ai works and how efficiently the pose can be performed. The user first needs to select the pose according to their benefit and need. The user should then face the camera. For proper functioning and to obtain good accuracy of the model, the user should practice yoga in a good lighting environment. Figure 3 depicts the Mobile-Net architecture details.

After the correct imitation of the pose, the counter begins. The sound of the counter timer motivates the yogi or the user to perform yoga for a longer period of time. If any difficulty is faced by the user while performing the pose, the user can visit the guidelines section for seeing the solution through the FAQ. Table 1 shows the various layers involved in the Y_PN-MSSD implementation. The outcome of the Pose-Net is to locate the position of various joints, while the role of Mobile-Net will be to determine the specified pose and validate the orientation and pose by connecting the joints. Figure 4 shows the role and block specific output of Pose-Net and Mobile-Net networks in determining the right yoga posture.

Table 1. Layer details of the proposed Y_PN-MSSD model.

Layer	Size
Input	224 × 224 × 3 (RGB)
Conv 1	64 × 64
Conv 2	192 × 192
Inception (1)	256 × 256
Inception (2)	480 × 480
Inception (3)	512 × 512
Inception (4)	512 × 512
Inception (5)	512 × 512
Inception (6)	528 × 528
Inception (7)	832 × 832
Inception (8)	832 × 832
Inception (9)	1024 × 1024
M_Con 3	1024 × 1024
M_Con 3	1024 × 1024
Avg_pool	1024
FC	1024
Soft_max	1000
Output	Pose
Output	Orientation

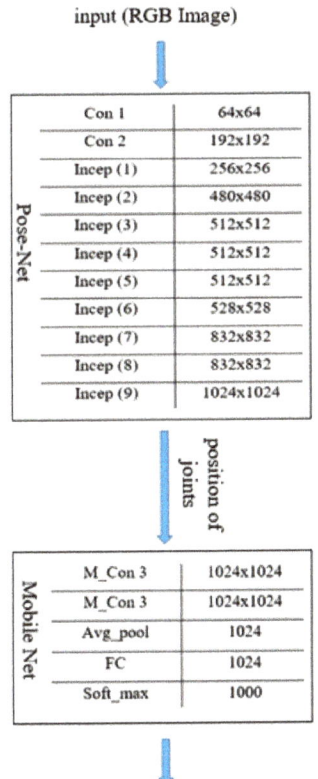

Figure 4. Block level details of Pose-Net and Mobile-Net and their respective output.

The input image is fed into the convolution layer (Con), where the edges are enhanced and embossed. Then, the output of convolution layer is fed to the Inception Layer (Incep), where a combination of layers (namely, 1 × 1 Convolutional layer, 3 × 3 Convolutional layer, 5 × 5 Convolutional layer) with their output filter banks concatenate the input into a single output vector (position of joints), which acts as an input of the next layer. The convolution layer of Mobile-Net (M_Con3) handles the output vector for depth-wise convolution. The separated output vectors are further fed into the average pooling layer (Avg_pool), which reduces the dimensions. In the fully connected layer, the output of the previous layer is multiplied with a weighted matrix and a bias vector is added. Finally, the soft max layer (soft_max) converts the vector scores based on the normalized probability distribution to detect the pose of the subject along with its orientation. Figure 5 shows the detailed structure of each layer in the proposed model for determining the right yoga posture.

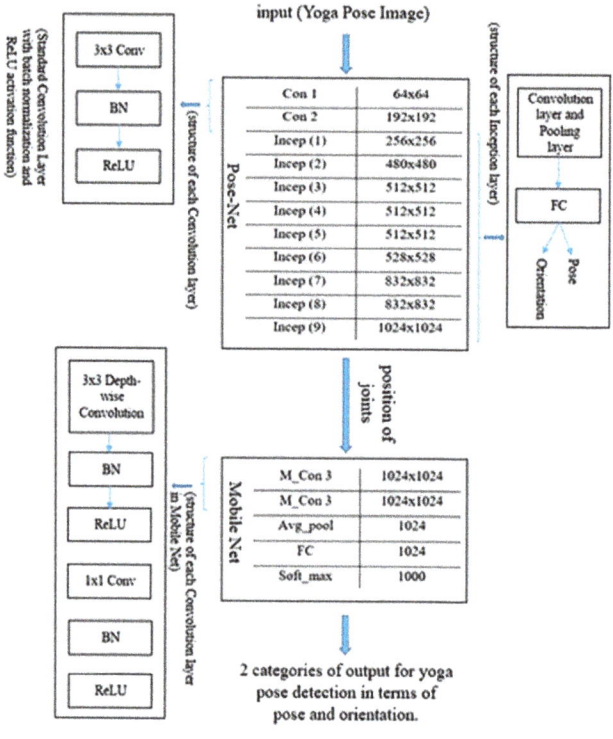

Figure 5. Block diagrammatic representation of the architectural flow of the proposed model.

The flow diagram pertaining to the working of this proposed Y_PN-MSSD approach has been depicted in Figure 6 and in terms of user interface has been depicted in Figure 7.

3.3. Dataset

The proposed model has been trained and tested on an open-source dataset [23] that is comprised of seven yoga poses: Cobra, Chair, Dog, Shoulder stand, Triangle, Tree, and Warrior. This dataset can be accessed using the following link: https://www.kaggle.com/datasets/amanupadhyay/yoga-poses (accessed on 2 January 2023). There are 70 total films of the seven positions, and 350 total instances of the seven poses. About 150 photographs were scraped for each posture from the internet and added to our own datasets. These films were recorded with a camera at a distance of four meters in a room with a frame rate of thirty frames per second. The frame rate can also be increased to a higher value and this will not have any significant bearing on the outcome. Four individuals performed these positions with little modifications in order to obtain robust trained models. Figure 8 shows an example of each pose. It must also be noted that no-pose data images are also taken into consideration. Here, the first image belongs to a no-pose dataset category and later the images of tree, shoulder stand and triangle are in the first row, whereas the image data of chair, cobra, dog and warrior postures are in the later row. Though some of the images in the dataset are in different orientations, they are adjusted by keeping the yoga mat as the reference and subsequent adjustments are made before subjecting the images for training. Images without regular orientation and not having a yoga mat are not considered for training and testing. When the yoga mat is not recognized, an automated alert prompting the user to practice yoga using a mat is generated by the model as shown in Figure 8. This way, the orientation of images is nullified. For both training and testing, the dataset has been used with 320 instances and 30 instances. For testing purpose, various films at different time intervals of data are built and called a secondary dataset. There are 30 examples in this unique dataset with 50 images for each posture.

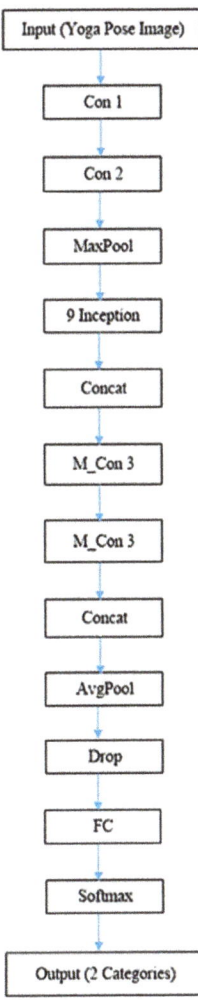

Figure 6. Block representation of the proposed Y_PN-MSSD model.

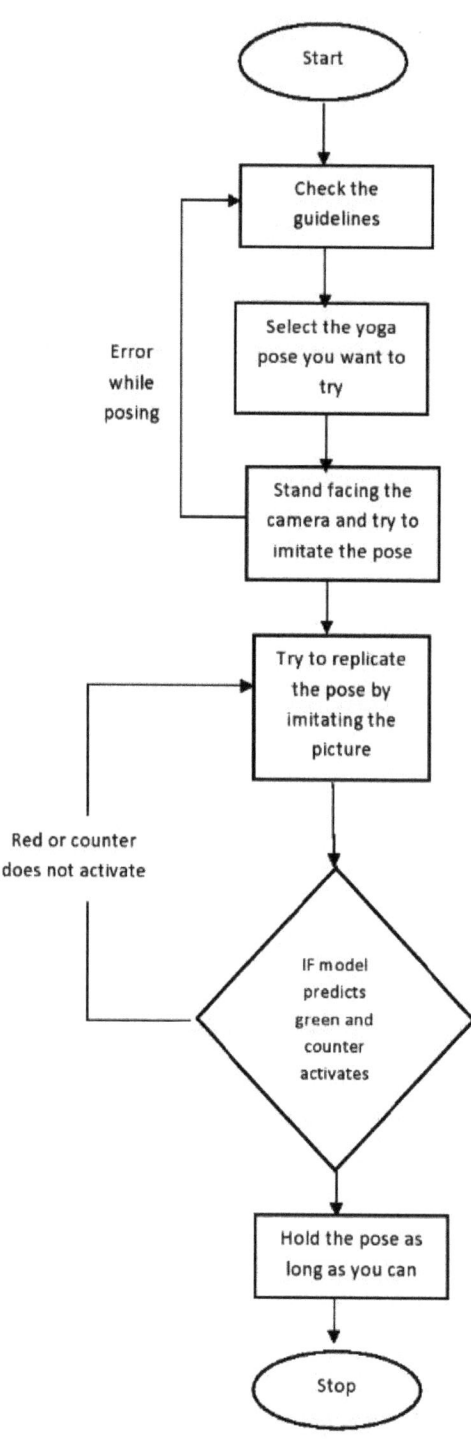

Figure 7. Working of Y_PN-MSSD model with respect to the user interface.

Figure 8. Various frames of yoga postures (cobra, chair, dog, tree, warrior, triangle, shoulder stand, neutral) in dataset used for training the Y_PN-MSSD model.

3.4. Real-Time Pose Estimation

In computer vision [24], human pose estimation is a major topic that has had significant progress in recent years from 2D-based single-person pose to multiple-person pose estimation. In general, algorithms [25] estimate poses by detecting key points of the body on an image or video and connecting them to generate output. Here, the x and y coordinates are considered as key points. Keras has been used for real-time multi-person pose estimation to extract critical points to estimate different postures. Each video is subjected to posture estimation. Each pose has been calculated for five consecutive frames, and each frame is taken every two seconds per video. It generates an array of 18 key points [26] for each pose with x and y coordinates. The pose estimation function extracts essential points from a frame, as shown in Figure 8.

The code generates a two-dimensional array in the form of a dictionary, which has keys that represent the coordinates of body parts and values. If several values are identified for a key in the dictionary, then all information has to be displayed, along with their respective levels of confidence. Even if the confidence level is low in comparison to other values, the first identified body point present in the dictionary is considered. As a result, the code must be updated based on a high level of confidence by choosing its value (example: in the tree or triangle pose) as depicted in Figure 9. Here, by using different confidence levels, six body points are detected, namely two arms, two leg joints, i.e., upper thighs, and two being the ends of the legs or the feet.

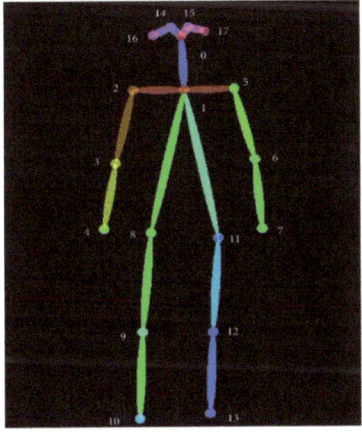
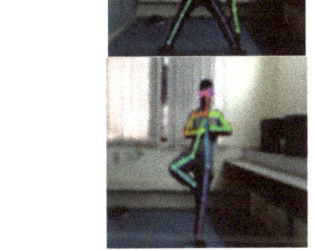

Figure 9. Feature extraction using Y_PN-MSSD model.

Many crucial points are observed with varying levels of confidence while detecting poses for a person. Keras pose estimation works by including the first key point found without taking confidence intervals into account. A few tweaks to the Keras posture estimation [27] were made in this work to account for important places with the highest confidence levels. The study used these x and y coordinates to extract characteristics such as angles between body joints and the ground, allowing models to be trained to reach high accuracy. These cases are given top consideration in order to ensure that no anomalous data are used as input. In this work, we have used cross-entropy as the cost function to gauge how well the model performs for the given dataset comprised of seven yoga poses. Cross-entropy can be defined as a measure of the difference between two probability distributions for a given random variable or set of events. The cross-entropy function between two probability distributions A and B can be stated formally as:

$$H(A, B) = -\text{sum } x \text{ in } X\, P(x) \times \log(Q(x)) \quad (1)$$

where $P(x)$ is the probability of the event x in P, and $Q(x)$ is the probability of event x in Q. A margin of error typically lets us know by how many percentage points the actual results obtained differ from the ideal value. In this case, the correct yogic pose is provided by the animated image and it is matched with the live pose frame. The joint positions are identified and lines are drawn. When comparing the joint positions of the live frame and animated image, a deviation of up to 5% is permitted. The Algorithm 1 for pose estimation is shown below:

Algorithm 1: Estimating correct pose

Input: Video, which is converted into frames and fed as images.
Step 1: Feed the input video. Convert it into N frames.
Step 2: Use the real-time multi-person pose estimation in Keras.
Step 3: Extract critical points to estimate posture.
Step 4: Set frame rate as 2 fps and estimate pose for five consecutive frames.
Step 5: Generate an array of 18 key points (p1, p2, ..., p18) for each pose with x and y coordinates.
Step 6: Form a dictionary D with these key points. {D} = {p1, p2, p3, ..., p18}
Step 7: Update the confidence values for the key points.
Step 8: Use these key points to detect arms, knees, joints, etc.
Step 9: Test the video frame with the model trained using these key points to estimate the correct posture.

- Let {Dv} be the dictionary generated by the input video. Let {Da} be the dictionary generated for a particular yoga pose (Yi) specified by the animated image (Ia).
- Compare {Dv} and {Da}.
    - If {Dv} = {Da} => Yi (Yoga pose recognized) => All connected lines become green.
    - If {Dv} ≠ {Da} => Yi not matched => The connected lines remain white.
- The user adjusts their pose according to the image until their pose matches with the yoga pose (Yi) specified by the animated image (Ia). [till {Dv} = {Da} situation is satisfied]

Output: Predicting whether the right pose is attained.
Step 10: Stop execution.

3.5. Interface

The interface was designed [28] with a view to help users see whether the yoga pose performed by them is correct or not. Therefore, a menu to select the yoga pose that the user would like to perform is provided. On selecting the required pose, the model will open an inbuilt webcam to analyze the pose. Here, the model Pose-Net in combination with Mobile-Net SSD are loaded in the background with much training performed on it. This recognizes the pose that the user performs through the webcam. In Figure 10, we can see that a tree pose (Figure 10a) has been performed by the user and similarly the chair pose (Figure 10b) has been enacted, and thus the model detects the pose. The model then checks whether the pose that user has enacted is right or not. If it matches with the training data of the required pose, then the connecting line goes green in color, otherwise it will remain in white. It will not turn green unless the user has performed the correct posture. Hence, the model can tell where the user goes wrong and simultaneously a timer is incremented until the pose is performed correctly. However, the timer stops if the user breaks the pose structure. This helps to record how much time the user has stood in the correct yoga pose. This will in turn help the user to improve next time. The user interface also shows the basic guidelines to perform the yoga pose. For performing any pose, this is the sequence of instruction the user should follow. If the user does not practice yoga on a mat, an automated alert is generated, prompting the user to practice yoga on a mat. This is depicted in Figure 9. The basic flow is as follows:

(a) When the app asks for permission to use the camera, allow it to access and capture the pose.
(b) Select what pose the user wants to perform using the dropdown.
(c) Read the instructions of that pose so that the user will know how to perform that pose.
(d) Click on Start pose and see the image of that pose on the right side and replicate that pose in front of the camera.
(e) If the user performs the correct pose, the skeleton over the video will turn green in color and a sound will be played.

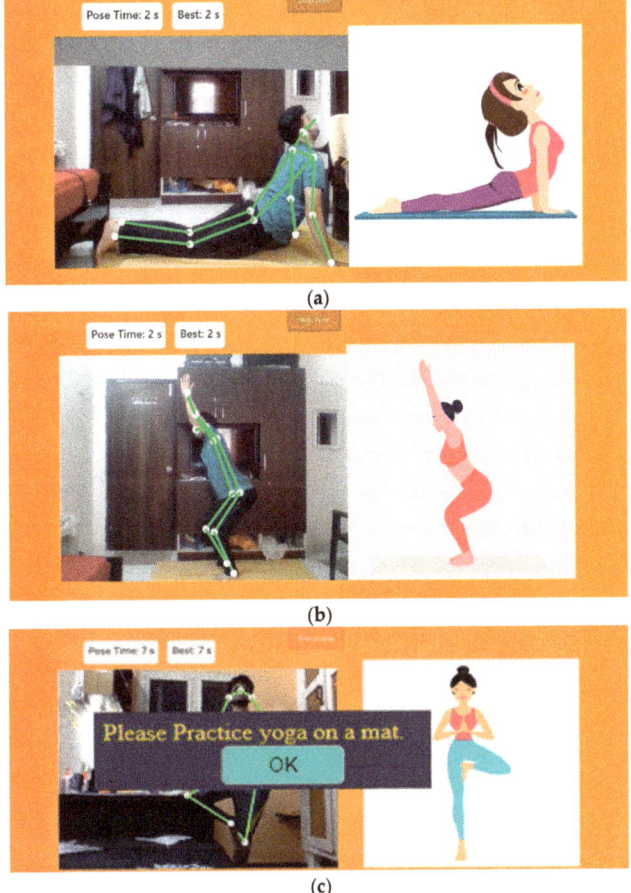

Figure 10. (**a**) Cobra pose recognition by Y_PN-MSSD model, (**b**) chair pose recognition by Y_PN-MSSD model, and (**c**) alert generated when a yoga mat is not recognized.

The described model uses Pose-Net and Mobile-Net for human pose prediction. For the ease of the user, the model has an easy-to-use graphical interface. The interface is made into an application with the help of JavaScript. The interface has been developed with ReactJS and an amalgamation of a plethora of json, php, and html scripts. Figure 11 shows the main home page that will be seen by the user. For users' convenience, the model divides the webpage into categories: 1. Working ("Lets hop in") and 2. Guide. The application also presents a guided menu wherein the user can select the yoga exercise of their choice. They can also see that each posture name has a guided figure attached within the menu display bar.

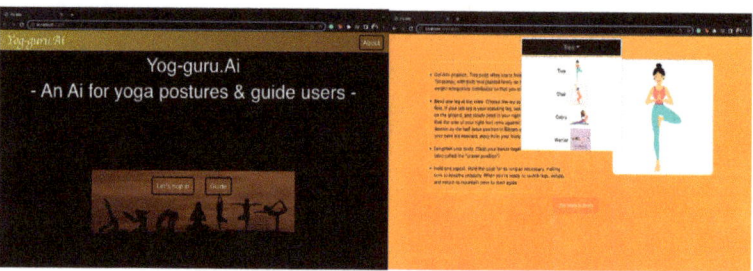

Figure 11. The home page and the user guide on various poses of the Y_PN-MSSD model.

Figure 12 shows tree and triangle yoga postures, respectively. Additionally, some facts and ways to perform certain yoga poses are also described in a side note. This will guide the user on how to carry out a certain pose even if any user is unaware of the yoga asana. In addition, there is a guide page on how to deal with hardware or camera issues on browser. Some of the potential issues faced during the experiment were occlusion and illumination changes. These challenges will be addressed in future work.

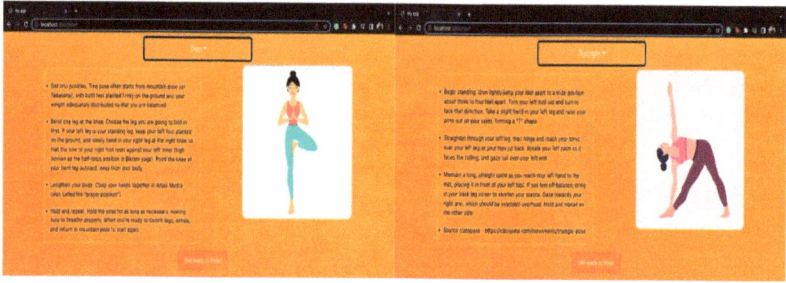

Figure 12. Tree posture and Triangle posture along with a proper guide on how to perform them.

4. Experimental Results

The proposed model provided in this paper uses the layer of a deep learning model to detect wrong yoga postures and correct them to improve. The vectors for nearby joints are used for estimating the angles. The extraction of feature points for pose estimation techniques are characterized in this work. These characteristics are then entered into categorization systems, which give feedback for the correctness of the yoga pose. As a result, the work is split into three parts: (1) feature extraction and time computation for every frame, (2) recognition, and (3) feedback generating time per frame for categorizing yoga poses. Each method's extraction and computation of features take the same amount of time. This experiment has been performed using NVIDIA GeForce GTX-1080 and Xeon(R) CPU E3-1240 v5. While training and test datasets consist of numerous ups and downs in accuracy until the 200 epoch, an accuracy of 0.9988 was finally attained after 200 epochs.

The loss of training and testing dropped gradually after 150 to 200 epochs. This results in coming up with a high confidence training model for classifying yoga postures. The accuracy and loss values have been depicted in Figure 13. It must be noted that the training was carried out using a dataset that contained frames taken from an open-source dataset and also data captured from four users performing seven yoga poses. These two entities constituted the training dataset. The testing was performed using the real time video captured by the user. Hence, the training accuracy is lower than the validation accuracy. The validation accuracy being higher than the training accuracy is a good indicator that the model performs very well in a real-time scenario. Additionally, another approach was carried out to validate the performance of the model in which the dataset was split into training and validation data in the ratio of 80:20. The training and validation sets were mutually exclusive. The accuracy plot for this approach is depicted in Figure 14. It can be seen that the validation accuracy was closely following the training accuracy and there were no major deviations between the curves. Additionally, the loss was decreasing significantly with increasing epochs, which once again confirmed the robust result of the model in terms of accuracy of classification.

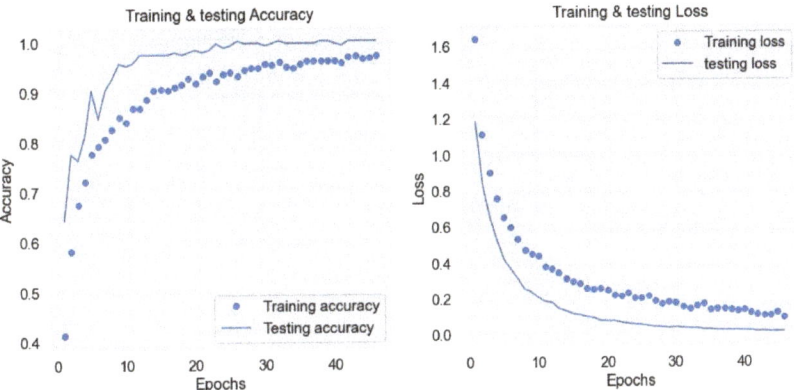

Figure 13. Accuracy and loss of Y_PN-MSSD model when tested in real time.

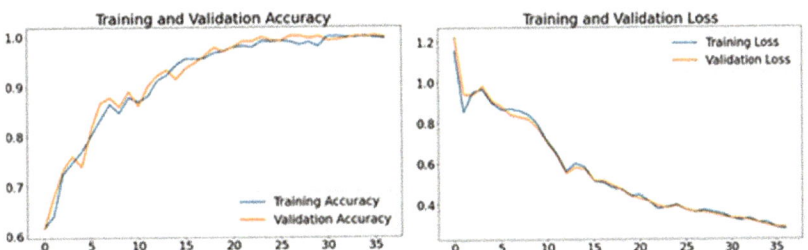

Figure 14. Accuracy and loss of Y_PN-MSSD model when tested with same dataset.

The loss of training and test datasets gradually decreased from epoch 0 to 200. The model does not appear to be over fitting based on the training and validation accuracy. Because the research is categorizing input features into one of seven labels, the loss function employed is categorical and the confusion matrix is depicted in Figure 15. This explains the obtained accuracy in a pictorial way, where in the last class-7, the accuracy is slightly fluctuating due to which the obtained accuracy is coming out to be 0.99885976. Table 2 depicts the details of hyperparameters used in the proposed model. An ablation study was carried out to choose the best possible optimizer. For obtaining the desired accuracy, hypermetric tuning was performed in which we have to change the optimizer, namely

Adam, AdaDelta, RMSProp, and Adagrad, along with activation function softmax. The comparison of optimizers along with loss and accuracy in 200 epochs is shown in below Table 3. From the table, it can be concluded that the proposed Y_PN-MSSD model with Adam optimizer outperforms the other optimizers.

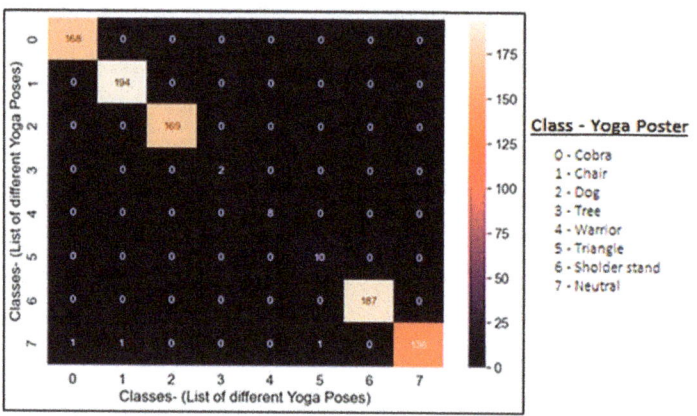

Figure 15. Confusion matrix of Y_PN-MSSD model with predicted labels on x-axis and true labels on y-axis.

Table 2. Hyperparameters of the proposed Y_PN-MSSD model.

Hyperparameters	Values
Image input shape	224 × 224
Channel	3
Batch size	64
Epochs	200
Optimizer	Adam
Loss function	cross-entropy
Learning rate	10^{-3}
Minimum learning rate	10^{-6}

Table 3. Performance of Y_PN-MSSD model based on different optimizers.

Sl. No	Optimizer	Epochs	Loss	Accuracy
1	Adam	200	0.00513306	0.99885976
2	AdaDelta	200	1.95524644	0.45952108
3	RMSProp	200	0.01959268	0.99087798
4	Adagrd	200	1.35576283	0.52793616

A confusion matrix is used to evaluate the performance of the proposed yoga recognition model. The frame-based metrics and even score are computed using the following metrics. They are True Positive Rate, False Positive Rate, precision, and recall, respectively. The overall performance is compared with the accuracy of the model. The mathematical model for the given metrics are shown below.

$$\text{False Positive Rate (FPR)} = FP/TN + FP \quad (2)$$

$$\text{Precision} = TP/TP + FP \quad (3)$$

$$\text{Recall, True Positive Rate} = TP/TP + FN \quad (4)$$

$$\text{Accuracy} = TP + TN/TP + TN + FP + FN \quad (5)$$

The proposed method is tested with the user pressing the "Get ready to pose" button in the user interface when the video is subsequently captured on the fly. The joints will be located and the connections between joints will be shown in white until the user's pose matches with that of the pose exhibited by the animated image shown. Once the pose is matched, the connections will turn from white to green, indicating to the user that the correct pose has been attained. The video capture is then turned off by hitting the "stop pose" button in the user interface. To explain this better, some additional sample output images indicating the transition from incorrect pose to correct pose on the fly have been included as Figure 16.

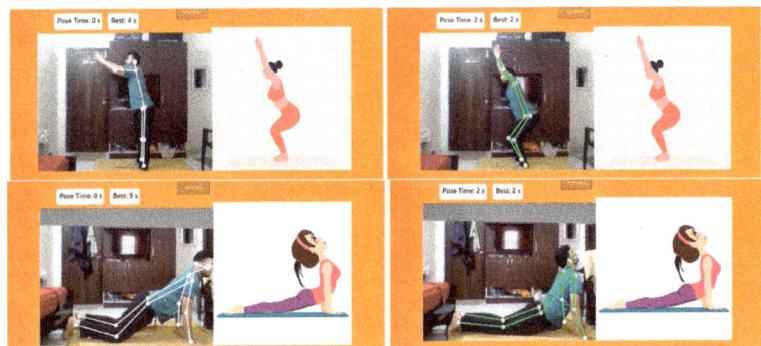

Figure 16. Sample output indicating incorrect pose (white connections) and user on the fly making adjustments until the correct pose (green connections) is attained.

The proposed model result is compared with the existing Pose-Net CNN model with respect to the derived metrics and it is observed that the proposed model outperforms the existing model and is tabulated in Table 4. It is observed that the false positive rate of the proposed model is less when compared to the existing model. Additionally, the precision, recall and accuracy of the proposed model show the best results, and both the models are trained and tested with the same seven yoga posture dataset. Table 5 compares the accuracy of the proposed model with other contemporary works involving yoga posture recognition. It can be seen that the proposed system yields the highest accuracy of 99.88%.

Table 4. The confusion metrics of the proposed model (Y_PN-MSSD) and existing model (Pose-Net CNN).

Model	Pose Net CNN [13]	Proposed Y_PN-MSSD
FPR	0.03	0.01
Precision	0.97	0.98
Recall, TPR	0.98	0.99
Accuracy	0.98	0.99

Table 5. Comparison of accuracy of proposed system and other works involving yoga posture recognition.

Model	Model Used	Best Accuracy Obtained in Each Work
[27]	Distributed CNN, 3D-CNN	96.31%
[26]	TL-VGG16-DA, TL-VGG19-DA, TL-MobileNet-DA, TL-MobileNetV2-DA, TL-InceptionV3-DA, and TL-DenseNet201-DA	98.43%
Proposed System	Y_PN-MSSD	99.88%

5. Conclusions and Future Work

Human posture estimation has been a subject of hot research in recent years. Human posture estimation varies from other computer vision problems in which key points of human body parts are tracked based on a previously defined human body structure. Yoga self-instruction systems have the potential to popularize yoga while also ensuring that it is properly performed. Deep learning models look to be promising because of substantial research in this field. In this paper, the proposed Y_PN-MSSD model is used to recognize seven yoga asanas with an overall accuracy of 99.88%. The accuracy of this model is based on Pose-Net posture assessment and Mobile-Net SSD. A Pose-Net layer takes care of feature point detection, whereas a Mobile-Net SSD layer performs human detection in each frame. This model has been categorized into three stages. In stage one, which is the data collection/preparation stage, the yoga posters are captured from the four users as well as an open-source dataset with seven yoga poses. Then, at the second stage, these collected data have been used for training the model where the feature extraction takes place by connecting key points of the human body. At last, the yoga posture has been recognized. This model will assist the user to perform yoga poses in a live tracking mode and they can correct the posture on the fly. When compared with the Pose-Net CNN model, the proposed model gives better results in terms of accuracy. Additionally, activity recognition is demonstrated in a real-world context and a model such as this could help with pose identification in sports, surveillance, and healthcare in the future. For self-training and real-time forecasting, this model can be used by augmenting the inputs. In future, the yoga posture recognition application can be trained with more number of yoga poses. Additionally, full-fledged training can be performed to build a fully adapted real time model for a real-time noise environment to act as a professional yoga trainer. Future work will be towards addressing other challenges faced during the implementation such as occlusion and illumination changes. Additionally, the proposed model will be enriched to detect more yoga postures. An audio-based alert can be included as part of the future scope to indicate a signal to the user when the correct posture is attained.

Author Contributions: A.U. and N.K.B. were involved in manuscript preparation. A.U., N.K.B. and B.A. were involved in the conceptualization, methodology, software, validation, formal analysis, investigation, resources, data curation, writing original draft, reviewing and editing of writing. A.U., N.K.B. and B.A. were involved in figures and tabulation of results. All authors were involved in collating the manuscript and formatting the same. All authors have read and agreed to the published version of the manuscript.

Funding: This research received no external funding.

Institutional Review Board Statement: Not applicable.

Informed Consent Statement: Not applicable.

Data Availability Statement: The data used to support the findings of this study are included and will be accessible using the following link https://www.kaggle.com/datasets/amanupadhyay/yoga-poses (accessed on 2 January 2023).

Acknowledgments: The authors express their sincere thanks to VIT management for their encouragement during this work.

Conflicts of Interest: The authors declare no conflict of interest.

References

1. Chu, X.; Ouyang, W.; Li, H.; Wang, X. Structured feature learning for pose estimation. In Proceedings of the IEEE Conference on Computer Vision and Pattern Recognition, Las Vegas, NV, USA, 27–30 June 2016; pp. 4715–4723.
2. Wu, Y.; Lin, Q.; Yang, M.; Liu, J.; Tian, J.; Kapil, D.; Vanderbloemen, L. A Computer Vision-Based Yoga Pose Grading Approach Using Contrastive Skeleton Feature Representations. *Healthcare* **2021**, *10*, 36. [CrossRef] [PubMed]
3. Chen, H.T.; He, Y.Z.; Hsu, C.C.; Chou, C.L.; Lee, S.Y.; Lin, B.S. Yoga posture recognition for self-training. In *Multimedia Modeling, MMM 2014*; Springer: Berlin/Heidelberg, Germany, 2014; pp. 496–505.

4. Islam, M.U.; Mahmud, H.; Bin Ashraf, F.; Hossain, I.; Hasan, M.K. Yoga posture recognition by detecting human joint points in real time using Microsoft Kinect. In Proceedings of the 2017 IEEE Region 10 Humanitarian Technology Conference (R10-HTC), Hyderabad, India, 21–23 December 2017; pp. 668–673.
5. Kendall, A.; Grimes, M.; Cipolla, R. PoseNet: A Convolutional Network for Real-Time 6-DOF Camera Relocalization. *Healthcare* **2021**, *35*, 36.
6. Chen, Y.; Shen, C.; Wei, X.; Liu, L.; Yang, J. Adversarial PoseNet: A Structure-aware Convolutional Network for Human Pose Estimation. *arXiv* **2017**, arXiv:1705.00389v2.
7. Zhang, Z.; Tang, J.; Wu, G. Simple and Lightweight Human Pose Estimation. *arXiv* **2019**, arXiv:1911.10346.
8. Soviany, P.; Ionescu, R.T. Continuous Trade-off Optimization between Fast and Accurate Deep Face Detectors. *arXiv* **2018**, arXiv:1811.11582.
9. Shavit, Y.; Ferens, R. Introduction to Camera Pose Estimation with Deep Learning. *arXiv* **2019**, arXiv:1907.05272.
10. Chen, H.; Feng, R.; Wu, S. 2D Human pose estimation: A survey. *Multimed. Syst.* **2022**. [CrossRef]
11. Belagiannis, V.; Zisserman, A. Recurrent human pose estimation. *arXiv* **2016**, arXiv:1605.02914.8.
12. Buehler, P.; Everingham, M.; Huttenlocher, D.P.; Zisserman, A. Upper body detection and tracking in extended signing sequences. *Int. J. Comput. Vision.* **2019**, *95*, 180–197. [CrossRef]
13. Chiddarwar, G.G.; Ranjane, A.; Chindhe, M.; Deodhar, R.; Gangamwar, P. AI-based yoga pose estimation for android application. *Int. J. Innov. Sci. Res. Technol.* **2020**, *5*, 1070–1073. [CrossRef]
14. Chaudhari, A.; Dalvi, O.; Ramade, O.; Ambawade, D. Yog-guru: Real-time yoga pose correction system using deep learning methods. In Proceedings of the International Conference on Communication Information and Computing Technology (ICCICT), Mumbai, India, 25–27 June 2021; pp. 1–6.
15. Qiao, S.; Wang, Y.; Li, J. Real-time human gesture grading based on OpenPose. In Proceedings of the 10th International Congress on Image and Signal Processing Bio-Medical Engineering, and Informatics (CISP-BMEI), Shanghai, China, 14–16 October 2017; pp. 1–6. [CrossRef]
16. Zhang, W.; Wang, Z.; Li, X. Blockchain-based decentralized federated transfer learning methodology for collaborative machinery fault diagnosis. *Reliab. Eng. Syst. Saf.* **2023**, *229*, 108885. [CrossRef]
17. Li, X.; Xu, Y.; Li, N.; Yang, B.; Lei, Y. Remaining Useful Life Prediction With Partial Sensor Malfunctions Using Deep Adversarial Networks. *IEEE/CAA J. Autom. Sin.* **2023**, *10*, 121–134. [CrossRef]
18. Yadav, S.K.; Agarwal, A.; Kumar, A.; Tiwari, K.; Pandey, H.M.; Akbar, S.A. YogNet: A two-stream network for realtime multiperson yoga action recognition and posture correction. *Knowl.-Based Syst.* **2022**, *250*, 109097. [CrossRef]
19. Long, C.; Jo, E.; Nam, Y. Development of a yoga posture coaching system using an interactive display based on transfer learning. *J. Supercomput.* **2022**, *78*, 5269–5284. [CrossRef]
20. Qian, J.; Cheng, X.; Yang, B.; Li, Z.; Ren, J.; Olofsson, T.; Li, H. Vision-Based Contactless Pose Estimation for Human Thermal Discomfort. *Atmosphere* **2020**, *11*, 376. [CrossRef]
21. Bulat, A.; Tzimiropoulos, G. Human pose estimation via convolutional part heatmap regression. In Proceedings of the Computer Vision–ECCV 2016: 14th European Conference, Amsterdam, The Netherlands, 11–14 October 2016; Volume 9911, pp. 717–732.
22. Carreira, J.; Agrawal, P.; Fragkiadaki, K.; Malik, J. Human pose estimation with iterative error feedback. In Proceedings of the IEEE Conference on Computer Vision and Pattern Recognition, Las Vegas, NV, USA, 27–30 June 2016; pp. 4733–4742.
23. Yu, N.; Huang, Y.T. Important Factors Affecting User Experience Design and Satisfaction of a Mobile Health APP: A Case Study of Daily Yoga APP. *Int. J. Environ. Res. Public Health* **2020**, *17*, 6967. [CrossRef]
24. Bilal, M.; Maqsood, M.; Yasmin, S.; Hasan, N.U.; Rho, S. A transfer learning-based efficient spatiotemporal human action recognition framework for long and overlapping action classes. *J. Supercomput.* **2021**, *78*, 2873–2908. [CrossRef]
25. Bukhari, M.; Bajwa, K.B.; Gillani, S.; Maqsood, M.; Durrani, M.Y.; Mehmood, I.; Ugail, H.; Rho, S. An efficient gait recognition method for known and unknown covariate conditions. *IEEE Access.* **2020**, *9*, 6465–6477. [CrossRef]
26. Verma, M.; Kumawat, S.; Nakashima, Y.; Raman, S. Yoga-82: A new dataset for fine-grained classification of human poses. *arXiv* **2020**, arXiv:2004.10362v1.
27. Chicco, D.; Jurman, G. The advantages of the Matthews correlation coefficient (MCC) over F1 score and accuracy in binary classification evaluation. *BMC Genom.* **2020**, *21*, 6. [CrossRef]
28. Bazarevsky, V.; Grishchenko, I.; Raveendran, K.; Zhu, T.; Zhang, F.; Grundmann, M. BlazePose: Ondevice real-time body pose tracking. *arXiv* **2020**, arXiv:2006.10204.

Disclaimer/Publisher's Note: The statements, opinions and data contained in all publications are solely those of the individual author(s) and contributor(s) and not of MDPI and/or the editor(s). MDPI and/or the editor(s) disclaim responsibility for any injury to people or property resulting from any ideas, methods, instructions or products referred to in the content.

Article

Human Activity Recognition Based on Embedded Sensor Data Fusion for the Internet of Healthcare Things

Mohamed E. Issa [1], Ahmed M. Helmi [1,2], Mohammed A. A. Al-Qaness [3,4,*], Abdelghani Dahou [5], Mohamed Abd Elaziz [6,7,8] and Robertas Damaševičius [9,*]

1. Computer and Systems Engineering Department, Faculty of Engineering, Zagazig University, Zagazig 44519, Egypt; Mamohamedali@eng.zu.edu.eg (M.E.I.); amhm162@gmail.com (A.M.H.)
2. College of Engineering and Information Technology, Buraydah Private Colleges, Buraydah 51418, Saudi Arabia
3. State Key Laboratory for Information Engineering in Surveying, Mapping and Remote Sensing, Wuhan University, Wuhan 430079, China
4. Faculty of Engineering, Sana'a University, Sana'a 12544, Yemen
5. LDDI Laboratory, Faculty of Science and Technology, University of Ahmed DRAIA, Adrar 01000, Algeria; dahou.abdghani@univ-adrar.edu.dz
6. Faculty of Computer Science and Engineering, Galala University, Suez 435611, Egypt; abd_el_aziz_m@yahoo.com
7. Artificial Intelligence Research Center (AIRC), College of Engineering and Information Technology, Ajman University, Ajman 346, United Arab Emirates
8. Department of Mathematics, Faculty of Science, Zagazig University, Zagazig 44519, Egypt
9. Department of Applied Informatics, Vytautas Magnus University, 44404 Kaunas, Lithuania
* Correspondence: alqaness@whu.edu.cn (M.A.A.A.-Q.); robertas.damasevicius@vdu.lt (R.D.)

Abstract: Nowadays, the emerging information technologies in smart handheld devices are motivating the research community to make use of embedded sensors in such devices for healthcare purposes. In particular, inertial measurement sensors such as accelerometers and gyroscopes embedded in smartphones and smartwatches can provide sensory data fusion for human activities and gestures. Thus, the concepts of the Internet of Healthcare Things (IoHT) paradigm can be applied to handle such sensory data and maximize the benefits of collecting and analyzing them. The application areas contain but are not restricted to the rehabilitation of elderly people, fall detection, smoking control, sportive exercises, and monitoring of daily life activities. In this work, a public dataset collected using two smartphones (in pocket and wrist positions) is considered for IoHT applications. Three-dimensional inertia signals of thirteen timestamped human activities such as Walking, Walking Upstairs, Walking Downstairs, Writing, Smoking, and others are registered. Here, an efficient human activity recognition (HAR) model is presented based on efficient handcrafted features and Random Forest as a classifier. Simulation results ensure the superiority of the applied model over others introduced in the literature for the same dataset. Moreover, different approaches to evaluating such models are considered, as well as implementation issues. The accuracy of the current model reaches 98.7% on average. The current model performance is also verified using the WISDM v1 dataset.

Keywords: Internet of Healthcare Things; human activity recognition; smart technologies for healthcare; m-Health; mobile devices; digital healthcare

Citation: Issa, M.E.; Helmi, A.M.; Al-Qaness, M.A.A.; Dahou, A.; Abd Elaziz, M.; Damaševičius, R. Human Activity Recognition Based on Embedded Sensor Data Fusion for the Internet of Healthcare Things. *Healthcare* 2022, 10, 1084. https://doi.org/10.3390/healthcare10061084

Academic Editors: Giner Alor-Hernández, Jezreel Mejía-Miranda, José Luis Sánchez-Cervantes and Alejandro Rodríguez-González

Received: 24 May 2022
Accepted: 9 June 2022
Published: 10 June 2022

Publisher's Note: MDPI stays neutral with regard to jurisdictional claims in published maps and institutional affiliations.

Copyright: © 2022 by the authors. Licensee MDPI, Basel, Switzerland. This article is an open access article distributed under the terms and conditions of the Creative Commons Attribution (CC BY) license (https://creativecommons.org/licenses/by/4.0/).

1. Introduction

1.1. Motivation

Smart solutions for Internet of Healthcare Things (IoHT) [1], also known as Healthcare Internet of Things [2], Internet of Medical Things [3], or Medical Internet of Things [4], systems have extensively emerged since the Industry 4.0 revolution [5], making use of digital devices, in particular wearable sensors and smart handheld devices. In the new phase of the industrial revolution, termed Industry 5.0, collaborative interaction between machines and people is coming back to the forefront [6]. Unlike aiming to find the best ways to connect devices together—in the first place—which was a goal of Industry 4.0, there is great interest in moving toward personalization in Industry 5.0. This means that creative thinking and smart usage of the entities of smart systems are expected to increase the productivity and benefits of emerging IoT-based solutions [7]. The guidelines of Industry 5.0—under the umbrella of IoT—open up a new window to the development and enhancement of existing smart IoHT systems, in particular, during present-day circumstances, such as the spread of COVID-19, and ehealth and telehealth services can be provided without in-person visits [8,9], while decision support provided by artificial intelligence methods can facilitate doctors' decisions [10]. Numerous applications are categorized under IoHT applications. For example, indoor localization and IoT applications inside smart buildings such as keeping social distances have been used since the COVID-19 pandemic began [11]. In addition, such applications are used for traditional tasks such as the monitoring of daily life activity [5,12,13], fall detection [14] and assisted living [15–18], bad habits (such as smoking) detection and control [19], monitoring of industrial workers' activity [20], monitoring the heart rate of vehicle drivers [21], using wearable sensors to monitor heart activity [22,23], mHealth Apps for Self-Management [24], gait detection for people with Parkinson's disease [25,26], and many others.

The implementation of IoHT systems starts with data acquisition, followed by a preprocessing and feature-extraction phase, and finally arrives at the decision-making stage. Most known approaches in the literature can be categorized as video-based, WiFi-based, and sensory-based. Video-based human activity monitoring approaches may provide rich information via videos and images for indoor activities when there are no ad hoc cameras in outdoor environments such as walking tracks, parks, traditional malls, and swimming pools. Conversely, both wearable sensors and smart handheld devices are very suitable for the environment-invariant Human Activity Recognition (HAR) models. Another concern is that maintaining the privacy of individuals is questionable in vision-based approaches [27], while dealing with data fusion from sensors presents no such compromise. However, WiFi-based recognition of activities of daily life [28–30] has the advantage of using the fixed WiFi devices, but such approach has no applicability in outdoor environments.

A great interest is devoted to employing wearable sensors (e.g., accelerometer units), embedded sensors in smart devices (e.g., accelerometer, gyroscope, and magnetometer), and Kinect sensors [31] to develop HAR models [12,32]. Currently, smart devices such as smartphones and smartwatches are receiving much attention in such IoHT applications for obvious reasons [5]. On the other hand, a special-purpose real-time health monitoring device may have concerns regarding the efficient implementation in terms of power consumption [33]. When data acquisition is performed through many sensors and/or devices, there is a need for a suitable IoT framework to be able to move to the preprocessing stage. In preprocessing stage, the tri-axial activity signals registered by the sensors usually first need noise filtration, then segmentation in window length that ranges from <1 to 30 s [5,34] with more focus on reasonable small lengths (e.g., 2–10 s) in order to simulate real-time situations. Furthermore, feature extraction can follow the traditional approach of handcrafting a set of fine features selected in the time domain (mean, standard deviation, min, max, Pearson coefficients, etc.) and the frequency domain (energy, entropy, FFT coefficients, etc.), or they may follow the modern trend of deep learning networks [16,34]. In the latter approach, features are implicitly extracted as the encodings of hidden layers of the network, while outer layers such as fully connected layers together with softmax layer are responsible for the decision-making (i.e., classification and recognition). Following the feature engineering approach, the Random Forest (RF) algorithm [35], Multilayer Perceptron (MLP) [36] (one variant of artificial neural networks), Support Vector Machines (SVMs) [37], and Naive Bayes (NB) [38] are among the well-known shallow classifiers.

However, deep learning models perform well for many available human activity datasets in the literature [34], but the RF algorithm, for example, performs better than a single LSTM classifier for a specific dataset addressed in [16]. In addition, the recent studies in [17,39–41] in IoT applications depend on shallow classifiers. Recently, hybrid ensemble approaches that make use of shallow classifiers in addition to deep convolutional layers are significantly bullish [28].

The limitation of existing approaches concerning a dataset collected by two smartphone units (in pocket and wrist positions) of human activities and gestures introduced by Shoaib et al. [42] motivates improving the state-of-the-art results. In this paper, an interesting and challenging dataset of thirteen activities is addressed. Activities are divided into two groups: the first group consists of hand gestures such as eating, smoking, drinking coffee, typing, and writing, and the other group consists of biking, jogging, standing, sitting, walking, walking upstairs, and walking downstairs. As a classification problem, the whole dataset is handled at a time in the training and testing processes. Using a feature set that is adequate to sensors' positions on the human body, an impartial comparison between the aforementioned shallow classifiers is conducted. The RF algorithm shows outstanding performance compared to previous models in the literature according to both subject-dependent and stratified k-fold cross-validation evaluation metrics. Furthermore, for testing the model generalization, another dataset, namely WISDM v1 [43], is used to examine the applied model performance.

1.2. Related Work

In the literature, numerous human activity datasets were collected from smartphones and/or smartwatches, e.g., WISDM v1 and v2, UCI–HAR, and UniMiB SHAR; see the survey by Demrozi et al. [44] for complete details. Shoaib et al. published a public dataset in [42] using two smartphone units. Below, we shed light on some closely related studies that addressed this dataset. In [42], a simple feature set of mean, standard deviation, median, min, max, semi-quartile, and the sum of the first ten FFT coefficients were extracted from each sensor stream, and the magnitude of its 3-dimensional signal was applied to the NB classifier. Since the readings of the accelerometer, linear accelerometer, gyroscope, and magnetometer sensors in both smartphones were registered, the focus in [42] was to evaluate the combination of sensors and device positions on the body, besides determining the effect of the window length from 2 to 30 s. The accelerometer and the gyroscope from both devices' positions gave the best performance. Baldominos et al. [45] performed a comparative study between different machine learning techniques (deep and shallow). Readings of the four sensors mentioned above were used. For shallow techniques, handcrafted features such as the mean and the standard deviation of raw signals and skewness, kurtosis, and the lower and upper quartiles of real coefficients of FFT of each dimension were obtained. The ensemble of randomized decision trees (ET) outperformed both shallow classifiers such as RF, MLP, NB, and K-nearest neighbors and convolutional neural networks (CNN). Alo et al. [46] examined two deep learning models, namely deep-stacked autoencoders (DSAE) and deep belief neural networks (DBNN). Only signals of the accelerometer are considered in both devices. Besides raw signals, the magnitude vector and the vectors of pitch and roll angles are used for training the models. The DSAE showed notable performance over both DBNN and the shallow classifiers (with the time-domain features in [42]) such as SVM, NB, and linear discriminant analysis. There are also deep learning models proposed for HAR using wearable sensors. For example, in [47], a combination of long short-term memory (LSTM) and a conventional neural network (CNN) was proposed to solve the HAR problem. In [48], a new HAR model was developed based on convolutional and LSTM recurrent units. In [49], a new model called iSPLInception was developed based on the Inception-ResNet framework from Google. It showed acceptable performance using different HAR datasets. In [50], the authors studied the applications of several deep learning methods. They found that the hybrid CNN-BiGRU showed the best results. Among the aforementioned studies, stratified k-fold evaluation criteria were applied by Shoaib et al. [42], while dataset samples were divided into train/test sets with a subject-dependent measure in [45,46]. Moreover, there is a variance between the different studies about the most suitable sensors for this task. Finally, there is some confusion about the superiority of conventional machine learning approaches versus deep learning models for this specific dataset.

To solve such conflicts, this paper proposes an individual model that proves superior according to both evaluation criteria. In addition, an impartial comparison between previous approaches and the current one has been performed.

1.3. Contribution of Current Work

- Presenting a light human-activity-recognition system using wearable sensors.
- Implementing a robust real-time model based on the Random Forest algorithm that outperforms other known classifiers and deep learning models.
- Handling a complex dataset of thirteen different human activities and gestures and improving the state-of-the-art results according to both subject-dependent and stratified k-fold cross-validation measures and using a different dataset, namely WISDM v1, for verifying model performance.
- Conducting sensitivity analysis for the applied model parameters (Random Forest size and depth).

1.4. Paper Organization

This document is organized as follows: Section 2 introduces the applied IoHT system framework. Section 3 presents the experimental results within the discussion. Section 4 handles the effect of important parameters on model performance. Section 5 provides a comparison with previous related studies. A different dataset is used to verify model performance in Section 6. The discussion of obtained results is given in Section 7. Section 8 includes conclusions, limitations, and future extensions of this work.

2. The Applied Approach

2.1. Dataset

Table 1 presents the generic information of dataset addressed here. Activity signals were recorded at a frequency of 50 Hz from the accelerometer, linear accelerometer, gyroscope, and magnetometer sensors of two Samsung Galaxy S2 smartphones. One device was put in the right pocket, and the other was placed on the right wrist. Ten subjects were asked to perform thirteen activities following a protocol; see Table 2 for the duration of each activity performed for each subject. This data set comprises six activities involving hand gestures, namely eating, smoking, drinking coffee, typing, and writing, and seven activities involving full-body motions, namely biking, jogging, standing, sitting, walking, walking upstairs, and walking downstairs. The total number of observations was 1,170,000. Activity signals were successfully registered, and there were no missing values. More details about the settings of collecting activities can be reviewed in [42].

Table 1. Dataset collection configuration.

Parameter	Information
# Subjects	10
# Activities	13
Total # Observations	1,170,000
Missing values	NO
Device	Two Samsung Galaxy S2 smartphones
Position on Body	Right pocket and right wrist
Sensors	Accelerometer, Linear Accelerometer, Gyroscope and Magnetometer
Frequency	50 Hz

Table 2. Dataset activities.

Activity	Abbreviation	Duration (min)
Biking	BK	3
Having Coffee	CO	5
Walking Downstairs	DS	3
Eating	ET	5
Jogging	JO	3
Sitting	ST	3
Smoking	SM	5
Standing	SN	3
Giving a Talk	TK	5
Typing	TP	5
Walking Upstairs	UP	3
Walking	WK	3
Writing	WR	5

2.2. Sensory Data Processing

The applied model makes use of the readings of accelerometer and gyroscope sensors, where the acceleration and angular velocity of body limbs are sufficient for characterizing the activities performed. This point of view coincides with the well-known study of Anguita et al. [51]. Figure 1 clarifies the sensors' positions on the human body in order to acquire activity signals. Figure 2 shows the signal separation into body and gravity components using the Butterworth filter. Figure 3 presents the IoHT framework applied here. When applying the model, it is suggested to connect devices through Bluetooth technology. Then, the processing takes place at one central point (i.e., smartphone) as shown in Figure 3.

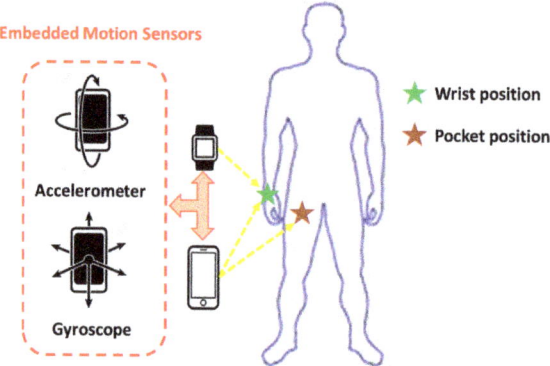

Figure 1. Activity signal acquisition from handheld smart devices.

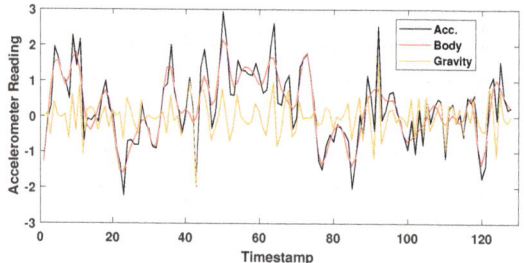

Figure 2. Accelerometer signal separation into body and gravity components using the Butterworth filter with a corner frequency of 20 Hz.

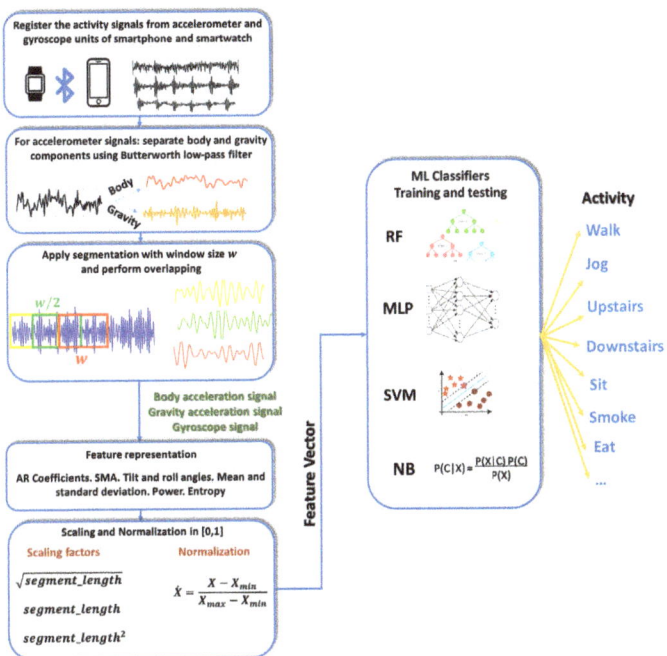

Figure 3. The composition of the applied IoHT system.

Activity Signal Preprocessing. According to previous studies, e.g., [51–53], it is preferred separate body and gravity components of accelerometer signals using, for example, a fourth-order Butterworth low-pass filter with a corner frequency of 20 Hz to filter out the body-acceleration component, since signals were collected at 50 Hz–. For real-time considerations, signals were segmented using a window length of 2.56 s (i.e., 128 data points) with an overlap of 50% [51]. Figure 2 presents an illustrating example of acceleration signal separation for the walking activity in a time interval of 2.56 s. Thus, there is a fusion of six time-series signals: body acceleration, gravity acceleration, and gyroscope readings of both devices.

Feature Representation. The features for smartphone-based activity signals (with the numerical participation in the feature vector in parenthesis) are listed as follows:

- (F1-12) Mean and standard deviation (STD) of each of the acceleration signal (AS) and its jerk signal (JS)
- (F13-24) Autoregressive (AR) model coefficients for AS
- (F25) Signal magnitude area (SMA)
- (F26) Tilt angle (TA)
- (F27-30) Roll angle (RA) Equation (1): mean, STD, entropy of JS, and power
- (F31) Angle of x-component of AS Equation (2)
- (F32-34) Entropy of JS
- (F35-37) Power of AS

$$\text{Roll angle} = arctan(-BA_z, -BA_y) \quad (1)$$

where BA_y and BA_z are body acceleration in y and z dimensions, respectively.

$$\text{Angle of x-component of AS} = real\left(arccos\left(max\left(min\left(\frac{B_x \cdot G_m}{||B_x|| * ||G_m||}, 1\right), -1\right)\right)\right) \quad (2)$$

where the only real part of the resulting quantity is used; B_x and G_m are body acceleration in the x-axis and the mean of gravity component in 3D, respectively; and the denominator represents the multiplication of the 2-norm of each vector. For the rest of the features, the readers can review [51]. Such a feature set is sensitive to body kinematics (e.g., wrist and leg motion in action). Thus, the 3D signals of each of the four operating sensors are represented by 37 features. Furthermore, combining the extracted features results in a 222-dimensional feature vector where the separation of body and gravity components of the accelerometer is performed.

Scaling and Normalization. The numerical values of the feature vector have a great variance in magnitude; e.g., SMA can reach a value that is a few hundred times that of the power of AS and the STD of acceleration JS. In order to eliminate the negative effect on the classification task, scaling is performed in terms of the segment length ($slen$). The coefficients of the AR model, TA, mean, and STD of AS, mean of JS, mean of RA, and power of RA are scaled by $\sqrt{slen}$, while the angle of the x-component of AS is scaled by $slen$, and finally the scaling factor $slen^2$ is applied for SMA. The rest of the features are used without scaling. This treatment is heuristically examined. After that, the whole feature vector is normalized in $[0, 1]$ as illustrated in Figure 3.

Classification Layer. Commonly applied classification algorithms in human-activity-recognition tasks are referred to here as RF, MLP, SVM, adn NB. RF [35] is a voting-based classifier where a decision tree is created for each sample inside a random subset of features. Then, the decision is taken for the sake of the class that is the most voted for. Thus, the most important parameters of the RF classifier are the number of decision trees and the maximum depth of the tree. MLP [36] contains interconnected processing units called neurons in one or more layers. Each neuron is characterized by its activation function, that is, a function of the weights of the preceding layer. The training algorithm, which is responsible for finding the best weights, plays a vital role in the network performance. In addition, the number of layers, number of neurons, and type of activation function are the most important parameters for the MLP. SVM [37] depends on finding the best hyperplanes that achieve the maximal margin between the nearest examples in high-dimensional spaces of two different classes. For a multiclass problem, $n * (n-1)/2$ binary SVM models are generated to distinguish n classes. NB [38] is a simple classifier that makes use of Bayes' rule to determine the class with the highest posterior probability.

3. Experimental Results and Analysis

3.1. Setup

Well-known machine learning (ML) classifiers in the IoT area, namely RF, MLP, SVM, and NB, are examined in an impartial comparison in order to clarify the most suitable one for this specific application. Since subject-dependent evaluation is usually easier than k-fold cross-validation in human-activity-recognition applications [54], the outstanding classifier according to the first mentioned criteria is examined in the later one. ML algorithms are referred to under the Scikit-learn framework in Python. Table 3 illustrates the parameters of each classifier during the experiments conducted here.

Table 3. Classifiers settings and parameter values.

Classifier	Function Call	Settings and Parameters
RF	RandomForestClassifier()	# estimators = 200, max. depth = 25, min. samples split = 2
MLP	MLPClassifier()	solver: quasi-Newton method, # hidden neurons = 75, activation function: tanh, max. # iterations = 1000, momentum = 0.9, initial learning rate is 0.01, validation ratio = 15%
SVM	svm.SVC()	kernel: radial basis function, polynomial degree is 3
NB	GaussianNB()	μ and σ parameters of Gaussian distribution are estimated using maximum likelihood

Performance of the examined ML algorithms is evaluated according to four metrics, namely the classification accuracy (Equation (3)); the F1-measure, which is the average of precision and recall of classification; (Equations (4) and (5)); execution time; and size on the disk.

$$\text{Accuracy} = \frac{TP}{TP + TN + FP + FN} \quad (3)$$

$$\text{Precision} = \frac{TP}{TP + FP} \quad (4)$$

$$\text{Recall} = \frac{TP}{TP + FN} \quad (5)$$

where TP represents the true-positive, TN is the true-negative, FP is the false-positive and FN is the false-negative classification rate. The best settings for each classifier are used in experiments after examining various training options. Experiments run on a computer machine with 10 GB RAM and 2.60 GHz i5 CPU.

3.2. Subject-Dependent Evaluation

The samples of each class are randomly separated, with 70% in the training and validation set and 30% in the testing set. The test samples are never introduced training any of examined classifiers, but samples of the same subject may appear in both training and testing sets. For impartial comparison, the simulation procedure was repeated by 10 independent runs, where each time, the same training/testing data are provided to each classifier. The average classification rates for activity recognition are presented in Figure 4.

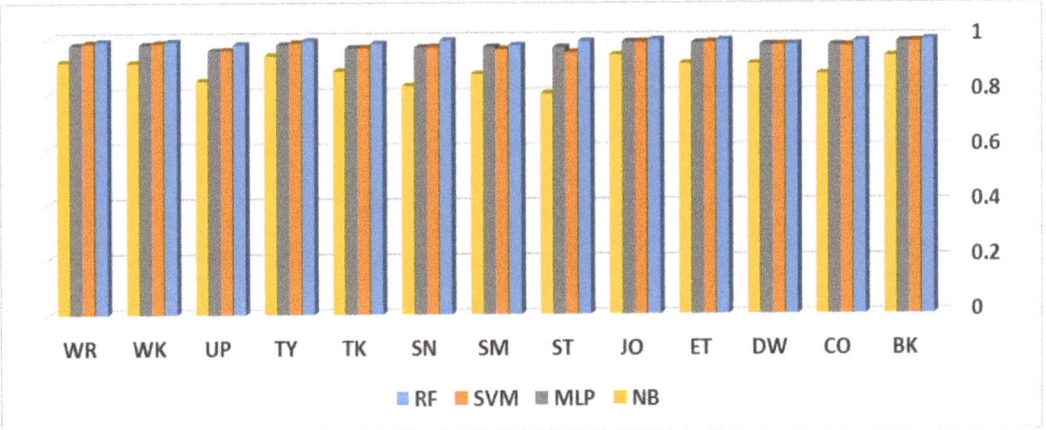

Figure 4. Average F-measure scores of the activities: Bike (BK), Coffee (CO), Downstairs (DW), Eat (ET), Jog (JO), Sit (ST), Smoke (SM), Stand (SN), Talk (TK), Type (TY), Upstairs (UP), Walk (WK) and Write (WR). Compared classifiers are evaluated under subject-dependent criteria.

Figure 4 shows the average classification rates for different activities per classifier. RF has the highest rate for each activity. Biking, eating, jogging, sitting, typing, and writing activities are successfully recognized with a rate > 99%. The activities walking downstairs, walking upstairs, and smoking are the least recognized by the RF classifier with a rate slightly less than 98%. Such behavior can be justified by reading the confusion matrix shown in Figure 5. On average, eight examples of walking downstairs were misclassified as walking upstairs, and vice versa for 11 examples of walking upstairs. Another notable conflict occurred for nine examples between smoking and giving a talk. It was noticed that conflicts occurred between very close activities, which is likely expected in such applications. However, the overall performance of the current model (employed sensors + preprocessing + features + classifier) is accepted, and it can be further improved by providing more training examples.

True Classes \ Predicted Classes	BK	CO	DW	ET	JO	ST	SM	SN	TK	TY	UP	WK	WR
BK	418	0	0	0	0	0	0	0	0	0	0	0	0
CO	0	413	0	0	0	4	0	0	0	0	0	0	1
DW	0	0	408	0	0	0	0	0	0	0	8	2	0
ET	0	0	0	417	0	0	0	0	0	1	0	0	0
JO	0	0	1	0	417	0	0	0	0	0	0	0	0
ST	0	1	0	1	0	415	0	0	0	0	0	0	1
SM	0	0	0	0	0	0	408	1	9	0	0	0	0
SN	0	0	0	0	0	0	7	410	1	0	0	0	0
TK	0	0	0	0	0	0	9	0	409	0	0	0	0
TY	0	0	0	0	0	0	0	0	0	417	0	0	1
UP	0	0	11	0	0	0	0	0	0	0	406	1	0
WK	0	0	2	0	0	0	0	0	0	0	3	413	0
WR	0	3	0	0	0	1	0	0	0	0	0	0	414

Figure 5. Confusion matrix for the RF classifier under subject-dependent evaluation criteria.

Table 4 provides a summary of comparing different ML algorithms, as well as important implementation issues. On average, the accuracy (and F-measure) of RF reaches 98.72%, which exceeds the accuracy of each of SVM, MLP, and NB by 1.3%, 1.27%, and 11.1%, respectively. MLP takes a notably long training time of 90.41 s, while NB training occurred quickly at less than one second, and RF needed about 29.3 s to announce its decisions. RF occupies about 22.68 MB of the disk, which is the largest size, while NB needs only 0.046 MB space. To improve the readability of comparative results of all classifiers, Figure 6 presents an illustrative radar plot.

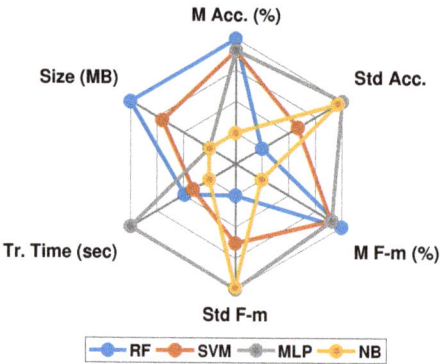

Figure 6. Radar plot for compared classifiers according to mean and standard deviation of accuracy (M Acc (%)) and (Std Acc), respectively; mean and standard deviation of F-measure (M F-m (%)) and (Std F-m), respectively; raining time in sec. (Tr. Time (sec)); and Size on disk in MB (Size (MB)).

3.3. *Stratified k-Fold Cross-Validation*

In the experimental settings of collecting this dataset, a controlled protocol was performed by each of the 10 participants. Each participating subject performed the same set of activities within the same permitted time duration. Thus, by chance, for this particular dataset, 10-fold cross-validation implicitly involved the stratified 10-fold validation followed in Shoaib et al. [42]. Moreover, the common evaluation criterion for human activity recognition models, i.e., leave-one-subject-out, can also be applied via the 10-fold cross-validation for this particular dataset. The latter measure criteria are of interest where the dataset provides subject-independent evaluation, and hence it examines the model's of generalization ability for newly introduced data. The average accuracy of the applied RF-based model here is equal to 92.54%.

Table 4. Performance of compared classifiers for subject-dependent evaluation.

		Accuracy		F-Measure	Training Time (sec)	Size on Disk (MB)
RF	Mean Std	98.72 0.1015	Mean Std	98.72 0.1015	29.3	22.683
SVM	Mean Std	97.43 0.2279	Mean Std	97.42 0.2398	19.69	13.593
MLP	Mean Std	97.47 0.3837	Mean Std	97.49 0.3736	90.41	0.143
NB	Mean Std	88.82 0.3693	Mean Std	88.87 0.3677	1	0.046

4. Sensitivity Analysis for Model Parameters

The performance of the RF algorithm is tremendously sensitive to both the number of decision trees (known as RF size) and the longest path from a tree head to the leaves (known as RF depth). For RF depth ≥ 15, with a suitable RF size ≥ 50, the applied RF-based model can provide notable recognition performance under subject-dependent evaluation measure; see Figure 7a. Moreover, increasing the RF size up to 400 trees has a slight improvement in the model accuracy. Conversely, under 10-fold cross-validation evaluation, the model accuracy grows by 1% when increasing both RF size and RF depth from (50, 10) to (15, 200); see Figure 7b. Moreover, increasing the RF size to 400, for example, will not enhance the model accuracy as much as the notable increment in processing time in this case. From Figure 7, we can conclude that with an RF depth between 15 and 25 and an RF size equal to 200, an efficient recognition model can be implemented for these kinds of IoHT systems that make use of sensory data from smartphones.

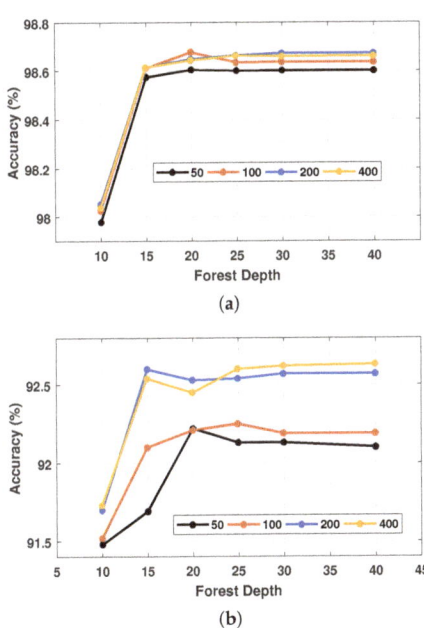

Figure 7. Model performance for different values of both forest size (50, 100, 200, 400) and forest depth (15, 20, 25, 30, 40) for (**a**) subject-dependent and (**b**) 10-fold cross validation criteria.

5. Comparison with Previous Studies

Different studies in the literature have addressed this dataset according to different evaluation measures. Table 5 provides the previous best recognition rates according to subject-dependent evaluation. Baldominos et al. [45] have tested shallow techniques against the deep CNN model. Only raw signals are used in 60 s segments. The ensemble of randomized decision trees (ET), with a set of handcrafted features, provides an average overall accuracy of 95.3%, while the accuracy of the CNN-based approach decreases to 85%. Stacked autoencoders provided better results than deep belief networks, where the accuracy reached 97.13% according to Alo et al. [46]. In a later study, besides raw activity signals, the magnitude vector and the vectors of pitch and roll angles were provided to deep networks in segments with a length of 2 s.

Table 5. Recognition rates of each activity for different models under subject-dependent validation criteria. ET: ensemble of randomized trees, FC: fully connected layer, AE: autoencoders, and DBN: deep belief networks.

Reference	Input Signals	Segment Length (s)	Feature Extraction	Classifier	Accuracy (%)
Baldominos et al. [45]	Raw signals	60	Handcrafted	ET	95.3
Baldominos et al. [45]	Raw signals	60	CNN hidden layers	FC layer	85
Alo et al. [46]	Raw signals, magnitude vector, pitch and roll vectors	2	Sparse AE layers	FC layer	97.13
Alo et al. [46]	Raw signals, magnitude vector, pitch and roll vectors	2	DBN hidden layers	DBN output layer	91.57
Current model	Raw signals	2.56	Handcrafted	RF	98.7

The proposed DL model was able to outperform the conventional classifiers such as support vector machines (SVM), Naive Bayes (NB), and linear discriminant analysis (LDA); however, the RF classifier was not included in this comparison. The current RF-based model presents the best recognition results among related studies. However, samples of the same person may appear in both the training and testing sets, but the experimental findings are still useful for seeking good models since registered data points occurred at different timestamps.

Moreover, the current model improves the recognition rates obtained by Shoaib et al. [42]. Table 6 shows the rates of each activity when stratified 10-fold cross-validation criteria are applied. Numerical values of Shoaib et al.'s model were computed from the confusion matrix in Figure 2c in [42]. The applied classifier was NB, but features were extracted from segments with a length of 5 s, and only accelerometer and gyroscope signals were used. Because of the suitable feature set used within the current model, the activities that directly depend on hand movement are well-recognized. The improvements in the rates of activities are as follows: having coffee (0.83 to 0.92), eating (0.89 to 0.99), smoking (0.82 to 0.95), giving a talk (0.86 to 0.97), typing (0.95 to 0.98), writing (0.89 to 0.97). For the other activities, the current model performs worse than or equal to Shoaib et al.'s model. In conclusion, the average overall accuracy is improved by 1.4%.

Table 6. Recognition rates of each activity for different models under 10-fold cross validation criteria.

Ref.	BK	CO	DW	ET	JO	ST	SM	SN	TK	TY	UP	WK	WR	Accuracy (%)
Shoaib et al. [42]	0.99	0.83	0.98	0.89	1	0.90	0.82	0.92	0.86	0.95	0.96	0.85	0.89	91.2
Current model	0.99	0.92	0.91	0.99	0.99	0.76	0.95	0.94	0.97	0.98	0.83	0.80	0.97	92.54

6. Applied Model Performance for WISDM Dataset

In this section, the validation of the applied framework is extended to the WISDM dataset [43]. It is one of the most addressed datasets in the HAR literature. WISDM v1 contains a total of 1,098,207 examples of activities that have been collected by 29 subjects. Six activities, namely walking (37.2%), jogging (29.2%), upstairs (12.0%), downstairs (10.2%), sitting (6.4%), and standing (5%), were registered via a smartphone in the front pants pocket (see Figure 1) of each subject. Walking and jogging activities were the most represented in this dataset. Activity signals were registered using the embedded accelerometer of the smartphone at a 20 Hz sampling rate. In the experimental settings, a window size of 10 s (according to the original study [43]) with 50% overlapping was applied to raw signals. The proposed feature set was generated for each activity segment, where the feature vector was 74 dimensions; since only the accelerometer signals are available, the RF classifier is called. Using the best settings, e.g., RF size and depth (200, 25), gave acceptable classification rates for this dataset. For 10-fold cross-validation criteria, the applied model gave an average accuracy of 94%, while for the subject-dependent evaluation (i.e., 70% training and 30% testing), the average accuracy reached 98.56%. This model performance regarding this dataset is comparable to many recent related studies in the literature, as summarized in Table 7.

Among the compared studies that appear in Table 7, using a window of 5 s for segments in [55] is more challenging than using longer segments, but a deep learning model was able to achieve 94.2% accuracy under 10-fold cross-validation. Moreover, an accuracy value of 98.85% was obtained in [56], but applying 95% overlapping when doing segmentation, and this is questionable in such a HAR study (i.e., overlapping usually ranges from 0 to 50%). In addition, for a 70%/30% split, using a more efficient RF such as (50, 20) gives an average accuracy of 98.34%, which is still close to the best performance obtained. However, under 10-fold cross-validation, using an RF with (50, 20) does not degrade the accuracy by more than 0.02%.

Summing up, the applied framework shows good performance for the WISDM v1 dataset under different evaluation criteria, while usually, only one of them is used in previous related studies. This model behavior reflects the robustness and suitability of both the feature set and the classifier algorithm for real-time HAR applications.

Table 7. Applied model results for WISDM dataset. MLP: multi-layer perceptron. LR: logistic regression. Stat. Feat.: statistical features. Att. M.: attention mechanism. R. B.: residual block. LSTM: Long short-term memory.

Evaluation	Reference	Segment Length (s)	Feature Extraction	Classifier	Accuracy (%)
10-fold cross validation	Kwapisz et al. [43]	10	Handcrafted	MLP	91.7
	Garcia-Ceja et al. [55]	5	CNN	FC layer	94.2
	Catal et al. [57]	10	Handcrafted	Ensemble of (LR, MLP, j48)	91.62
	Ignatov [58]	10	CNN + Stat. Feat.	FC layer	93.32
	Current model	10	Handcrafted	RF	94
70%/30% split	Gao et al. [56]	10	CNN + Att. M.	FC layer	98.85
	Suwannarat et al. [59]	8	CNN	FC layer	95
	Abdel-Basset et al. [60]	10	CNN + R. B. + LSTM + Att. M.	MLP	98.90
	Zhang et al. [61]	11.2	CNN	FC layers	96.4
	Zhang et al. [62]	10	CNN + Att.	FC layer	96.4
	Current model	10	Handcrafted	RF	98.56

7. Discussion

The applied framework introduces one example of an IoHT system that is examined using two datasets with different settings. Shoaib's dataset contains thirteen activities gathered by 10 subjects at a sampling rate of 50 Hz, while WISDM v1 has six activities collected by 29 subjects at a sampling rate of 20 Hz. Such a variety of activity signal resources constitutes a strong test for any proposed HAR model. Applying the different common evaluation criteria of HAR models in the same study is highly recommended to ensure its superiority. Later observation is missing in most studies in the literature.

More evidence is needed for the use of the dense production of deep learning models in the HAR field. Such models have thousands of parameters learned during training (tremendous computational load). However, they should at least outperform the conventional shallow approaches. Classical handcrafted features are meaningful and interoperable to a great extent, while the interpretation of most deep models, in particular in the HAR field, is still in its infancy.

In [46], the applied DL model required the help of extra inputs such as magnitude and pitch and roll signals, together with the raw 3D acceleration signals, in order to improve the performance. One the other hand, features extracted implicitly from DL models may need refinement via feature selection approaches in order to eliminate illusive features of classifiers. Recent studies such as [63] and others have emphasized the role of applying feature selection with DL models. On the other hand, the RF algorithm performs feature selection as one of the steps performed to achieve its classification result. One important observation is the degradation of accuracy when moving from the subject-dependent to 10-fold cross-validation criteria. For the WISDM v1 dataset, the misclassification is relatively high between upstairs and downstairs in comparison to other activities, in addition to the difficulty when applying 10-fold cross-validation (i.e., different subjects are used for training and testing). The later result has also been reported by different previous models such as [43,55,60], which cn probably be attributed to the sensor position on subjects' bodies. A similar notation also holds for Shoaib's dataset, where in Figure 5, the confusion matrix shows that the majority of false predictions take place between the activities of walking upstairs and walking downstairs .

8. Conclusions and Future Trends

In this work, an efficient model for an IoHT system is introduced through a set of carefully handcrafted features and a shallow classifier such as Random Forest for the dataset of Shoaib et al. [42]. Participants used to collect this dataset followed a specific protocol, which may be called a controlled environment. Similarly to related studies, using accelerometers and gyroscope sensors in smartphones is convenient for such applications. Moreover, inducing features (e.g., statistics of the roll angle vector and the angle of the x–component of body acceleration with a gravity vector) that depend on body kinematics (e.g., wrist and leg motion) improve the model performance. The presented model provides state-of-the-art results under both subject-dependent and 10-fold cross-validation criteria. Moreover, the current model performance was verified by another dataset, namely WISDM v1 [43] under both aforementioned evaluation criteria.

Author Contributions: All authors contributed equally to this paper. All authors have read and agreed to the published version of the manuscript.

Funding: This work was supported by National Natural Science Foundation of China (Grant No. 62150410434).

Institutional Review Board Statement: Not applicable.

Informed Consent Statement: Not applicable.

Data Availability Statement: The data are publicly available as described in the main text.

Conflicts of Interest: The authors declare no conflict of interest.

References

1. Baker, S.B.; Xiang, W.; Atkinson, I. Internet of Things for Smart Healthcare: Technologies, Challenges, and Opportunities. *IEEE Access* **2017**, *5*, 26521–26544. [CrossRef]
2. Qadri, Y.A.; Nauman, A.; Zikria, Y.B.; Vasilakos, A.V.; Kim, S.W. The Future of Healthcare Internet of Things: A Survey of Emerging Technologies. *IEEE Commun. Surv. Tutor.* **2020**, *22*, 1121–1167. [CrossRef]
3. Kamruzzaman, M.M.; Alrashdi, I.; Alqazzaz, A. New Opportunities, Challenges, and Applications of Edge-AI for Connected Healthcare in Internet of Medical Things for Smart Cities. *J. Healthc. Eng.* **2022**, *2022*, 2950699. [CrossRef] [PubMed]
4. Dimitrov, D.V. Medical internet of things and big data in healthcare. *Healthc. Inform. Res.* **2016**, *22*, 156–163. [CrossRef] [PubMed]
5. Wang, Y.; Cang, S.; Yu, H. A survey on wearable sensor modality centred human activity recognition in health care. *Expert Syst. Appl.* **2019**, *137*, 167–190. [CrossRef]
6. Pillai, S.G.; Haldorai, K.; Seo, W.S.; Kim, W.G. COVID-19 and hospitality 5.0: Redefining hospitality operations. *Int. J. Hosp. Manag.* **2021**, *94*, 102869. [CrossRef] [PubMed]
7. Demir, K.A.; Döven, G.; Sezen, B. Industry 5.0 and human–robot co-working. *Procedia Comput. Sci.* **2019**, *158*, 688–695. [CrossRef]
8. Vanagas, G.; Engelbrecht, R.; Damaševičius, R.; Suomi, R.; Solanas, A. eHealth Solutions for the Integrated Healthcare. *J. Healthc. Eng.* **2018**, *2018*, 3846892. [CrossRef] [PubMed]
9. Hernández-Chan, G.S.; Ceh-Varela, E.E.; Sanchez-Cervantes, J.L.; Villanueva-Escalante, M.; Rodríguez-González, A.; Pérez-Gallardo, Y. Collective intelligence in medical diagnosis systems: A case study. *Comput. Biol. Med.* **2016**, *74*, 45–53. [CrossRef]
10. Rodríguez-González, A.; Torres-Niño, J.; Mayer, M.A.; Alor-Hernandez, G.; Wilkinson, M.D. Analysis of a multilevel diagnosis decision support system and its implications: A case study. *Comput. Math. Methods Med.* **2012**, *2012*, e0148991. [CrossRef]
11. Barsocchi, P.; Calabrò, A.; Crivello, A.; Daoudagh, S.; Furfari, F.; Girolami, M.; Marchetti, E. COVID-19 & privacy: Enhancing of indoor localization architectures towards effective social distancing. *Array* **2021**, *9*, 100051.
12. Kiran, S.; Khan, M.A.; Javed, M.Y.; Alhaisoni, M.; Tariq, U.; Nam, Y.; Damaševičius, R.; Sharif, M. Multi-Layered Deep Learning Features Fusion for Human Action Recognition. *Comput. Mater. Contin.* **2021**, *69*, 4061–4075. [CrossRef]
13. Şengül, G.; Ozcelik, E.; Misra, S.; Damaševičius, R.; Maskeliūnas, R. Fusion of smartphone sensor data for classification of daily user activities. *Multimed. Tools Appl.* **2021**, *80*, 33527–33546. [CrossRef]
14. Şengül, G.; Karakaya, M.; Misra, S.; Abayomi-Alli, O.O.; Damaševičius, R. Deep learning based fall detection using smartwatches for healthcare applications. *Biomed. Signal Process. Control.* **2022**, *71*, 103242. [CrossRef]
15. Khan, M.F.; Ghazal, T.M.; Said, R.A.; Fatima, A.; Abbas, S.; Khan, M.; Issa, G.F.; Ahmad, M.; Khan, M.A. An IoMT-Enabled Smart Healthcare Model to Monitor Elderly People Using Machine Learning Technique. *Comput. Intell. Neurosci.* **2021**, *2021*, 2487759. [CrossRef]
16. Farsi, M. Application of ensemble RNN deep neural network to the fall detection through IoT environment. *Alex. Eng. J.* **2021**, *60*, 199–211. [CrossRef]
17. Moualla, S.; Khorzom, K.; Jafar, A. Improving the performance of machine learning-based network intrusion detection systems on the UNSW-NB15 dataset. *Comput. Intell. Neurosci.* **2021**, *2021*, 5557577. [CrossRef]
18. Maskeliunas, R.; Damaševicius, R.; Segal, S. A review of internet of things technologies for ambient assisted living environments. *Future Internet* **2019**, *11*, 259. [CrossRef]
19. Agac, S.; Shoaib, M.; Incel, O.D. Context-aware and dynamically adaptable activity recognition with smart watches: A case study on smoking. *Comput. Electr. Eng.* **2021**, *90*, 106949. [CrossRef]
20. Patalas-maliszewska, J.; Halikowski, D.; Damaševičius, R. An automated recognition of work activity in industrial manufacturing using convolutional neural networks. *Electronics* **2021**, *10*, 2946. [CrossRef]
21. Bharti, R.; Khamparia, A.; Shabaz, M.; Dhiman, G.; Pande, S.; Singh, P. Prediction of heart disease using a combination of machine learning and deep learning. *Comput. Intell. Neurosci.* **2021**, *2021*, 8387680. [CrossRef]
22. Girčys, R.; Kazanavičius, E.; Maskeliūnas, R.; Damaševičius, R.; Woźniak, M. Wearable system for real-time monitoring of hemodynamic parameters: Implementation and evaluation. *Biomed. Signal Process. Control.* **2020**, *59*, 101873. [CrossRef]
23. Olmedo-Aguirre, J.O.; Reyes-Campos, J.; Alor-Hernández, G.; Machorro-Cano, I.; Rodríguez-Mazahua, L.; Sánchez-Cervantes, J.L. Remote Healthcare for Elderly People Using Wearables: A Review. *Biosensors* **2022**, *12*, 73. [CrossRef]
24. Cruz-ramos, N.A.; Alor-hernández, G.; Colombo-mendoza, L.O.; Sánchez-cervantes, J.L.; Rodríguez-mazahua, L.; Guarneros-nolasco, L.R. mHealth Apps for Self-Management of Cardiovascular Diseases: A Scoping Review. *Healthcare* **2022**, *10*, 322. [CrossRef]
25. Yang, Z. An Efficient Automatic Gait Anomaly Detection Method Based on Semisupervised Clustering. *Comput. Intell. Neurosci.* **2021**, *2021*, 8840156. [CrossRef]
26. Priya, S.J.; Rani, A.J.; Subathra, M.S.P.; Mohammed, M.A.; Damaševičius, R.; Ubendran, N. Local pattern transformation based feature extraction for recognition of parkinson's disease based on gait signals. *Diagnostics* **2021**, *11*, 1395. [CrossRef]
27. Bokhari, S.M.; Sohaib, S.; Khan, A.R.; Shafi, M. DGRU based human activity recognition using channel state information. *Measurement* **2021**, *167*, 108245. [CrossRef]
28. Cui, W.; Li, B.; Zhang, L.; Chen, Z. Device-free single-user activity recognition using diversified deep ensemble learning. *Appl. Soft Comput.* **2021**, *102*, 107066. [CrossRef]
29. Al-qaness, M.A. Device-free human micro-activity recognition method using WiFi signals. *Geo-Spat. Inf. Sci.* **2019**, *22*, 128–137. [CrossRef]

30. Al-Qaness, M.A.; Abd Elaziz, M.; Kim, S.; Ewees, A.A.; Abbasi, A.A.; Alhaj, Y.A.; Hawbani, A. Channel state information from pure communication to sense and track human motion: A survey. *Sensors* **2019**, *19*, 3329. [CrossRef]
31. Ryselis, K.; Petkus, T.; Blažauskas, T.; Maskeliūnas, R.; Damaševičius, R. Multiple Kinect based system to monitor and analyze key performance indicators of physical training. *Hum.-Centric Comput. Inf. Sci.* **2020**, *10*, 51. [CrossRef]
32. Al-Qaness, M.A.; Dahou, A.; Abd Elaziz, M.; Helmi, A. Multi-ResAtt: Multilevel Residual Network with Attention for Human Activity Recognition Using Wearable Sensors. *IEEE Trans. Ind. Inform.* **2022**. [CrossRef]
33. Siam, A.I.; Almaiah, M.A.; Al-Zahrani, A.; Elazm, A.A.; El Banby, G.M.; El-Shafai, W.; El-Samie, F.E.A.; El-Bahnasawy, N.A. Secure Health Monitoring Communication Systems Based on IoT and Cloud Computing for Medical Emergency Applications. *Comput. Intell. Neurosci.* **2021**, *2021*, 8016525. [CrossRef] [PubMed]
34. Sousa Lima, W.; Souto, E.; El-Khatib, K.; Jalali, R.; Gama, J. Human Activity Recognition Using Inertial Sensors in a Smartphone: An Overview. *Sensors* **2019**, *19*, 3213. [CrossRef]
35. Svetnik, V.; Liaw, A.; Tong, C.; Culberson, J.C.; Sheridan, R.P.; Feuston, B.P. Random forest: A classification and regression tool for compound classification and QSAR modeling. *J. Chem. Inf. Comput. Sci.* **2003**, *43*, 1947–1958. [CrossRef]
36. Priddy, K.L.; Keller, P.E. *Artificial Neural Networks: An Introduction*; SPIE Press: Bellingham, WA, USA, 2005; Volume 68.
37. Wang, L. *Support Vector Machines: Theory and Applications*; Springer Science & Business Media: Berlin/Heidelberg, Germany, 2005; Volume 177.
38. Murphy, K.P. Naive bayes classifiers. *Univ. Br. Columbia* **2006**, *18*, 1–8.
39. Khan, R.; Kumar, S.; Srivastava, A.K.; Dhingra, N.; Gupta, M.; Bhati, N.; Kumari, P. Machine Learning and IoT-Based Waste Management Model. *Comput. Intell. Neurosci.* **2021**, *2021*, 5942574. [CrossRef]
40. Alsamhi, S.H.; Almalki, F.A.; Al-Dois, H.; Ben Othman, S.; Hassan, J.; Hawbani, A.; Sahal, R.; Lee, B.; Saleh, H. Machine learning for smart environments in B5G networks: Connectivity and QoS. *Comput. Intell. Neurosci.* **2021**, *2021*, 6805151. [CrossRef]
41. Atalaa, B.A.; Ziedan, I.; Alenany, A.; Helmi, A. Feature Engineering for Human Activity Recognition. *Int. J. Adv. Comput. Sci. Appl.* **2021**, *12*, 160–167 [CrossRef]
42. Shoaib, M.; Bosch, S.; Incel, O.D.; Scholten, H.; Havinga, P.J. Complex human activity recognition using smartphone and wrist-worn motion sensors. *Sensors* **2016**, *16*, 426. [CrossRef]
43. Kwapisz, J.R.; Weiss, G.M.; Moore, S.A. Activity recognition using cell phone accelerometers. *ACM SigKDD Explor. Newsl.* **2011**, *12*, 74–82. [CrossRef]
44. Demrozi, F.; Pravadelli, G.; Bihorac, A.; Rashidi, P. Human activity recognition using inertial, physiological and environmental sensors: A comprehensive survey. *IEEE Access* **2020**, *8*, 210816–210836. [CrossRef]
45. Baldominos, A.; Cervantes, A.; Saez, Y.; Isasi, P. A comparison of machine learning and deep learning techniques for activity recognition using mobile devices. *Sensors* **2019**, *19*, 521. [CrossRef]
46. Alo, U.R.; Nweke, H.F.; Teh, Y.W.; Murtaza, G. Smartphone Motion Sensor-Based Complex Human Activity Identification Using Deep Stacked Autoencoder Algorithm for Enhanced Smart Healthcare System. *Sensors* **2020**, *20*, 6300. [CrossRef]
47. Xia, K.; Huang, J.; Wang, H. LSTM-CNN architecture for human activity recognition. *IEEE Access* **2020**, *8*, 56855–56866. [CrossRef]
48. Ordóñez, F.J.; Roggen, D. Deep convolutional and lstm recurrent neural networks for multimodal wearable activity recognition. *Sensors* **2016**, *16*, 115. [CrossRef]
49. Ronald, M.; Poulose, A.; Han, D.S. iSPLInception: An inception-ResNet deep learning architecture for human activity recognition. *IEEE Access* **2021**, *9*, 68985–69001. [CrossRef]
50. Mekruksavanich, S.; Jitpattanakul, A. Deep convolutional neural network with rnns for complex activity recognition using wrist-worn wearable sensor data. *Electronics* **2021**, *10*, 1685. [CrossRef]
51. Anguita, D.; Ghio, A.; Oneto, L.; Parra, X.; Reyes-Ortiz, J.L. A public domain dataset for human activity recognition using smartphones. *Esann* **2013**, *3*, 437–442.
52. Hämäläinen, W.; Järvinen, M.; Martiskainen, P.; Mononen, J. Jerk-based feature extraction for robust activity recognition from acceleration data. In Proceedings of the 2011 11th International Conference on Intelligent Systems Design and Applications, Córdoba, Spain, 22–24 November 2011; pp. 831–836.
53. Quiroz, J.C.; Banerjee, A.; Dascalu, S.M.; Lau, S.L. Feature selection for activity recognition from smartphone accelerometer data. *Intell. Autom. Soft Comput.* **2017**, 1–9. [CrossRef]
54. Dehghani, A.; Sarbishei, O.; Glatard, T.; Shihab, E. A quantitative comparison of overlapping and non-overlapping sliding windows for human activity recognition using inertial sensors. *Sensors* **2019**, *19*, 5026. [CrossRef] [PubMed]
55. Garcia-Ceja, E.; Uddin, M.Z.; Torresen, J. Classification of recurrence plots' distance matrices with a convolutional neural network for activity recognition. *Procedia Comput. Sci.* **2018**, *130*, 157–163. [CrossRef]
56. Gao, W.; Zhang, L.; Teng, Q.; He, J.; Wu, H. DanHAR: Dual attention network for multimodal human activity recognition using wearable sensors. *Appl. Soft Comput.* **2021**, *111*, 107728. [CrossRef]
57. Catal, C.; Tufekci, S.; Pirmit, E.; Kocabag, G. On the use of ensemble of classifiers for accelerometer-based activity recognition. *Appl. Soft Comput.* **2015**, *37*, 1018–1022. [CrossRef]
58. Ignatov, A. Real-time human activity recognition from accelerometer data using Convolutional Neural Networks. *Appl. Soft Comput.* **2018**, *62*, 915–922. [CrossRef]
59. Suwannarat, K.; Kurdthongmee, W. Optimization of Deep Neural Network-based Human Activity Recognition for a Wearable Device. *Heliyon* **2021**, *7*, e07797. [CrossRef]

60. Abdel-Basset, M.; Hawash, H.; Chakrabortty, R.K.; Ryan, M.; Elhoseny, M.; Song, H. ST-DeepHAR: Deep learning model for human activity recognition in IoHT applications. *IEEE Internet Things J.* **2020**, *8*, 4969–4979. [CrossRef]
61. Zhang, Y.; Zhang, Z.; Zhang, Y.; Bao, J.; Zhang, Y.; Deng, H. Human activity recognition based on motion sensor using u-net. *IEEE Access* **2019**, *7*, 75213–75226. [CrossRef]
62. Zhang, H.; Xiao, Z.; Wang, J.; Li, F.; Szczerbicki, E. A novel IoT-perceptive human activity recognition (HAR) approach using multihead convolutional attention. *IEEE Internet Things J.* **2019**, *7*, 1072–1080. [CrossRef]
63. Fan, C.; Gao, F. Enhanced human activity recognition using wearable sensors via a hybrid feature selection method. *Sensors* **2021**, *21*, 6434. [CrossRef]

Article

Geriatric Care Management System Powered by the IoT and Computer Vision Techniques

Agne Paulauskaite-Taraseviciene [1], Julius Siaulys [1], Kristina Sutiene [2,*], Titas Petravicius [1], Skirmantas Navickas [1], Marius Oliandra [1], Andrius Rapalis [3,4] and Justinas Balciunas [5]

[1] Faculty of Informatics, Kaunas University of Technology, Studentu 50, 51368 Kaunas, Lithuania
[2] Department of Mathematical Modeling, Kaunas University of Technology, Studentu 50, 51368 Kaunas, Lithuania
[3] Biomedical Engineering Institute, Kaunas University of Technology, K. Barsausko 59, 51423 Kaunas, Lithuania
[4] Faculty of Electrical and Electronics Engineering, Kaunas University of Technology, Studentu 48, 51367 Kaunas, Lithuania
[5] Faculty of Medicine, Vilnius University, Universiteto 3, 01513 Vilnius, Lithuania
* Correspondence: kristina.sutiene@ktu.lt

Citation: Paulauskaite-Taraseviciene, A.; Siaulys, J.; Sutiene, K.; Petravicius, T.; Navickas, S.; Oliandra, M.; Rapalis, A.; Balciunas, J. Geriatric Care Management System Powered by the IoT and Computer Vision Techniques. *Healthcare* **2023**, *11*, 1152. https://doi.org/10.3390/healthcare11081152

Academic Editors: Giner Alor-Hernández, Jezreel Mejía-Miranda, José Luis Sánchez-Cervantes and Alejandro Rodríguez-González

Received: 22 February 2023
Revised: 3 April 2023
Accepted: 13 April 2023
Published: 17 April 2023

Copyright: © 2023 by the authors. Licensee MDPI, Basel, Switzerland. This article is an open access article distributed under the terms and conditions of the Creative Commons Attribution (CC BY) license (https://creativecommons.org/licenses/by/4.0/).

Abstract: The digitalisation of geriatric care refers to the use of emerging technologies to manage and provide person-centered care to the elderly by collecting patients' data electronically and using them to streamline the care process, which improves the overall quality, accuracy, and efficiency of healthcare. In many countries, healthcare providers still rely on the manual measurement of bioparameters, inconsistent monitoring, and paper-based care plans to manage and deliver care to elderly patients. This can lead to a number of problems, including incomplete and inaccurate record-keeping, errors, and delays in identifying and resolving health problems. The purpose of this study is to develop a geriatric care management system that combines signals from various wearable sensors, noncontact measurement devices, and image recognition techniques to monitor and detect changes in the health status of a person. The system relies on deep learning algorithms and the Internet of Things (IoT) to identify the patient and their six most pertinent poses. In addition, the algorithm has been developed to monitor changes in the patient's position over a longer period of time, which could be important for detecting health problems in a timely manner and taking appropriate measures. Finally, based on expert knowledge and a priori rules integrated in a decision tree-based model, the automated final decision on the status of nursing care plan is generated to support nursing staff.

Keywords: geriatric care; IoT; vital parameters; posture recognition; image recognition; deep learning; non-contact monitoring

1. Introduction

Geriatric care is a field of healthcare that focuses on the physical, mental, and social needs of older adults. As people age, they may experience physical, cognitive, and social changes that require special care and support. Geriatric care is based on the specific needs of older adults and aims to improve their health and well-being as well as manage age-related diseases and conditions so that they can maintain their independence, quality of life, and overall comfort. Such care often involves a multidisciplinary approach with care provided by a team of health care professionals, including physicians, nurses, therapists, and social workers, who are trained in gerontology and geriatrics [1,2]. The estimated number of dependent people in need of some form of long-term care in Europe is 30.8 million, and this is expected to increase to 38 million by 2050. Furthermore, the expected shortage of nurses will reach 2.3 million in 2030. By 2080, the population aged over 80 years and older in Europe will have multiplied by 2.5. It should be noted that the majority of dependent patients suffer from Alzheimer's and chronic diseases, such as past myocardial infarction, congestive heart failure, cardiac arrhythmia, renal failure, and chronic pulmonary disease, have an increased risk of mortality in nursing homes [3].

Currently, the main problems are caused by the absence of tools to design automated care plans. The problems identified are related to the lack of digital evidence-based protocols for different situations and the nonadherence to existing protocols by nursing staff. Typically, an individualised nursing care plan is developed for the elderly patient upon admission to meet their needs. This plan is developed based on a thorough assessment of the person's medical history and evidence-based care practices. As elderly individuals reside in nursing homes, it is common for their health to decline, which makes it crucial to monitor their health status while they are there. Thus, caregivers must regularly check important biometric data, such as blood pressure, heart rate, body temperature, and respiratory rate. Collecting and documenting patient vital signs data manually is a relatively slow and therefore inefficient process. Depending on the types of vital signs, it usually takes up to five minutes to assess three to six vital signs [4]. Moreover, this information is usually documented in paper form separately from the nursing care plans, and therefore, the whole process takes up to 13 min per patient [5]. Furthermore, care plans have to be regularly re-evaluated by comparing current and historical health records to look for abnormalities and changes that could have clinical significance. However, biometric data are documented separately from nursing care plans and records of doctors. With such fragmented data sources, the process is human-dependent, highly inefficient, and cumbersome and can take up to 37 min per patient [5,6]. Moreover, in the absence of a systematic approach in geriatric care management, it becomes challenging to quickly capture monitoring data and act on them. This can cause caregivers to miss any unusual changes in the biometric data, leading to delays in administering treatment.

During the course of our research, several hospices from Latvia, Estonia, and Poland (e.g., Orpea) were contacted, and it was concluded that geriatric care management systems with a digital care plan and remote monitoring solutions are currently not available in these markets. Facilities rely on outdated software that was developed for inpatient hospital services without taking into account the nursing care plan. In particular, in Scandinavian and UK markets (e.g., Appva), some tools have been developed that include a simple digitised nursing care plan without remote monitoring or decision support capabilities; however, none of these companies have shown an interest in providing the service in the Baltic States. Therefore, in many countries, including the Baltic States, nurses use paper-based care plan templates and manually prepare time-consuming documents. Consequently, data loss and missing information in care plans are common problems. Based on the problems identified during oral interviews and discussions with various stakeholders in Lithuania, the following needs for long-term care at home and in specialised institutions have been narrowed down as the most recurrent and yet relatively possible to complete with limited funding: (1) easily create nursing care plans for new patients with action protocols for nursing staff; (2) ensure adherence to and traceability of the execution of the protocols; (3) automate patient monitoring; (4) reduce manual paper documentation; (5) easily adapt nursing care plans according to changes in the health of the patient; and (6) enable a transition from reactive care to proactive care.

Digitalised care systems could be a solution to meet the multidimensional need to monitor whether elderly patients in geriatric care facilities are receiving optimal care, thus monitoring patients more efficiently and providing personalised care. Digitalisation also helps a relatively small number of healthcare workers to reduce the need for repetitive manual work and use the collected data for proactive decision making. Furthermore, the combination of Internet of Things (IoT) and artificial intelligence (AI) technologies can aid in the analysis of data and ensure continuous monitoring of elderly patients to positively impact their care and outcomes [7–10]. By collecting data on patient activity and health, advanced AI algorithms can analyse patterns and detect deviations from normal behaviour, allowing caregivers to respond in a timely manner.

In this study, we propose an intelligent geriatric care management system based on AI and IoT to track and detect changes in the health status of elderly patients, thus ensuring efficient digitalisation of personalised care plans. The proposed solution can be used to tackle two of the most urgent problems in the area: nursing staff shortages and the costly and inefficient long-term care process. Although home care for dependent and elderly people is becoming more and more popular, it is still not a viable option for everyone due to the expensive infrastructure required and the difficulties in gaining access to their homes in an emergency. Even if people choose to live in a nursing home, it is still difficult to monitor, care, and treat elderly residents on a regular basis. With the growing demand for healthcare nurses, fragmented remote health monitoring tools, and lack of existing solutions for real-time modifications of nursing care plans, it is crucial to have a cost-effective and semi-autonomous solution available in the market.

2. Related Works

In recent years, there has been a growing interest in the development of digital health solutions to support older people and promote healthy ageing [11,12]. However, elderly individuals are more likely to develop diseases such as dementia, diabetes, and cataracts, suffer from physical and cognitive impairments, and have low levels of physical activity, all of which lead to a continuous decline in their health. This makes it difficult for staff to keep track of elderly people, to monitor changes in their health, to record and store all readings systematically, and to always react quickly and appropriately to the changes and adjust the care plan. Furthermore, as life expectancy continues to increase, the need for nurses working in geriatrics is also increasing. As such, remote monitoring and wearable devices can be used to measure vital signals, evaluate physical activity, and inform caregivers or physicians about changes in their health, which aids in the early detection of health risks [13,14].

2.1. The Use of Wearable Devices

Recently, wearable technology has benefited from technological progress, as the size of devices has significantly reduced, while the efficiency of energy consumption has improved simultaneously [15]. In particular, wearable technology can be used for a variety of purposes, ranging from keeping track of physical activity to monitoring clinically important health and safety data. Wearable devices provide real-time monitoring of the wearer's walking speed, respiratory rate, measuring sleep, energy expenditure, blood oxygen and pressure, and other related parameters [16]. Such devices can also be useful tools for people living with heart failure to facilitate exercise and recovery [17,18]. Comparatively, a study demonstrated the strong potential for improvement in healthcare through the use of wearable activity monitors in oncology trials [19]. The use of wearable technology to identify gait characteristics is another intriguing example [20], where lower limb joint angles and stride length were measured simultaneously with a prototype wearable sensor system. The study [21] investigated how a wearable device could help physicians to optimize antiepileptic treatment and prevent patients from sudden unexpected death due to epilepsy. For particular groups of individuals that suffer from chronic disease such as diabetes mellitus, cardiac disease, or chronic obstructive pulmonary disease, wearables may be used to monitor changes in health symptoms during treatment and may contribute to the personalisation of healthcare [22–24]. The use of wearables within a group of elderly population brings additional challenges. For example, it is very important to detect falls, which has already become a topic of particular importance in this field. For example, in [25], a framework was proposed for edge computing to detect individual's falls using real-time monitoring by cost-effective wearable sensors. For this purpose, an IoT-based system that makes the use of big data, cloud computing, wireless sensor networks, and smart devices was developed and integrated with an LSTM model, showing very promising results for the detection of falls by elderly people in indoor circumstances. The validity and reliability of wearables have been addressed by many studies focusing on different classes of devices used to measure activity or biometric data [26–29]. Apparently, there is no consensus among researchers, as findings depend on the manufacturer, device type, and the purpose for which it was used. This is also true because devices are constantly being upgraded to new models, which suggests that their validity and reliability will improve with time.

2.2. Contactless Measurement of Vital Signs

There are still some concerns regarding the reliability and accuracy of wearables to detect physical activity and evaluate health-related outcomes within elderly individuals, as they are generally designed primarily to collect biometric information during activities of daily living in the general population [30–33]. First, the ability of older people to recognise the need for wearables and properly use them poses new challenges. Second, the high prevalence of different diseases in this population and the heterogeneity associated with their lifestyle, needs, preferences, and health point to the need for wearable devices that are valid and reliable and that can accurately measure and monitor important signals. Additionally, taking into account the problems associated with time-inefficient work in care homes, contactless monitoring of vital signs may be beneficial for healthcare [34–36]. In particular, contactless measurement techniques can be applied to measure the respiratory rate and monitor the heart rate variability, which is one of the fourth most important vital parameters [37]. Monitoring the respiratory rate is useful for the recognition of psychophysiological conditions, the treatment of chronic diseases of the respiratory system, and the recognition of dangerous conditions [38,39]. Combining respiratory rate and heart rate data provides even more useful information on the condition of the cardiovascular system [40,41]. The most promising method of noncontact monitoring of the respiratory process is through infrared and near-infrared cameras [42,43]. An infrared camera is a device that can capture small temperature changes on the surface of an object and/or in the environment. This device can record the temperature fluctuations of airflow from the mouth or nose. Infrared cameras can successfully measure the respiration rate if advanced computer vision algorithms that are insensitive to constantly varying lighting and temperature conditions are applied.

2.3. Benefits of Computer Vision Techniques

Image recognition is one of the main methods used to determine an individual's pose and activity. The use of pose estimation technology in geriatric care offers several advantages, including the continuous monitoring of patients, early detection of potential health problems, essential data on the patient's movements, and, in particular, the detection of extra situations (e.g., the person is lying on the ground and not moving) [44]. Pose estimation algorithms vary in complexity and accuracy, ranging from simple rule-based algorithms to more complex deep-learning-based algorithms. Simple algorithms may be faster and easier to implement but typically they are not as accurate as more complex ones. Deep-learning-based algorithms, on the other hand, may provide more accurate results but may be more computationally intensive and require large amounts of training data. Comparatively, deep-learning-based methods have shown great potential for improving the accuracy of human posture recognition, for both single individuals [45,46] and multiple individuals [47,48] in images or videos. In particular, methods such as the multisource deep model [49], the position refinement model [50,51], and the stacked hourglass network [52] have demonstrated the effectiveness of deep learning in human posture recognition. These methods use convolutional neural networks to extract features from input images and estimate the positions of human joints. However, the early detection of falls [53–55] is one of the most important functions of the geriatric care system as it allows prompt medical assistance to be provided and can prevent further injuries. Human fall detection systems can help to identify when a fall has occurred and alert caregivers or emergency services immediately. Therefore, various types of fall detection and prediction systems suggested in the field not only rely on image recognition techniques [42,56,57] but also employ other information sources, for example, biological factors or signals obtained by wearable devices that are more commonly used for fall risk assessments [58,59]. Although computer vision techniques have been used widely and very successfully in medicine, the monitoring and identification of patients in nursing homes should take into account the fact that image capture devices cannot always be used to track patients (e.g., hygiene rooms) according to privacy and ethical requirements [60,61]. In addition, capturing certain information with cameras may not always be possible due to changes in the environment or, for instance, in cases when the person reappears or is partially obscured by other objects, which poses the additional challenge of re-identifying the same individual. Therefore, it is important to determine which factors may be automatically recorded and tracked over time utilising image processing technology. It is also crucial that the solution is quick. As such, it is essential to carefully assess the trade-off between precision and speed in order to choose a solution that meets the specific requirements of the application.

3. Materials and Methods

The proposed solution includes (1) an IoT module with integrated wearable and contactless devices; (2) an AI module that utilises deep learning architectures for the image recognition of patient posture and activity; and (3) a decision support module for generating the patient-personalised nursing care plan.

An IoT module has been developed to monitor and transfer data in real time. It consists of sensors connected to an Arduino microprocessor to monitor the patient's vital signs. This module integrates not only body-worn devices that are networked but also a number of remote devices for monitoring health data. In general, such devices can collect and transmit the collected data, such as heart rate, body temperature, and physical activity, to a remote system or application, usually through wireless connectivity (e.g., Bluetooth, Wi-Fi). Some wearable health devices also have built-in sensors and algorithms that can perform basic health assessments, such as tracking sleep patterns, counting steps taken, and estimating calorie expenditure.

In this study, four IoT devices, a Fitbit wristband, smart scale, smart blood pressure device, and a camera, were used to monitor the health of elderly patients in a nursing home (Table 1). Data collected from these devices were sent to the server and processed to obtain the final decision (Figure 1).

Table 1. Types of IoT devices used in the research.

IoT Device Type	Device Name
Camera	EZVIZ CS-C3TN 1920 × 1080
Wrist band	Fitbit Charge 5
Blood Pressure	Withings BPM Connect
Scales	Withings Body+

A patient room in a hospital for the elderly was equipped with cameras to continuously monitor the status of the patients in real time. The video footage from these cameras was sent directly to a server where it was stored, processed, and analysed using image processing algorithms. This was necessary to monitor patients' motor activity, changes, or progress in movement and consequently make the necessary changes to the care plan or react in emergency situations such as falls, pressure sores, etc. In parallel to the cameras, the patients were also given Fitbit wristbands for the additional monitoring of physiological parameters. These wristbands were equipped with sensors to monitor the patient's vital signs, such as heart rate and respiratory rate. The data from the Fitbit bracelets were sent to Google Cloud and then to a server using APIs. The geriatric nurse also used specialised equipment to monitor the patients' weight and blood pressure. Withings body+ connected scales make it easy to monitor weight, BMI, fat, water, and body mass, which is later automatically synchronised with the smartphone via Wi-Fi or Bluetooth. In particular, monitoring the following parameters is important for patients at risk of complications such as high blood pressure, diabetes, etc.

One of the main limitations is that off-the-shelf IoT devices do not offer the option of sending data directly to a third-party server. As a result, all data must first pass through the provider's cloud services and use their API. This also leads to software limitations, such as only allowing one IoT device of a certain type per account, making the data collection pipeline more complex than is necessary.

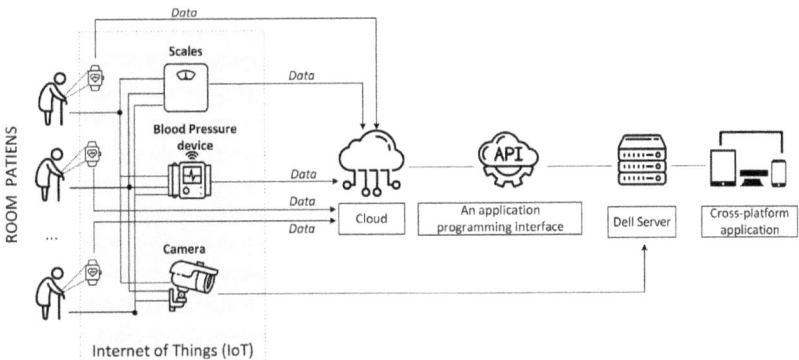

Figure 1. Data collection pipeline of the GCM system.

Data captured by all smart devices not only digitalise the tracking of key physiological parameters but also enable the investigation of dependencies between these indicators and a patient's health status or its change, but only when a statistically reliable sample is collected. If computer-vision-based health monitoring is involved, real-time visual information collection must include data storage and analysis [44,62,63]. For the experiment, data collection started on 15 September 2022 and data were uploaded to the server Dell PowerEdge R7525 (AMD EPYC 7452 32-Core Processor/2350 MHz; 512 GB RAM; NVIDIA GA100 [A100 PCIe 40 GB], 2 × 450 GB SSD; 2 × 25 Gbps LAN MT27800 Family [ConnectX-5] 2 × 100 GBps [ConnectX-6]). In total, 1.412 TB of data were accumulated during the observation period between 15 September 2022 and 28 December 2022. In addition to the data collected from the IoT devices, the system also allowed manual input from healthcare personnel. This included additional parameters that were not captured by the IoT devices, such as bedsores, changes in eating habits, changes in bowel movements, etc. These data were entered into an Excel spreadsheet by healthcare professionals and then automatically uploaded to the database.

By continuously monitoring a patient, wearable health devices can provide a more comprehensive view of a patient's health status. However, it is important to ensure that the system is secure, respects the patient's privacy, and complies with relevant regulations and standards [64,65]. However, it has been observed that wearable gadgets are frequently taken off and thrown away for either purposeful or unintentional reasons, so a balance needs to be struck between functionality, dependability, and cost. This is a common issue with wearable health monitoring devices, particularly among patients with dementia, who may forget where they have placed their device or may not understand the importance of wearing it consistently.

Non-contact monitoring of vital signs using cameras and image recognition techniques is a promising area of development in healthcare technology and has the potential to improve the accessibility, efficiency, and cost-effectiveness of vital sign monitoring. The use of AI-based image recognition algorithms, mainly deep learning architectures, allows images to be automatically analysed to assess vital signs.

YOLOv3 (You Only Look Once, Version 3) [66] is a real-time object detection algorithm that allows specific objects to be identified in videos. YOLOv3 uses a variant of the Darknet neural network architecture, specifically Darknet-53 as its backbone network. The architecture consists of 53 convolutional layers, which was trained on the ImageNet dataset, which was designed for computer vision research [67]. YOLOv3 also contains several key features that help to improve the detection accuracy and performance, including residual skip connections, upsampling, and multiscale detection. The most important feature of the algorithm is that it performs detection at three different scales by downsampling the dimensions of the input image by factors of 32, 16, and 8, respectively (see Figure 2).

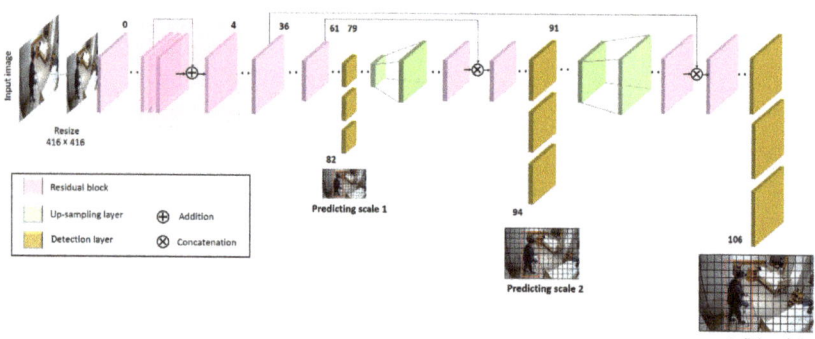

Figure 2. The architecture of YOLOv3 algorithm.

The AlphaPose algorithm allows us to detect keypoints in the bodies of several people with high accuracy in real-time video or images. The 17 keypoints detected by AlphaPose include the nose, eyes, ears, shoulders, elbows, wrists, hips, knees, and ankles (see Figure 3). As the Figure 3 shows, the algorithm can successfully detect the following keypoints in video footage of a patient in a movement position. All of these keypoints are used to construct a human body skeleton representation, which can be used for various applications such as activity estimation [68], process recognition [69], and human fall detection in different environments [54,55,70,71].

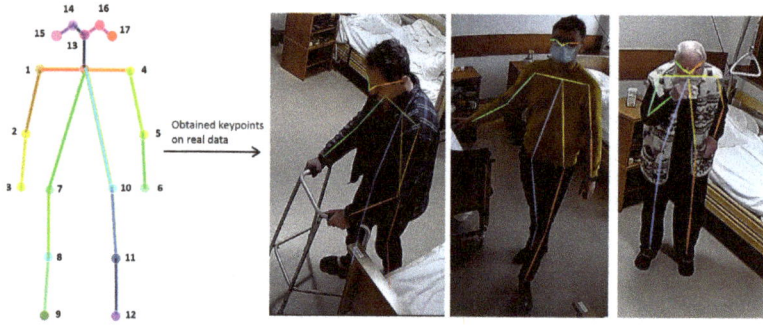

Figure 3. AlphaPose algorithm illustration: keypoints on patients' bodies in video footage.

In particular, a decision support system relies primarily on the expert knowledge of geriatric staff nurses who are experienced in developing nursing plans for patients with different health problems. Their expertise has been used to create the rules that guide the decisions made by nursing professionals which, in this case, are mapped into the output of how to proceed with the nursing plan. Individual experts suggest different decisions based on critical factors in certain cases, so it would seem reasonable to use Fuzzy logic or Neuro-fuzzy models, which are more similar to human thinking. However, given that most of the input variables are of the verbal and integer type, the use of such models will not be efficient. In addition, we do not have enough statistical data to create mappings between numerical values and verbal estimates (e.g., Breathing: Increased $\rightarrow$ X breaths per minute) and to create fuzzy sets based on this. Therefore, we decided to rely on the Decision Tree supervised learning approach which can handle both numeric and non-numeric values, has fast decision times, enables parameter optimisation, and has the possibility of refinement if the accuracy of the result is not satisfactory (e.g., Random Forest). In the decision support module, a Decision Tree with a Gini impurity value was used, and a prepruning process was applied to prevent overfitting. The Gini impurity value is given by

$$Gini = 1 - \sum_{j=1}^{c} p_j^2,$$

where p_j is a proportion of observations that belong to class c for a particular node.

The fine-tuning of Decision Tree hyperparameters involves a depth limited to a maximum of 3 and a minimum number of samples equal to 6 in a finite node. An average classification error of 92% was achieved.

For patient reidentification, the study made use of the Bag of Visual Words (BOVW), since it has been proven to be successful in a number of computer vision tasks, including human reidentification and human action classification [72–74]. With the BOVW approach, local features (such as SIFT descriptors) from images are first extracted and then grouped into a visual vocabulary. Each image is then represented as a histogram of visual words, which may be used for classification or retrieval tasks using machine learning algorithms. More specifically, the K-means algorithm was trained using the final list of features that were retrieved from patient images. As a result, the features were grouped into visual words. Finally, a ML-based classifier was used to generate a categorisation of images based on a newly created vocabulary.

Performance Metrics

The F1 score is a metric that is widely used to evaluate the performance of a classification model. For a multiclass classification, the F1 score for each class is calculated using the one-against-rest (OvR) method. In this approach, the metric for each class is determined separately. However, rather than assigning multiple F1 scores to each class, it is more common to take an average and obtain a single value to measure the overall performance. Three types of averaging methods are commonly used to calculate F1 scores in a multiclass classification, but only two of them are recommended for unbalanced data, as in our case. More specifically, macroaveraging calculates the F1 score for each class separately and derives an unweighted average of these scores. This means that each class is treated equally, regardless of the number of samples it contains. The macroaveraging F1 score is given by

$$Macro_{avg}\ F1 = \frac{\sum_{i=1}^{n} F1_i}{n},$$

where n is the number of classes. In contrast, a weighted averaging calculates the F1 score for each class separately and then takes the weighted average of these scores, where the weight for each class is proportional to the number of samples in that class. In this case, the F1 result is biased towards the larger classes, i.e.,

$$Weighted_{avg}\ F1 = \frac{\sum_{i=1}^{n} w_i \times F1_i}{n},$$

where $w_i = \frac{k_i}{N}$ is the weight of the class i, N is the total number of samples, and k_i is the number of samples in the class i.

4. Results

4.1. Implementation of the Geriatric Care System

The geriatric care plan system for end-users, i.e., nursing home staff, was created using C# programming language and the ASP.NET Core 6.0 framework for the back-end. The front-end was built using Node.js version 19 and the Angular framework, while testing was carried out using Karma. PostgreSQL was used as an open-source relational database management system. The use of these technologies allowed developers to create a robust and scalable system that was able to handle the large amounts of data generated by IoT devices. In addition, Docker was used to containerise the software for deployment by combining the system and all of its dependencies into a single container that could be quickly deployed on any platform that is compatible with Docker. The architecture of the system is demonstrated in Figure 4.

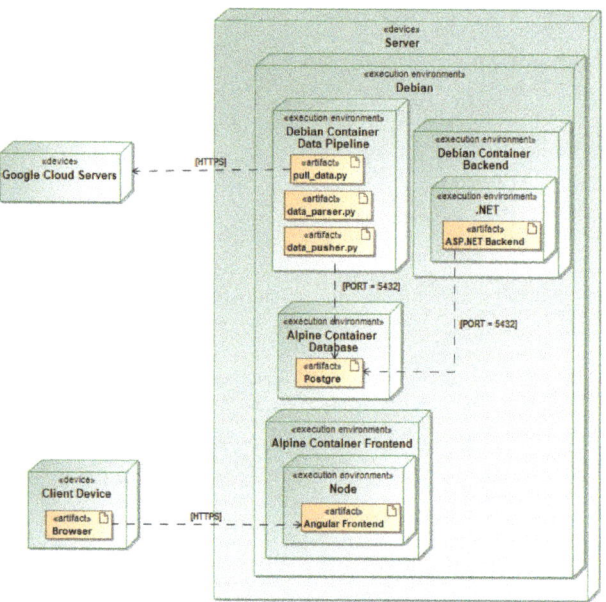

Figure 4. UML deployment diagram of the geriatric care system architecture.

Wearable gadgets synchronise the data with cloud servers, since the data they generate needs to be processed and analysed. Once the data have been received by the cloud servers, the company's server pulls the data from the Google cloud servers using API and then parses the files and saves information in the Postgre database. In contrast, the data captured by the cameras are sent directly to the server. This dataset is then processed in the back-end and analysed alongside the wearable data in order to provide a more comprehensive view of a patient's health status. The main purpose of .NET backend is to act as a bridge, passing data between the Angular front-end and the Postgre database. The back-end is written using REST API methodology to provide a standardised way for different applications or devices to communicate.

4.2. AI-Based Data Analytics and Decision Making

To prepare a nursing care plan, a rich set of data is collected about the patient, as summarised in Table 2. Then, the recommendations for the actions to be taken in a nursing plan are generated from the geriatric care management system.

For demonstration purposes, the collected data were analysed to detect possible dependencies. The radar graph below (see Figure 5) is a single patient's chart of selected vital signs over a 50-day observation period, displaying SBP (systolic blood pressure), DBP (diastolic blood pressure), HR (heart rate), SPO2 (oxygen saturation), sleeping hours, and weight measurements. The data analysis was carried out on three patients on the ward, but no significant dependencies between variables were identified. It may be assumed that that some trends could be determined if the data were gathered over a longer period of time and additional variables, such as pain level, temperature, and even verbal type indicators, were included.

Table 2. Patient information.

No	Variable	Definition	Instances of Possible Values/Range
1	FN	First name	-
2	LN	Last name	-
3	BD	Birth date	yyyy/mm/dd
4	HE	Height	1.20 m–2.20 m
		Input data	
1	MoveC	Movement capabilities	Lying; sitting in a wheelchair; with assistive devices; etc.
2	RiskC	Risk of collapse	None; low; medium; high
3	Bedsores	Bedsores	Yes; no
4	Diseases	All patient's diseases	Heart failure; Alzheimer; dementia; Cancer; etc.
5	Med	Taken medications	Antibiotics; antihypertensives; antidepressants; etc.
6	BMI	BMI unit change per week	<0.5 plus; <0.5 minus; 0.5–1 plus; etc.
7	MoveH	Movement habits	Unchanged; slowed down; increased; falling on the ground
8	EatH	Eating habits	Parenteral nutrition; fed by another person; independent eating; etc.
9	EatC	Eating capabilities	Swallows solid food; swallows only mashed food; swallows only liquids; etc.
10	Bowel	Bowel habits	Regular bowel movements; diarrhoea; constipation; faecal incontinence
11	Sleep	Sleeping	<4 h; 4–6 h; 6–8 h; >8 h; apnoea
12	Breath	Breathing	Increased; slowing down; with apnoeas
13	PL	Pulse	Normal; bradycardia; tachycardia
14	BP	Blood Pressure	Normotension; hypotension; hypertension mild; hypertension moderate; hypertension severe; etc.
15	Temp	Temperature	<36.0 °C; 36.0–37.4 °C; etc.
16	Sat	Saturation	≥94%; <94%
17	Urine	Daily urine output	Concentrated urine; very frequent; etc.
18	Fluid	Fluid tracking	<500 mL; ≥500 mL
19	Gly	Glycaemia	<2.5 mmol/l; ≥2.5 mmol/l
20	Con	Consciousness	Unchanged; changed; unconscious
21	Pain	Perceived level of pain	None; mild; moderate; severe; unbearable
		Output data	
1	Plan	Nursing plan	Continue current plan; monitor; adjust; extra situation

The geriatric care personnel was responsible for writing the rules for the care plans. These rules were based on best practise and experience in the field and were designed to ensure that patients receive the most appropriate and effective care. More specifically, care plans were tailored to the specific needs of current and future patients, taking into account their medical history, current condition, and other relevant factors. The variables listed in Table 2 were included in the care plan, as they indicate the patient's medical history, current health status, and other relevant factors that can influence treatment. On the basis of this information, the initial set of rules covered a wide range of scenarios and options, but after optimisation, the patient care plan eventually consisted of 61 rules with the four possible outputs of the care plan: "Continue current treatment", "Monitor", "Adjust", or

"Extra situation" (see Figure 6). All remaining cases that were not included in the rules were assigned to the care plan "Continue current treatment" by default.

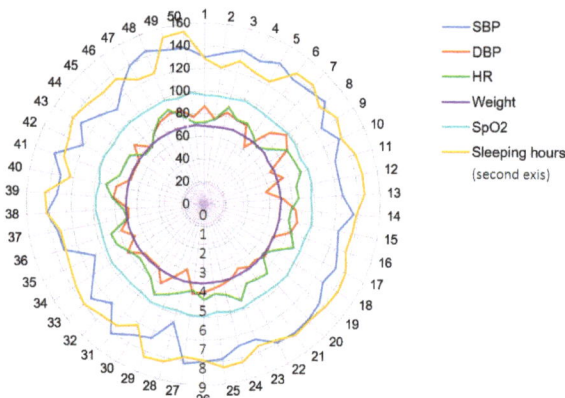

Figure 5. Six different health parameters collected for a single patient.

```
IF MoveH is slowed down AND EatH is vomiting        IF Diseases is dementia AND MoveH is increased
   AND Breath is Increased AND Pulse is tachycardia    AND Sleep is 4-6 hrs. AND BMI is 0.5 - 1 minus
   AND Saturation <94%                                 THEN Plan is Adjust
   AND Con is changed THEN Plan is Extra Situation

IF Temp is >38 AND Urine is very frequent           IF Med is BZD hypnotics AND Sleep is >8 hrs.
   AND Pain is moderate and Pulse is tachycardia       THEN Plan is Monitor
   THEN Plan is Adjust
```

Figure 6. Examples of different care plan with IF-THEN rules defined by the staff.

In particular, the nursing care plan was designed to be flexible and adaptable to allow healthcare professionals to adjust the patient's geriatric care according to his or her health status and changing needs. Those rules and the output generated by the geriatric care management system help healthcare personnel to respond more quickly to changes in a patient's health, shape the patient-personalised geriatric care, reduce the risk of human error, and make better use of staff time by concentrating more on essential social support.

Figure 7 shows a schema for an AI-based decision support system. Four of the 21 variables (see Table 2) are automatically registered; that is, three of them were retrieved from IoT devices and one (change in movement) was obtained from the camera. The value of the latter variable was generated from the AI-based image recognition module. The remaining variables were taken from the MS Excel spreadsheet file, where all data were entered manually.

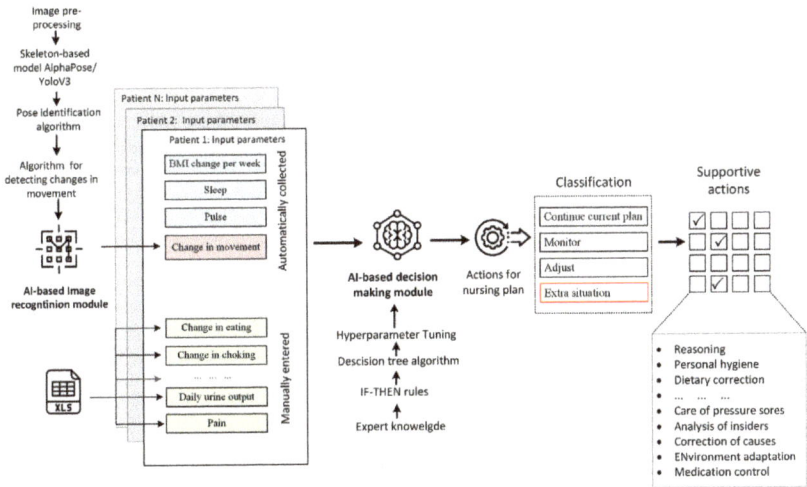

Figure 7. Schematic diagram of proposed geriatric care management systems

Image Recognition Solution

An AI-based image recognition module is a block consisting of several sequential algorithms that detects changes in the movement of a patient. In this project, we used the camera to film nursing home patients, that is, one room with three patients. The video was recorded at 1920 × 1080 pixel resolution with a frequency of 10 FPS, therefore storing 10 unique images per second to obtain 10FPS × 60 = 600 images per minute. An image was analysed every five seconds with the assumption that no significant changes in motion would be detected in that time period.

The image processing included
- Brightening: to increase the overall luminosity of the image, improve visibility, and increase the clarity of the image during low light conditions;
- Cropping: to keep only regions of interest in the image;
- Denoising: to remove noise from the image, typically by applying a low-pass filter. It also improved the quality and clarity of the image by removing noise, which could be especially useful if the image was taken under poor conditions or with a low-quality camera.
- Edge detection: to identify edges in the image by finding points of a rapid intensity change. It can also be used to identify and extract features or objects in the image, such as lines, shapes, or boundaries.

After image processing, the algorithm integrating YOLOv3 and AlphaPose [75] was used to detect human poses. The algorithm includes the three main components [76]. First, the Symmetric Spatial Transformer Network (SSTN) takes the detected bounding boxes to generate pose proposals. The SSTN allows the spatial context and correlations between the keypoints to be captured, leading to more accurate pose estimates. Second, the Parametric Pose Non-Maximum-Suppression (NMS) is a component that is used to remove redundant pose detections and improve the overall accuracy of the pose estimation. Finally, the Pose-guided Proposals Generator is used to create a large sample of training proposals with the same distribution as the output of the human detector.

The next step is the problem of identifying and classifying patient postures, which in this case, included the following six postures: "walking", "standing", "sitting", "fallen on the ground", "lying in bed", and "sleeping". For the verification of all poses, a sequence of three images was taken for a period of 15 s, except for the last two poses. The poses of "sleeping" and "laying in bed" correlate with the parameters of the smart bracelet (sleep time and heart rate), so these parameters were also assessed. If the patient was found to be lying in bed, the assessment time was extended by up to one minute to identify whether the patient was "lying in bed" or "sleeping". In particular, the pose was assessed every minute until a new pose was captured.

A pose change algorithm was developed to detect differences between adjacent images, that is, to identify that a person was walking rather than standing or that a person was just lying on a bed rather than sleeping. Figure 8 illustrates the example of three iterations of assessment frames of the "walking" pose taken every five seconds. Comparing the images taken every five seconds, we can see that the pose remained the same, although the frames were not identical and the patient's coordinates varied.

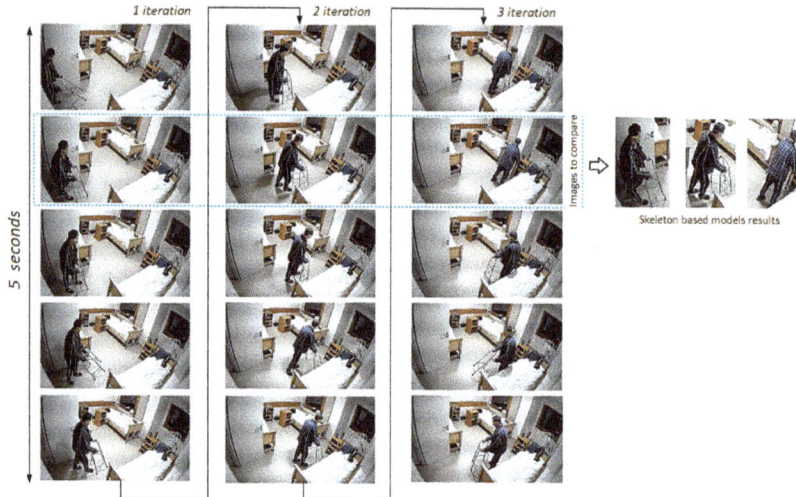

Figure 8. Three iterations of the assessment frames of the patient in the "walking" pose taken every five seconds.

In order to define changes in movement habits, an additional algorithm (see Algorithm 1) was created to evaluate movement changes over a longer period of time t_{m-n}, where m is a current time moment, n is a number of days before t_m, $1 \leq n \leq 3$. This algorithm calculates the duration (hours) in each pose per day. The percentage change is then evaluated, compared with threshold value k_{th} and a response is generated that includes three possible values: "Unchanged", "Slowed down", or "Increased". The pseudocode of the algorithm is provided below.

Algorithm 1 Evaluation of changes in movement habits

$Act_H = Walking(hrs) + Sitting(hrs) + Standing(hrs)$
$k_{th} = 12.5\%$
$Diff(a, b) = ((a - b)/b) * 100$
if $Diff(Act_H(t_{m-1}), Act_H(t_{m-2})) \leq -k_{th} \wedge Diff(Act_H(t_{m-2}), Act_H(t_{m-3})) \leq -k_{th}$ **then**
    $MoveH$ is slowed down
else if $Diff(Act_H(t_{m-1}), Act_H(t_{m-2})) \geq k_{th} \wedge Diff(Act_H(t_{m-2}), Act_H(t_{m-3})) \geq k_{th}$ **then**
    $MoveH$ is increased
else $MoveH$ is unchanged
end if

For demonstration purposes, the identification of tough poses observed in the real-world environment is shown below. For instance, Figure 9 shows a skeleton-based posture recognition in various lighting environments. In well-illuminated areas, patients can be detected by identifying all skeleton keypoints (Figure 9c). It has been observed that at night or at twilight/night, walking patients can be identified quite accurately with all keypoints (Figure 9a,b), but when patients are sleeping with their blankets, few keypoints were successfully detected (Figure 9d) or keypoints were not detected at all (Figure 9a).

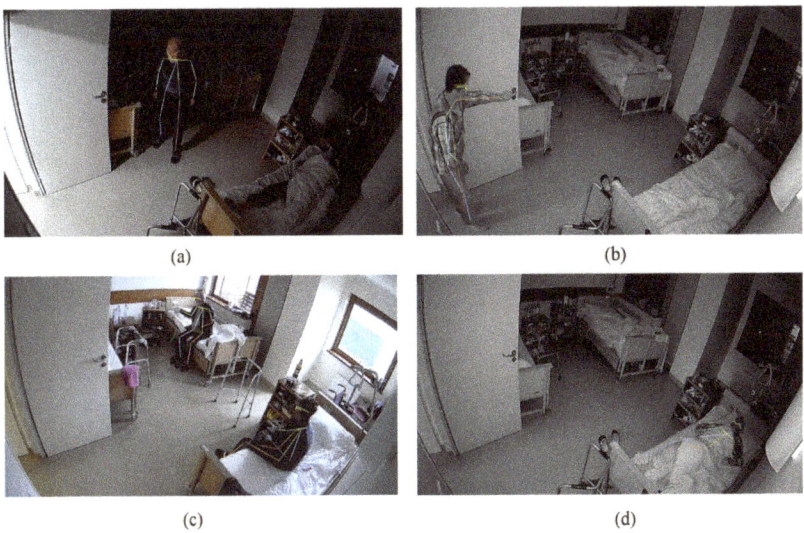

Figure 9. Examples of skeleton-based posture recognition in various ambient light conditions: (**a**) patient walks in a semi-lit environment; (**b**) patient walks during the night; (**c**) two patients sit in a fully-lit environment; (**d**) patient is lying down at night.

Another example demonstrates a skeleton-based posture recognition for two different scenarios. In Figure 10a, the keypoints in the patient's body were detected when the patient was lying on the ground, which refers to the status "falling on the ground". To correctly recognise this pose, a training dataset with artificially simulated falling poses was created. Comparatively, Figure 10b shows that the keypoints in the body were identified for all persons located in the ward, but the nursing personnel needed to be the exception. Therefore, additional data were collected to train a deep learning algorithm to distinguish staff from patients. Consequently, the nursing personnel was identified by their clothes, more specifically, white trousers and a blue top, which they had to always wear.

Figure 10. Examples of skeleton-based posture recognition in different scenarios: (**a**) the patient is lying on the ground; (**b**) patients are visited by nursing staff.

4.3. Experimental Results

A posture detection algorithm of captured video material was tested to identify six different poses. The results are summarised in a confusion matrix to evaluate the performance of the algorithm. More specifically, the confusion matrix provides a visual representation of the number of correct and incorrect predictions made by the classifier: the rows represent the actual class labels, while the columns represent the predicted class labels. The diagonal elements show the number of correct predictions (see Figure 11).

	Walking	Standing	Sitting	Falling on the ground	Lying in bed	Sleeping	
Walking	944	52	11	0	0	0	93.74% / 6.26%
Standing	39	471	3	0	1	0	93.74% / 8.37%
Sitting	5	14	428	1	6	0	94.27% / 5.73%
Falling on the ground	0	2	4	145	14	9	91.77% / 8.23%
Lying in bed	0	1	6	5	760	84	88.79% / 11.21%
Sleeping	0	0	3	4	68	712	90.47% / 9.53%
	95.55% / 4.45%	87.22% / 12.78%	94.07% / 5.93%	93.55% / 6.45%	90.48% / 9.52%	89.22% / 10.78%	91.63% / 8.37%

Figure 11. Testing results: confusion matrix of posture classification.

The posture recognition algorithm was trained using 9300 labelled images and tested using 3792 images. An average posture recognition accuracy of 91.63% was achieved for the testing data set (Figure 11). Posture labelling was performed manually on the images obtained from the video stream for training and testing purposes. The Receiver Operating Characteristic (ROC) curve of the stratified testing dataset is provided in Figure 12.

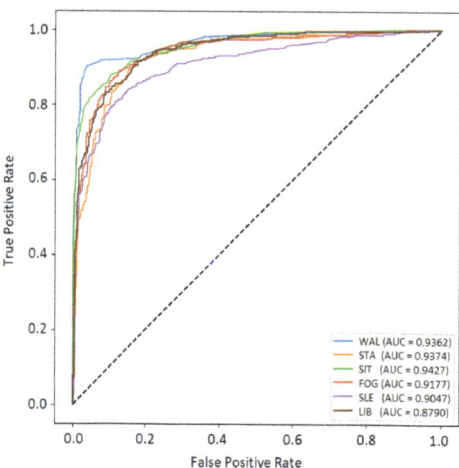

Figure 12. Testing results: ROC curve of the posture recognition algorithm.

The AUC values for each posture class ranged from 0.8790 to 0.9427, with the highest value obtained for the sitting posture class. The sleeping and lying in bed posture classes resulted in the lowest AUC values, with values of 0.9047 and 0.8790, respectively. These lower values suggest that it might be more difficult for the classifier to distinguish between these postures and others. Comparatively, the AUC value for the fallen on the ground posture class was 0.9177, which is slightly lower than those of the other more successfully recognised posture classes. This could be due to the lack of training data for this posture, which might have led to lower accuracy. Next, Table 3 summarises the estimated values of precision, recall, and F1 score for each class of interest, together with macro and weighted F1 scores for the evaluation of the overall performance of the posture recognition algorithm.

Table 3. Testing results: performance metrics of the posture recognition algorithm.

Class	Precision	Recall	F1 Score
Walking (WAL)	0.9554	0.9374	0.9463
Standing (STA)	0.8722	0.9163	0.8937
Sitting (SIT)	0.9406	0.9427	0.9416
Fallen on the ground (FOG)	0.9354	0.8333	0.8814
Lying in bed (LIB)	0.8951	0.8878	0.8914
Sleeping (SLE)	0.8844	0.9047	0.8944
		Macro F1 score	0.9082
		Weighted F1 score	0.9125

The patient re-identification testing results are summarised in Figure 13. The support vector machine (SVM) method was used to generate categories of images, providing labels for the patient classes. In our case, the maximum number of classes was set to four: three classes represented the maximum number of patients the ward can accommodate, while the separate class "None" referred to unauthorised individuals such as nursing staff, family members, doctors, or others. The class names for patients were labelled "First", "Second", and "Third" (see Figure 13).

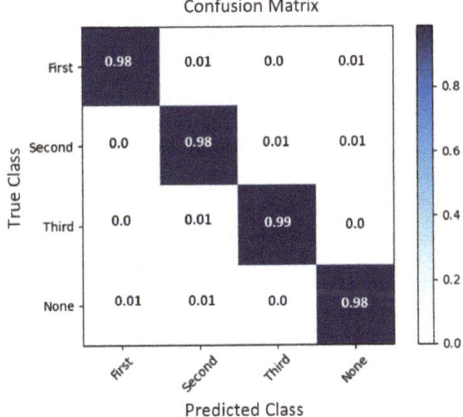

Figure 13. Confusion matrix showing the re-identification of three patients (referred to by the class labels "First", "Second", and "Third".)

The confusion matrix in Figure 13 summarises how successfully the algorithm identifies three ward patients in common areas. One can observe that an accuracy level of 90% was obtained for the "First" class, a value of 88% was obtained for the "Second" class, and a value of 91% was obtained for the "Third" class. Although the lowest accuracy level of 87% was achieved in the "None" class, considering that there can be around 6–13 people in a single tray, this is a pretty good accuracy level. It was observed that female patients and nursing home staff were more easily recognised, but other patients, nonmedical nursing home staff, and visiting relatives were the most confused with these patients.

Finally, to conduct a real-time experiment, patient positioning verification was carried out. This included 16 scenarios with diverse positions. The results are summarised in

Table 4. Two prediction errors were determined. More specifically, the patient was "lying in the bed", but he was detected as "sleeping", as he was covered up, his heart rate was reduced, he did not move for more than one minute, and it was night time. Another prediction error also related to the sleeping pose. The patient was lying in the bed without covering up; however, it was determined that he was not sleeping based on readings from the smart wristband. It should be noted that the prediction may also be impacted by ambient light conditions. From a technical perspective, the proposed system performed pose estimation with an average output time of 182 ms, including the algorithm used to predict the pose from the possible outcomes.

Table 4. Real-time scenario testing results of posture recognition.

No.	Actual Pose	Predicted Pose	Ambient Lighting	Confidence
1	Walking	Walking	Day time (well-lit)	98.0%
2	Sitting	Sitting	Day time (well-lit)	97.5%
3	Sitting	Sitting	Day time (well-lit)	98.2%
4	Lying in bed	Sleeping	Night time (poorly lit)	89.3%
5	Standing	Standing	Day time (perfect)	99.7%
6	Lying in bed	Lying in bed	Evening time (semi-lit)	87.9%
7	Sleeping	Lying in bed	Evening time (semi-lit)	88.6%
8	Standing	Standing	Day time (perfect)	99.1%
9	Sleeping	Sleeping	Night time (poorly lit)	85.4%
10	Walking	Walking	Day time (perfect)	93.6%
11	Lying in bed	Lying in bed	Day time (perfect)	94.2%
12	Standing	Standing	Day time (perfect)	99.3%
13	Walking	Walking	Night time (poorly lit)	96.0%
14	Sitting	Sitting	Day time (perfect)	98.5%
15	Sleeping	Sleeping	Day time (perfect)	91.0%
16	Fallen on the ground	Fallen on the ground	Day time (perfect)	99.8%

To test the correctness of the output of the geriatric care management system, different scenarios involving nursing home staff were developed. The results revealed that the system provided the correct output in all cases. The system was designed to generate changes to the treatment plan immediately after any changes are made. When a healthcare professional makes a change to the care plan, the system analyses the data from the patient's IoT devices and determines the appropriate course of action. The system then automatically updates the results of the action to be taken for the individual patient and alerts the healthcare professional. This allows healthcare professionals to stay up-to-date with the patient's condition and make any necessary adjustments to their treatment in a timely manner.

5. Discussion

There are a few areas for improvement, as the proposed geriatric management care system is still in its initial stage of functioning. Personality identification, which relates to the continuous contactless assessment of the patient, is the most challenging concern. Comparatively, wearable devices do not raise any questions at the moment; their purpose is clear, but elderly people have a problem with wearing them because they find them annoying. The creation of the nursing care plan itself could be fully automated later on, with a follow-up on what action should be taken when the situation changes. However, to fully automate it, a lot of statistical data are needed, including actions taken by nurses, from which the system could be learnt, that is, from the actions taken by the care worker on each individual situation. Taking into account the current data (Table 2), there are at least 20,155,392,000 possible combinations of parameters that define the health condition, which are likely to increase in the future due to the inclusion of additional parameters. For this purpose, a list of actions is provided in the geriatric care management system, from which the care worker must indicate (select from the list) what they intend to do. In this way, a dataset of situations and decisions with all the actions taken accordingly is continuously accumulated. Once a representative sample of data has been accumulated (say after at least one year), the correctness of the automated action is improved.

The challenge with consistent and accurate patient identification makes it reasonable to consider other methods of individual identification than BOVW. As patients usually stay in their own wards, the accuracy of identification is high when the patients are present and nursing staff visit them a certain number of times per day. However, the accuracy drops in common areas (e.g., resting, eating) because there are more patients and personnel present. Mainly because of their distinguishing clothes, nursing workers are simpler to identify (see column "None" in Figure 14). However, the elderly patients themselves are more likely to be confused with each other in common areas, with a best individual identification result of 0.914 achieved (see Figure 14).

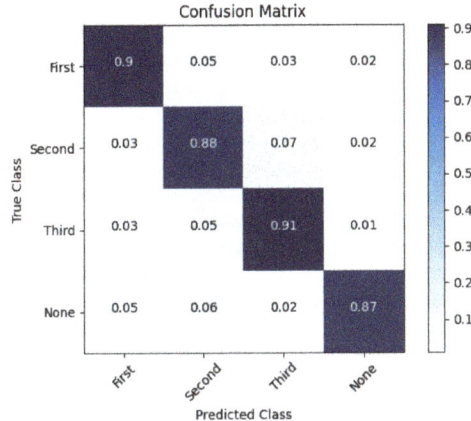

Figure 14. Confusion matrix of the re-identification of three patients (referred to class labels "First", "Second", and "Third") in the two common areas of nursing homes.

As an alternative method, gait recognition (GR) technology can be used for patient identification. This method examines the uniqueness of an individual's walking or running pattern using machine learning (ML) techniques [77]. More specifically, ML algorithms are trained to recognize subtle differences in a person's gait and thus can use this information to identify individuals even if their face is obscured in the image [78,79]. An additional benefit of GR technology is that gait information can be used not only for personal identification, but also for medical purposes, such as monitoring and for the diagnosis and treatment of various movement disorders [80,81]. For example, gait recognition technology can be used to identify and diagnose various types of neurodegenerative diseases (such as Parkinson's and Alzheimer's disease) or assess the course of disease [82–85]. This can help doctors and healthcare professionals to develop more effective nursing care plans and interventions as well as monitor the progression of these conditions over time. However, GR technology usually requires a variety of sources or capture devices to gather data about an individual's gait, including multiple video cameras, motion sensors, radars, and other specialized equipment [79,86]. In addition, the accuracy of gait recognition technology can be affected by a range of factors, including the angle at which the gait is captured.

Finally, it should be noted that elderly people are choosing to live independently at home for as long as possible. In such cases, intelligent geriatric care management system monitoring adapted to the individual home and operating remotely can be very helpful for ensuring that the elderly person is safe and providing faster reactions to emergencies (i.e., fall detection) and appropriate care. In the near future, we plan to develop the necessary software and hardware package (e.g., for the proper functioning of the system such as a stable internet connection) for the home care services and to test it in real-world environment with the possibility of transmitting the data to the responsible physician for monitoring.

6. Conclusions

In this study, a geriatric care management system based on IoT and AI algorithms was proposed to monitor some of the most important vital signs in a noncontact manner and to facilitate the adjustment of the care plan. The system provides an intelligent assistance function, which suggests how to proceed with the patient's care plan based on the available data and the decision support module.

A built-in posture recognition algorithm allows staff to react quickly to extreme situations, which are highly expected at night or during peak working hours. Another algorithm was developed to monitor changes in a patient's movement habits over a longer period of time, which can be important for detecting health problems more quickly and taking appropriate action. This is a value-added functionality of the system, as it is very difficult for nursing staff to do this in a natural way, as it is not possible to monitor every patient 24 h a day without smart technology. During this study, it was observed that the most confusing poses are "lying in bed" and "sleeping". Detecting the individual or pose when the patient is fully or partially occluded is also quite challenging. However, capturing the pulse and sleep mode and combining these indicators with the outputs of the image recognition algorithms resulted in better detection of the "sleeping" and "lying in bed" poses, i.e., the accuracy was improved by around 15.48% and 22.06%, respectively. Additionally, the system is resistant to data deficiencies; if certain data are not received at the current time, the value is taken from the last time of recording. In any case, the final decision is made by the human, and in case of error or incorrect output, one has the opportunity to correct it.

Other concerns are ensuring that smart health monitoring devices are worn and maintained at all times, as patients often want to remove devices (particularly patients with a dementia), and nursing staff do not always notice quickly when the devices need to be loaded. Therefore, the involvement of care specialists is crucial to ensure the system operates effectively and efficiently. In addition, it is equally important to make sure that patients feel comfortable and moreover that their privacy and trust in smart technologies are maintained at the appropriate level. By involving nursing staff in the implementation process, they can provide valuable feedback, suggestions, and ideas, leading to a better overall outcome.

Author Contributions: Conceptualization, A.P.-T., J.S. and J.B.; methodology, A.P.-T., J.B.; software, J.S., S.N., M.O. and T.P.; validation, J.S., M.O., S.N. and T.P.; formal analysis, A.P.-T., J.S., K.S. and A.R.; investigation, A.P.-T., J.S., K.S., A.R. and J.B.; resources, J.B., A.P.-T. and A.R.; data curation, A.P.-T., J.B., J.S., S.N. and A.R.; writing—original draft preparation, A.P.-T., J.S., K.S., S.N.; writing—review and editing, A.P.-T., K.S. and J.S.; visualization, A.P.-T. and K.S; supervision, A.P.-T. and J.B.; project administration, J.B.; funding acquisition, J.B. All authors have read and agreed to the published version of the manuscript.

Funding: This research was supported by the project EIT Regional Innovation Scheme (EIT RIS)-EIT Health-Nursing.AI, 2022, Project ID: 2021-RIS_Innovation-033.

Institutional Review Board Statement: The research study was reviewed and approved by the Kaunas Region Biomedical Research Ethics Committee (No. BE-2-24).

Informed Consent Statement: Informed consent was obtained from patients' relatives/caregivers involved in the study.

Data Availability Statement: The data are not publicly available due to privacy restrictions.

Conflicts of Interest: The authors declare no conflict of interest.

References

1. Ellis, G.; Sevdalis, N. Understanding and improving multidisciplinary team working in geriatric medicine. *Age Ageing* **2019**, *48*, 498–505. [CrossRef] [PubMed]
2. Elliott, M.N.; Beckett, M.K.; Cohea, C.; Lehrman, W.G.; Russ, C.; Cleary, P.D.; Giordano, L.A.; Goldstein, E.; Saliba, D. The hospital care experiences of older patients compared to younger patients. *J. Am. Geriatr. Soc.* **2022**, *70*, 3570–3577. [CrossRef] [PubMed]
3. Reber, K.C.; Lindlbauer, I.; Schulz, C.; Rapp, K.; König, H.H. Impact of morbidity on care need increase and mortality in nursing homes: A retrospective longitudinal study using administrative claims data. *BMC Geriatr.* **2020**, *20*, 439. [CrossRef] [PubMed]
4. Dall'Ora, C.; Griffiths, P.; Hope, J.; Briggs, J.; Jeremy, J.; Gerry, S.; Redfern, O. How long do nursing staff take to measure and record patients' vital signs observations in hospital? A time-and-motion study. *Int. J. Nurs. Stud.* **2021**, *118*, 103921. [CrossRef] [PubMed]
5. Tang, V.; Choy, K.; Ho, G.; Lam, H.; Tsang, Y.P. An IoMT-based geriatric care management system for achieving smart health in nursing homes. *Ind. Manag. Data Syst.* **2019**, ahead-of-print. [CrossRef]
6. Flores-Martin, D.; Rojo, J.; Moguel, E.; Berrocal, J.; Murillo, J.M.; Cai, Z. Smart Nursing Homes: Self-Management Architecture Based on IoT and Machine Learning for Rural Areas. *Wirel. Commun. Mob. Comput.* **2021**, *2021*. [CrossRef]
7. Lu, Z.X.; Qian, P.; Bi, D.; Ye, Z.W.; He, X.; Zhao, Y.H.; Su, L.; Li, S.L.; Zhu, Z.L. Application of AI and IoT in Clinical Medicine: Summary and Challenges. *Curr. Med Sci.* **2021**, *41*, 1134–1150. [CrossRef] [PubMed]
8. Mbunge, E.; Muchemwa, B.; Jiyane, S.; Batani, J. Sensors and healthcare 5.0: Transformative shift in virtual care through emerging digital health technologies. *Glob. Health J.* **2021**, *5*, 169–177. [CrossRef]
9. Khan, M.F.; Ghazal, T.M.; Said, R.A.; Fatima, A.; Abbas, S.; Khan, M.A.; Issa, G.F.; Ahmad, M.; Khan, M.A. An IoMT-Enabled Smart Healthcare Model to Monitor Elderly People Using Machine Learning Technique. *Comput. Intell. Neurosci.* **2021**, *2021*, 1–10. [CrossRef]
10. Alshamrani, M. IoT and artificial intelligence implementations for remote healthcare monitoring systems: A survey. *J. King Saud Univ.—Comput. Inf. Sci.* **2022**, *34*, 4687–4701. [CrossRef]
11. Ienca, M.; Schneble, C.; Kressig, R.W.; Wangmo, T. Digital health interventions for healthy ageing: A qualitative user evaluation and ethical assessment. *BMC Geriatr.* **2021**, *21*, 412. [CrossRef]
12. Andreoni, G.; Mambrettii, C. Privacy and Security Concerns in IoT-Based Healthcare Systems. In *Digital Health Technology for Better Aging*; Springer: Cham, Switzerland, 2021; p. 365. [CrossRef]
13. Kekade, S.; Hseieh, C.H.; Islam, M.M.; Atique, S.; Mohammed Khalfan, A.; Li, Y.C.; Abdul, S.S. The usefulness and actual use of wearable devices among the elderly population. *Comput. Methods Programs Biomed.* **2018**, *153*, 137–159. [CrossRef]
14. Chandrasekaran, R.; Katthula, V.; Moustakas, E. Too old for technology? Use of wearable healthcare devices by older adults and their willingness to share health data with providers. *Health Inform. J.* **2021**, *27*, 14604582211058073. [CrossRef]
15. Prieto-Avalos, G.; Cruz-Ramos, N.A.; Alor-Hernández, G.; Sánchez-Cervantes, J.L.; Rodríguez-Mazahua, L.; Guarneros-Nolasco, L.R. Wearable Devices for Physical Monitoring of Heart: A Review. *Biosensors* **2022**, *12*, 292. [CrossRef]
16. Lu, L.; Zhang, J.; Xie, Y.; Gao, F.; Xu, S.; Wu, X.; Ye, Z. Wearable health devices in health care: Narrative systematic review. *JMIR mHealth uHealth* **2020**, *8*, e18907. [CrossRef] [PubMed]
17. Singhal, A.; Cowie, M.R. The Role of Wearables in Heart Failure. *Curr. Heart Fail. Rep.* **2020**, *17*, 125–132. [CrossRef] [PubMed]
18. Alharbi, M.; Straiton, N.; Gallagher, R. Harnessing the Potential of Wearable Activity Trackers for Heart Failure Self-Care. *Curr. Heart Fail. Rep.* **2017**, *14*, 23–29. [CrossRef]
19. Gresham, G.; Schrack, J.; Gresham, L.M.; Shinde, A.M.; Hendifar, A.E.; Tuli, R.; Rimel, B.; Figlin, R.; Meinert, C.L.; Piantadosi, S. Wearable activity monitors in oncology trials: Current use of an emerging technology. *Contemp. Clin. Trials* **2018**, *64*, 13–21. [CrossRef] [PubMed]
20. Watanabe, T.; Saito, H.; Koike, E.; Nitta, K. A Preliminary Test of Measurement of Joint Angles and Stride Length with Wireless Inertial Sensors for Wearable Gait Evaluation System. *Comput. Intell. Neurosci.* **2011**, *2011*, 1–12. [CrossRef] [PubMed]
21. Ryvlin, P.; Ciumas, C.; Wisniewski, I.; Beniczky, S. Wearable devices for sudden unexpected death in epilepsy prevention. *Epilepsia* **2018**, *59* (Suppl. 1), 61–66. [CrossRef]
22. Takei, K.; Honda, W.; Harada, S.; Arie, T.; Akita, S. Toward flexible and wearable human-interactive health-monitoring devices. *Adv. Healthc. Mater.* **2015**, *4*, 487–500. [CrossRef] [PubMed]
23. Kamei, T.; Kanamori, T.; Yamamoto, Y.; Edirippulige, S. The use of wearable devices in chronic disease management to enhance adherence and improve telehealth outcomes: A systematic review and meta-analysis. *J. Telemed. Telecare* **2022**, *28*, 342–359. [CrossRef]
24. Yu, S.; Chen, Z.; Wu, X. The Impact of Wearable Devices on Physical Activity for Chronic Disease Patients: Findings from the 2019 Health Information National Trends Survey. *Int. J. Environ. Res. Public Health* **2023**, *20*, 887. [CrossRef] [PubMed]
25. Kulurkar, P.; kumar Dixit, C.; Bharathi, V.; Monikavishnuvarthini, A.; Dhakne, A.; Preethi, P. AI based elderly fall prediction system using wearable sensors: A smart home-care technology with IOT. *Meas. Sensors* **2023**, *25*, 100614. [CrossRef]
26. Cudejko, T.; Button, K.; Al-Amri, M. Validity and reliability of accelerations and orientations measured using wearable sensors during functional activities. *Sci. Rep.* **2022**, *12*, 14619. [CrossRef] [PubMed]
27. Fuller, D.; Colwell, E.; Low, J.; Orychock, K.; Tobin, M.A.; Simango, B.; Buote, R.; Heerden, D.V.; Luan, H.; Cullen, K.; et al. Reliability and Validity of Commercially Available Wearable Devices for Measuring Steps, Energy Expenditure, and Heart Rate: Systematic Review. *JMIR mHealth uHealth* **2020**, *8*, e18694. [CrossRef]

28. Patel, V.; Orchanian-Cheff, A.; Wu, R. Evaluating the Validity and Utility of Wearable Technology for Continuously Monitoring Patients in a Hospital Setting: Systematic Review. *JMIR mHealth uHealth* **2021**, *9*, e17411. [CrossRef]
29. Chan, A.; Chan, D.; Lee, H.; Ng, C.C.; Yeo, A.H.L. Reporting adherence, validity and physical activity measures of wearable activity trackers in medical research: A systematic review. *Int. J. Med Inform.* **2022**, *160*, 104696. [CrossRef] [PubMed]
30. Teixeira, E.; Fonseca, H.; Diniz-Sousa, F.; Veras, L.; Boppre, G.; Oliveira, J.; Pinto, D.; Alves, A.J.; Barbosa, A.; Mendes, R.; et al. Wearable Devices for Physical Activity and Healthcare Monitoring in Elderly People: A Critical Review. *Geriatrics* **2021**, *6*, 38. [CrossRef]
31. Moore, K.; O'Shea, E.; Kenny, L.; Barton, J.; Tedesco, S.; Sica, M.; Crowe, C.; Alamaki, A.; Condell, J.; Nordstrom, A.; et al. Older Adults' Experiences With Using Wearable Devices: Qualitative Systematic Review and Meta-synthesis. *JMIR mHealth uHealth* **2021**, *9*, e23832. [CrossRef]
32. Koerber, D.; Khan, S.; Shamsheri, T.; Kirubarajan, A.; Mehta, S. Accuracy of Heart Rate Measurement with Wrist-Worn Wearable Devices in Various Skin Tones: A Systematic Review. *J. Racial Ethn. Health Disparities* **2022**. [CrossRef]
33. Ferguson, C.; Hickman, L.D.; Turkmani, S.; Breen, P.; Gargiulo, G.; Inglis, S.C. "Wearables only work on patients that wear them": Barriers and facilitators to the adoption of wearable cardiac monitoring technologies. *Cardiovasc. Digit. Health J.* **2021**, *2*, 137–147. [CrossRef] [PubMed]
34. Kristoffersson, A.; Lindén, M. Wearable Sensors for Monitoring and Preventing Noncommunicable Diseases: A Systematic Review. *Information* **2020**, *11*, 521. [CrossRef]
35. Rohmetra, H.; Raghunath, N.; Narang, P.; Chamola, V.; Guizani, M.; Lakkaniga, R. AI-enabled remote monitoring of vital signs for COVID-19: Methods, Prospects and Challenges. *Computing* **2021**, *105*, 783–809. [CrossRef]
36. Guo, K.; Zhai, T.; Purushothama, M.H.; Dobre, A.; Meah, S.; Pashollari, E.; Vaish, A.; DeWilde, C.; Islam, M.N. Contactless Vital Sign Monitoring System for In-Vehicle Driver Monitoring Using a Near-Infrared Time-of-Flight Camera. *Appl. Sci.* **2022**, *12*, 4416. [CrossRef]
37. Guo, K.; Zhai, T.; Pashollari, E.; Varlamos, C.J.; Ahmed, A.; Islam, M.N. Contactless Vital Sign Monitoring System for Heart and Respiratory Rate Measurements with Motion Compensation Using a Near-Infrared Time-of-Flight Camera. *Appl. Sci.* **2021**, *11*, 10913. [CrossRef]
38. Jelinčić, V.; Diest, I.V.; Torta, D.M.; von Leupoldt, A. The breathing brain: The potential of neural oscillations for the understanding of respiratory perception in health and disease. *Psychophysiology* **2022**, *59*, e13844. [CrossRef] [PubMed]
39. Andrea, N.; Carlo, M.; Emiliano, S.; Massimo, S. The Importance of Respiratory Rate Monitoring: From Healthcare to Sport and Exercise. *Sensors* **2020**, *20*, 6396. [CrossRef]
40. Baumert, M.; Linz, D.; Stone, K.; McEvoy, R.D.; Cummings, S.; Redline, S.; Mehra, R.; Immanuel, S. Mean nocturnal respiratory rate predicts cardiovascular and all-cause mortality in community-dwelling older men and women. *Eur. Respir. J.* **2019**, *54*, 1802175. [CrossRef]
41. Fox, H.; Rudolph, V.; Munt, O.; Malouf, G.; Graml, A.; Bitter, T.; Oldenburg, O. Early identification of heart failure deterioration through respiratory monitoring with adaptive servo-ventilation. *J. Sleep Res.* **2023**, *32*, e13749. [CrossRef]
42. Scebba, G.; Da Poian, G.; Karlen, W. Multispectral Video Fusion for Non-Contact Monitoring of Respiratory Rate and Apnea. *IEEE Trans. Biomed. Eng.* **2021**, *68*, 350–359. [CrossRef]
43. Nakagawa, K.; Sankai, Y. Noncontact Vital Sign Monitoring System with Dual Infrared Imaging for Discriminating Respiration Mode. *Adv. Biomed. Eng.* **2021**, *10*, 80–89. [CrossRef]
44. Yacchirema, D.C.; de Puga, J.S.; Palau, C.E.; Esteve, M. Fall detection system for elderly people using IoT and ensemble machine learning algorithm. *Pers. Ubiquitous Comput.* **2019**, *23*, 801–817. [CrossRef]
45. Esmaeili, B.; AkhavanPour, A.; Bosaghzadeh, A. An Ensemble Model For Human Posture Recognition. In Proceedings of the 2020 International Conference on Machine Vision and Image Processing (MVIP), Teheren, Iran, 18–20 February 2020; pp. 1–7. [CrossRef]
46. Artacho, B.; Savakis, A.E. UniPose: Unified Human Pose Estimation in Single Images and Videos. *CoRR* **2020**, abs/2001.08095.
47. Insafutdinov, E.; Pishchulin, L.; Andres, B.; Andriluka, M.; Schiele, B. DeeperCut: A Deeper, Stronger, and Faster Multi-Person Pose Estimation Model. *CoRR* **2016**, abs/1605.03170.
48. Li, J.; Wang, C.; Zhu, H.; Mao, Y.; Fang, H.; Lu, C. CrowdPose: Efficient Crowded Scenes Pose Estimation and a New Benchmark. In Proceedings of the IEEE Conference on Computer Vision and Pattern Recognition, CVPR 2019, Long Beach, CA, USA, 16–20 June 2019; pp. 10863–10872. [CrossRef]
49. Ouyang, W.; Chu, X.; Wang, X. Multi-source Deep Learning for Human Pose Estimation. In Proceedings of the 2014 IEEE Conference on Computer Vision and Pattern Recognition, Columbus, OH, USA, 23–28 June 2014; pp. 2337–2344. [CrossRef]
50. Moon, G.; Chang, J.; Lee, K.M. PoseFix: Model-Agnostic General Human Pose Refinement Network. In Proceedings of the 2019 IEEE/CVF Conference on Computer Vision and Pattern Recognition (CVPR), Long Beach, CA, USA, 15–20 June 2019; pp. 7765–7773. [CrossRef]
51. Nie, X.; Feng, J.; Xing, J.; Xiao, S.; Yan, S. Hierarchical Contextual Refinement Networks for Human Pose Estimation. *IEEE Trans. Image Process.* **2019**, *28*, 924–936. [CrossRef]
52. Newell, A.; Yang, K.; Deng, J. Stacked Hourglass Networks for Human Pose Estimation. In Proceedings of the Computer Vision–ECCV 2016, Amsterdam, The Netherlands, 11–14 October 2016; pp. 483–499.

53. Núñez-Marcos, A.; Azkune, G.; Arganda-Carreras, I. Vision-Based Fall Detection with Convolutional Neural Networks. *Wirel. Commun. Mob. Comput.* **2017**, *2017*, 1–16. [CrossRef]
54. Xu, C.; Xu, Y.; Xu, Z.; Guo, B.; Zhang, C.; Huang, J.; Deng, X. Fall Detection in Elevator Cages Based on XGBoost and LSTM. In Proceedings of the 2021 26th International Conference on Automation and Computing (ICAC), Portsmouth, UK, 2–4 September 2021; pp. 1–6. [CrossRef]
55. Ren, X.; Zhang, Y.; Yang, Y. Human Fall Detection Model with Lightweight Network and Tracking in Video. In Proceedings of the 2021 5th International Conference on Computer Science and Artificial Intelligence, CSAI 2021, Beijing, China, 4–6 December 2021; pp. 1–7. [CrossRef]
56. De Miguel, K.; Brunete, A.; Hernando, M.; Gambao, E. Home Camera-Based Fall Detection System for the Elderly. *Sensors* **2017**, *17*, 2864. [CrossRef]
57. Sadreazami, H.; Bolic, M.; Rajan, S. Contactless Fall Detection Using Time-Frequency Analysis and Convolutional Neural Networks. *IEEE Trans. Ind. Informatics* **2021**, *17*, 6842–6851. [CrossRef]
58. Butt, F.S.; La Blunda, L.; Wagner, M.F.; Schafer, J.; Medina-Bulo, I.; Gomez-Ullate, D. Fall Detection from Electrocardiogram (ECG) Signals and Classification by Deep Transfer Learning. *Information* **2021**, *12*, 63. [CrossRef]
59. Bhattacharya, A.; Vaughan, R. Deep Learning Radar Design for Breathing and Fall Detection. *IEEE Sensors J.* **2020**, *20*, 5072–5085. [CrossRef]
60. Martinez-Martin, N.; Luo, Z.; Kaushal, A.; Adeli, E.; Haque, A.; Kelly, S.S.; Wieten, S.; Cho, M.K.; Magnus, D.; Fei-Fei, L.; et al. Ethical issues in using ambient intelligence in health-care settings. *Lancet Digit. Health* **2021**, *3*, e115–e123. [CrossRef]
61. Esteva, A.; Chou, K.; Yeung, S.; Naik, N.; Madani, A.; Mottaghi, A.; Liu, Y.; Topol, E.; Dean, J.; Socher, R. Deep learning-enabled medical computer vision. *NPJ Digit. Med.* **2021**, *4*, 5. [CrossRef] [PubMed]
62. Babar, M.; Rahman, A.; Arif, F.; Jeon, G. Energy-harvesting based on internet of things and big data analytics for smart health monitoring. *Sustain. Comput. Inform. Syst.* **2018**, *20*, 155–164. [CrossRef]
63. Syed, L.; Jabeen, S.; Manimala, S.; Elsayed, H.A. Data Science Algorithms and Techniques for Smart Healthcare Using IoT and Big Data Analytics. In *Smart Techniques for a Smarter Planet: Towards Smarter Algorithms*; Springer International Publishing: Cham, Switzerland, 2019; pp. 211–241. [CrossRef]
64. Tawalbeh, L.; Muheidat, F.; Tawalbeh, M.; Quwaider, M. IoT Privacy and Security: Challenges and Solutions. *Appl. Sci.* **2020**, *10*, 4102. [CrossRef]
65. Awotunde, J.B.; Jimoh, R.G.; Folorunso, S.O.; Adeniyi, E.A.; Abiodun, K.M.; Banjo, O.O. Privacy and Security Concerns in IoT-Based Healthcare Systems. In *The Fusion of Internet of Things, Artificial Intelligence, and Cloud Computing in Health Care*; Springer International Publishing: Cham, Switzerland, 2021; pp. 105–134. [CrossRef]
66. Redmon, J.; Farhadi, A. YOLOv3: An Incremental Improvement. *arXiv* **2018**, arXiv:1804.02767.
67. Russakovsky, O.; Deng, J.; Su, H.; Krause, J.; Satheesh, S.; Ma, S.; Huang, Z.; Karpathy, A.; Khosla, A.; Bernstein, M.S.; et al. ImageNet Large Scale Visual Recognition Challenge. *Int. J. Comput. Vis.* **2015**, *115*, 211–252. [CrossRef]
68. Pan, C.; Cao, H.; Zhang, W.; Song, X.; Li, M. Driver activity recognition using spatial-temporal graph convolutional LSTM networks with attention mechanism. *IET Intell. Transp. Syst.* **2020**, *15*, 297–307. [CrossRef]
69. Vasconez, J.; Admoni, H.; Cheein, F.A. A methodology for semantic action recognition based on pose and human-object interaction in avocado harvesting processes. *Comput. Electron. Agric.* **2021**, *184*, 106057. [CrossRef]
70. Zhang, C.; Yang, X. Bed-Leaving Action Recognition Based on YOLOv3 and AlphaPose. In Proceedings of the 2022 the 5th International Conference on Image and Graphics Processing (ICIGP), ICIGP 2022, Beijing, China, 7–9 January 2022; pp. 117–123. [CrossRef]
71. Zhao, X.; Hou, F.; Su, J.; Davis, L. An Alphapose-Based Pedestrian Fall Detection Algorithm. In Proceedings of the Artificial Intelligence and Security, Qinghai, China, 15–20 July 2022; pp. 650–660.
72. Cortés, X.; Conte, D.; Cardot, H. A new bag of visual words encoding method for human action recognition. In Proceedings of the 2018 24th International Conference on Pattern Recognition (ICPR), Beijing, China, 20–24 August 2018; pp. 2480–2485. [CrossRef]
73. Aslan, M.; Durdu, A.; Sabanci, K. Human action recognition with bag of visual words using different machine learning methods and hyperparameter optimization. *Neural Comput. Appl.* **2020**, *32*, 8585–8597. [CrossRef]
74. Nazir, S.; Yousaf, M.H.; Velastin, S.A. Evaluating a bag-of-visual features approach using spatio-temporal features for action recognition. *Comput. Electr. Eng.* **2018**, *72*, 660–669. [CrossRef]
75. Fang, H.S.; Li, J.; Tang, H.; Xu, C.; Zhu, H.; Xiu, Y.; Li, Y.L.; Lu, C. AlphaPose: Whole-Body Regional Multi-Person Pose Estimation and Tracking in Real-Time. *arXiv* **2022**, arXiv:2211.03375.
76. Fang, H.S.; Xie, S.; Tai, Y.W.; Lu, C. RMPE: Regional Multi-person Pose Estimation. *arXiv* **2016**, arXiv:1612.00137.
77. Wan, C.; Wang, L.; Phoha, V.V. A Survey on Gait Recognition. *ACM Comput. Surv.* **2018**, *51*, 1–35. [CrossRef]
78. Semwal, V.B.; Mazumdar, A.; Jha, A.; Gaud, N.; Bijalwan, V. Speed, Cloth and Pose Invariant Gait Recognition-Based Person Identification. In *Machine Learning: Theoretical Foundations and Practical Applications*; Springer Singapore: Singapore, 2021; pp. 39–56. [CrossRef]
79. Elharrouss, O.; Almaadeed, N.; Al-ma'adeed, S.; Bouridane, A. Gait recognition for person re-identification. *J. Supercomput.* **2021**, *77*, 3653–3672. [CrossRef]
80. Sun, F.; Zang, W.; Gravina, R.; Fortino, G.; Li, Y. Gait-based identification for elderly users in wearable healthcare systems. *Inf. Fusion* **2020**, *53*, 134–144. [CrossRef]

81. Liu, X.; Zhao, C.; Zheng, B.; Guo, Q.; Duan, X.; Wulamu, A.; Zhang, D. Wearable Devices for Gait Analysis in Intelligent Healthcare. *Front. Comput. Sci.* **2021**, *3*, 661676. [CrossRef]
82. Zhao, A.; Li, J.; Dong, J.; Qi, L.; Zhang, Q.; Li, N.; Wang, X.; Zhou, H. Multimodal Gait Recognition for Neurodegenerative Diseases. *IEEE Trans. Cybern.* **2022**, *52*, 9439–9453. [CrossRef]
83. Din, S.; Elshehabi, M.; Galna, B.; Hobert, M.; Warmerdam, E.; Sünkel, U.; Brockmann, K.; Metzger, F.; Hansen, C.; Berg, D.; et al. Gait analysis with wearables predicts conversion to Parkinson disease. *Ann. Neurol.* **2019**, *86*, 357–367. [CrossRef] [PubMed]
84. Rucco, R.; Agosti, V.; Jacini, F.; Sorrentino, P.; Varriale, P.; De Stefano, M.; Milan, G.; Montella, P.; Sorrentino, G. Spatio-temporal and kinematic gait analysis in patients with Frontotemporal dementia and Alzheimer's disease through 3D motion capture. *Gait Posture* **2017**, *52*, 312–317. [CrossRef]
85. de Oliveira Silva, F.; Ferreira, J.V.; Plácido, J.; Chagas, D.; Praxedes, J.; Guimarães, C.; Batista, L.A.; Laks, J.; Deslandes, A.C. Gait analysis with videogrammetry can differentiate healthy elderly, mild cognitive impairment, and Alzheimer's disease: A cross-sectional study. *Exp. Gerontol.* **2020**, *131*, 110816. [CrossRef]
86. Yamada, H.; Ahn, J.; Mozos, O.; Iwashita, Y.; Kurazume, R. Gait-based person identification using 3D LiDAR and long short-term memory deep networks. *Adv. Robot.* **2020**, *34*, 1201–1211. [CrossRef]

Disclaimer/Publisher's Note: The statements, opinions and data contained in all publications are solely those of the individual author(s) and contributor(s) and not of MDPI and/or the editor(s). MDPI and/or the editor(s) disclaim responsibility for any injury to people or property resulting from any ideas, methods, instructions or products referred to in the content.

Article

Are Health Information Systems Ready for the Digital Transformation in Portugal? Challenges and Future Perspectives

Leonor Teixeira [1,*], Irene Cardoso [2], Jorge Oliveira e Sá [3] and Filipe Madeira [4]

1. Department of Economics, Management, Industrial Engineering and Tourism (DEGEIT), Institute of Electronics and Informatics Engineering of Aveiro (IEETA)/Intelligent Systems Associate Laboratory (LASI), University of Aveiro, 3810-193 Aveiro, Portugal
2. Associação Portuguesa de Sistemas de Informação (APSI), 4800-058 Guimarães, Portugal
3. Department of Information Systems, Centro ALGORITMI, University of Minho, 4800-058 Guimarães, Portugal
4. Department of Informatics and Quantitative Methods, Research Centre for Arts and Communication (CIAC)/Pole of Digital Literacy and Social Inclusion, Polytechnic Institute of Santarém, 2001-904 Santarem, Portugal
* Correspondence: lteixeira@ua.pt

Citation: Teixeira, L.; Cardoso, I.; Oliveira e Sá, J.; Madeira, F. Are Health Information Systems Ready for the Digital Transformation in Portugal? Challenges and Future Perspectives. *Healthcare* **2023**, *11*, 712. https://doi.org/10.3390/healthcare11050712

Academic Editors: Giner Alor-Hernández, Jezreel Mejía-Miranda, José Luis Sánchez-Cervantes and Alejandro Rodríguez-González

Received: 21 December 2022
Revised: 19 February 2023
Accepted: 24 February 2023
Published: 28 February 2023

Copyright: © 2023 by the authors. Licensee MDPI, Basel, Switzerland. This article is an open access article distributed under the terms and conditions of the Creative Commons Attribution (CC BY) license (https://creativecommons.org/licenses/by/4.0/).

Abstract: Purpose: This study aimed to reflect on the challenges of Health Information Systems in Portugal at a time when technologies enable the creation of new approaches and models for care provision, as well as to identify scenarios that may characterize this practice in the future. Design/methodology/approach: A guiding research model was created based on an empirical study that was conducted using a qualitative method that integrated content analysis of strategic documents and semi-structured interviews with a sample of fourteen key actors in the health sector. Findings: Results pointed to the existence of emerging technologies that may promote the development of Health Information Systems oriented to "health and well-being" in a preventive model logic and reinforce the social and management implications. Originality/value: The originality of this work resided in the empirical study carried out, which allowed us to analyze how the various actors look at the present and the future of Health Information Systems. There is also a lack of studies addressing this subject. Research limitations/implications: The main limitations resulted from a low, although representative, number of interviews and the fact that the interviews took place before the pandemic, so the digital transformation that was promoted was not reflected. Managerial implications and social implications: The study highlighted the need for greater commitment from decision makers, managers, healthcare providers, and citizens toward achieving improved digital literacy and health. Decision makers and managers must also agree on strategies to accelerate existing strategic plans and avoid their implementation at different paces.

Keywords: digital transformation; health information systems; emerging technologies; Health 4.0; empirical study

1. Introduction

Globalization, associated with many rapidly evolving factors, such as the COVID-19 pandemic, leads to an ever-increasing need to share health data outside of the physical space where they are generated [1,2]. Demographic changes, increased chronic diseases, rising health spending, and fairer healthcare access are global challenges [3], which, if associated with the increase in people's average life expectancy and the growth in their literacy, show the greater importance of new Health Information Systems (HISs) that allow efficient communication between the Health Systems (HS) and their stakeholders. This has resulted in recent years in a new generation of emerging technologies that offer new opportunities for healthcare delivery and the practice of medicine while also ensuring greater efficiency of HIS and more responsive communication channels.

HIS includes mechanisms for capturing, processing, analyzing, and transmitting any needed information in health services whilst also having an important role in care planning, management, and even in research for public health [4]. With a growing need for decentralized and remote work, HIS also plays a key role since they support communication among geographically dispersed actors, promote solutions to value chain management, and support new business models. In addition, digital health offers a valuable opportunity to handle health issues, such as the pandemic situation, with near real-time responsiveness [5]. This trend is transversal to other areas of knowledge, with different types of applications, as demonstrated in the study of Epizitone et al. [6]. Information and Communication Technologies (ICT) applied to the health context have been the subject of several research works, with a greater incidence in the Digital Transformation (DT) of this sector [1,7], such as processes related to healthcare delivery models and medical practices [8]. Some studies point to digitalization as a future priority in the health and public sector, reinforcing the need to adopt intelligent technological applications [9–12] and connectivity mechanisms [13,14]. Cavallone and Palumbo [15], for example, stated that Industry 4.0 (I4.0), Artificial Intelligence (AI) and Digitalization are revolutionizing the design and the delivery of care.

Despite the progress observed in recent decades, the gap that emphasizes the need to create solutions with responses to extreme events is demonstrated by some authors [16,17]. In addition, Gehring and Eulenfeld [18] argued that there is still a pressing need to significantly improve the infrastructure and functionalities of the HIS for the benefit of users and also for research in areas such as biomedical sciences, health sciences, and also computer and information sciences.

Knowledge of the current HISs' development state is a requirement when researching future directions. An analysis of current HISs [19] identifies five different groups of obstacles that limit these systems' application and development—(i) technical problems, (ii) usage problems, (iii) quality problems, (iv) operational functionality, and issues related to (v) maintenance and support. In addition, the study of Khubone et al. [20] discussed a set of challenges that should not be ignored when adopting HIS, which are mainly related to the lack of technical consensus, poor leadership and limited human resource, staff resistance and lack of management, and non-engagement of the users. In the area of telemedicine, Tabaeeian et al. [21] compiled a set of barriers that should not be ignored when implementing such solutions, concluding that "the future of telemedicine depends on consistency in system usage and minimizing problems, increasing system compatibility with users and learning how to use". In turn, the development of a HIS requires several connections between local systems, which can be small county or regional systems or national platforms, a connection between different sectors (such as public, private, and other), and can even need the articulation at an international level (e.g., between European countries, or USA and Canada).

There is a generally recognized importance of HISs by health policymakers. In Portugal, this is a reality reflected by the creation of the National Strategy for the Health Information Ecosystem 2020 (ENESIS-2020) [22]. This working framework elected the main goal to have more efficient information processes that would (i) increase the general sharing of information and knowledge between all actors to promote the literacy and general health of citizens; (ii) offer greater efficiency for healthcare providers; (iii) offer greater rationalization of resources with an impact on global efficiency and health management; and (iv) offer an alignment of HIS strategies with other European countries (i.e., standards and interoperability).

Considering the impact that these systems have on healthcare management and the medical practice, whilst also considering the current technological trends currently reported in the literature, this study aimed to understand the current state of HIS in Portugal and its main challenges, as well as foresee future trends about the use and impact of this type of systems that can benefit with emerging I4.0 technologies. The authors adopted a two-phase methodological approach. First, the authors carried out a literature review, covering topics related to the role of HIS with existing technologies, namely the ones that could enhance the development of these systems. A focus on digitalization technologies was pursued. Secondly, the results from the literature review were compared to the current Portuguese HIS situation. This was achieved by conducting an empirical study based on interviews with some representative HIS actors in Portugal while also considering some official documents, legislation, and strategies/policies currently in force in Portugal.

Theoretically, this work aimed to advance some likely future trends and scenarios for the Portuguese's HIS based on the Industry 4.0 drivers. In practical terms, it was expected to provide some recommendations to practitioners and decision makers on the opportunities of DT and the expected main impacts on the Portuguese HIS. As an additional contribution, it was also intended to identify some aspects that would enable societal health empowerment through the adoption of emerging technologies.

This article is structured into six sections. The current section presents the gaps, the motivations for this study, and the main objectives to be achieved. In Section two, a literature review is conducted. The third section describes the methodology adopted in this study. Section four presents the main results, challenges, and future scenarios for HISs in Portugal obtained from the point of view of the study participants (i.e., the interviewees) and from documents, legislation, and strategies/policies analyzed. In Section five, the authors present their views on future challenges and scenarios for the Portuguese HISs. Finally, the last section is devoted to some final remarks.

2. Theoretical Framework

2.1. Evolution of HIS to Date

The practice of medicine has changed significantly in the last 50 years, with ICT making a strong contribution to this change. There have also been changes in the information paradigm itself, moving from institution-centered information to a more patient-centered approach [8].

Due to globalization and other circumstantial factors, such as the COVID-19 pandemic, there is an increasing need to share health data outside of the physical space where they are generated [1,2,20]. Furthermore, the change in the clinical information consumption patterns, where the citizen assumes an increasingly significant role, is also a reason that highlights the importance of HIS. The final use given to the data, which before was focused on responses to clinical practice, has gradually assumed greater importance also in health planning and clinical and epidemiological research.

Over the past decades, HIS has had various stages, as briefly described in Figure 1, which presents HIS usage in health in four digital eras—from Health 1.0 to Health 4.0.

The first era, also called Health 1.0, began in the 1960s and was associated with the introduction of patients' records. During this period, HIS was exclusively designed to store patients' information locally, in paper or digital format. Access was limited and only available in each service, department, or institution.

Afterward, in the Health 2.0 era, which started in the late 1980s, HIS was further developed, so to allow the grouping of patients' data in digital repositories with private access to authorized users and necessary services, materializing the Electronic Health Record (EHR). The information was then citizen-centered [23], increasing the role of patients/citizens in HISs as they started to have limited access to the information recorded by health professionals.

The Health 3.0 era, which began in the early 2000s, was characterized by the development of Personal Health Records (PHR). During this time, the main goal of HISs was to support the citizen's life cycle, with data introduced by both healthcare providers and the citizen himself. This further advancement of HIS allowed patients to engage proactively and collaboratively in their care. Thus, data were co-created and maintained by both providers and patients [24], and society moved from institution-centered information to a more patient-centered approach [8]. PHR quickly became attractive, as it allowed the centralization of each citizen's health data in digital platforms widely available, engaging both providers and receivers in healthcare deliverance.

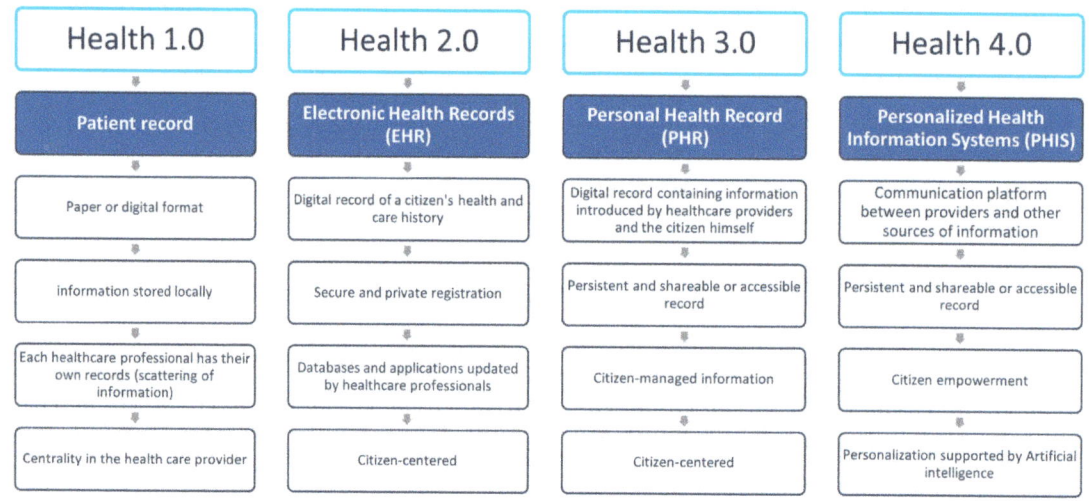

Figure 1. Evolution of HIS in health according to technological "eras".

Nevertheless, most countries and providers still felt that there was a need to have a connected decentralized health system. This was made possible by new communication platforms and emerging technologies, as well as the use of Artificial Intelligence (AI), developed during this past decade. Personalized Health Information Systems prevail, but with a vastly improved connectivity between healthcare actors, in what is called the Health 4.0 era.

2.2. The Fourth Industrial Revolution and the HIS

The creation of digital ecosystems through a set of tools that enable the connection between the digital and the physical worlds is paramount in Industry 4.0 (I4.0) technologies. This last concept, which is a product of the fourth Industrial Revolution (4IR), is intricately related to principles such as interoperability, decentralization, real-time responsiveness, data-based services, virtualization, and modularity [25]. I4.0 uses, for the technologies developed, a variety of concepts such as automation and data exchange through cloud computing, Big Data, Internet of Things (IoT), Robotics, 5G Technologies, Virtual and Augmented Reality, Additive Manufacturing, Cyber-Physical Systems, and AI, among others [25–27].

With the promise of empowering the creation of better healthcare services, 4IR enhances the personalization and individualization of the services provided, the optimization of resources associated with the practice of medicine, as well as the promotion of health based on preventive models [28,29]. Concepts include Health 4.0 [25,30,31] Healthcare 4.0 [32], Medicine 4.0 [33] or Care 4.0 [34], Hospital 4.0 [35,36], or more specific applications such as Surgery 4.0 [37], represent only some approaches/applications that make use of I4.0 emerging technologies to create new models of medical practice and health promotion. These concepts, in their different terminologies, represent just some extensions of the I4.0 principles applied to the medical/health area.

Several authors considered that Health 4.0 could not be dissociated from the Digital Transformation (DT) concept, as the latter is used not only in the deliverance of care but also in the governance processes of all the value chains [38]. Health 4.0 makes possible the future virtualization of healthcare delivery and medical practice [31].

Connectivity and computing power, enhanced by emerging technologies, are crucial factors in the deliverance of need-oriented care, considering individualized approaches based on preventive and predictive models. The emerging technologies of I4.0, when applied to healthcare, can greatly enhance the productivity of the providers, as well as promote the creation of preventive care models since they allow for early detection of health-related anomalous situations [28]. It is then possible to avoid future health issues (and costs) for both citizens and society alike.

Some other applications include, for example, additive manufacturing or 3D printing, which allows faster and more personalized creation of health products, such as implants, tools, and specific devices, according to the different needs and requirements of each patient [39]; or robotics, which can be used in surgery and physiotherapy services, fostering improvements in performance, movement, and control. IoT allows connectivity with mobile and other devices, enabling the automatic collection of human data. Other examples include Big Data which, in addition to storing a large amount of data, allows, through Data Analytics, the identification of patterns and trends, enabling the decision-making on predicted problems of future health events. Finally, AI can help manage and analyze data, make decisions, and identify and forecast upcoming health trends or issues [40].

Recently, Large Language Models (LLMs) with billions of parameters have brought a big boost to the base models of deep learning, and several architectures have emerged, such as ChatGPT and BERT (which are among the best known). The research and interest around these deep learning technologies are huge and promising. ChatGTP, launched in late 2022, presents both great potential and challenges for the Medicine and Healthcare sectors.

AI-assisted technologies have for some time (and with different degrees of success) been employed in several aspects' areas of healthcare. For example, in 2016, IBM launched Watson for Oncology [41], an AI clinical decision-support for cancer treatment, that achieved moderate success before being discontinued for failure to achieve the same clinical marks as real-time physicians [42]. Other examples come from radiology or pathology, where AI-assisted tools are being employed to identify tumors, differentiate between healthy and abnormal tissue samples, and provide clinicians with diagnostic suggestions, which lead to faster and more efficient results and prompt treatment [43–45].

So, if AI is not new to the healthcare sector, it has mostly been used in specific areas and as support for clinical decisions. ChatGTP is different! Firstly, it was not designed with a specific medical intention in mind. Secondly, it is widely available. Finally, it has shown a considerable level of clinical accuracy [46], e.g., achieving the passing threshold of the USMLE (the United States Medical Licensing Examination). These characteristics can lead to important innovations in the healthcare sector, namely in developed countries, where this sector is struggling to deliver good healthcare in an aging society:

- Triage of patients—LLMs, such as ChatGTP, can be used as primary points of contact between the patient and the healthcare system, triaging patients and decreasing the burden on the healthcare system. Moreover, these tools can also reduce clinical biases, providing a standard of consideration to every patient, independent of personal characteristics.
- Medical scribe functions—modern healthcare systems require the input by the physicians or their assistants of large amounts of data. LLMs can be used to help or reduce this workload, performing note-taking tasks and writing brief patient summaries and presentations. One such example is a recent Microsoft announcement that Teams would provide note-taking features for meetings [47].
- Diagnosis assistance—LLMs can become important tools to help clinicians to make an evidence-based differential diagnosis as unbiased tools that can be trained not only with large amounts of medical information but which can also be updated with the latest relevant data, including innovative academic studies or clinical trials.

LLMs can also drive innovation and competition in the healthcare sector. Medical Sciences are, by nature, an uneven market field, where the provider (physicians) have all the knowledge, and the user (patient) does not know what he/she is getting before the service is complete. The digital revolution of the past few decades has reduced this gap, both empowering patients with information and promoting health literacy in general. Just in Europe, it is estimated that half of the patients look for health information online [48]. LLMs, such as ChatGTP, can increase this trend, promoting more informed choices and leading to more demanding customers (i.e., patients). This has the chance to drive innovation and promote excellence across the medical field.

There are, nevertheless, some challenges. ChatGTP, for example, remains an imperfect tool, with the CEO of OpenAI, the company behind this tool, recently twittered that ChatGPT remains "incredibly limited", and that "It's a mistake to rely on it for anything important right now" [49]. Obstacles such as misinformation, artificial hallucination, data protection, or ethical questions remain relevant and should, if not limit, at least warrant some cautions in the use and dissemination of these tools.

In conclusion, LLMs present significant opportunities for the healthcare sector, but a careful approach involving practitioners, patients, policymakers, and other relevant field professionals are needed before they become mainstream. With all these developments, the Health 4.0 concept is closer than humanity imagines.

Thus, the main pillars of Health 4.0 (and its derivations) are framed within digital ecosystems and are focused on people, technologies, and co-design because it presupposes a change in hospital business models to an ever-increasing citizen-centered care provision. Technology insofar represents the drives that are at the basis of the Health 4.0 concept itself, and without which its implementation would not be possible. Finally, to co-design patients' involvement is a requirement, not only in the HISs as active actors but also in the design and development of these systems, to allow their future participation [29].

3. Materials and Methods

Starting from the objective that supported this research (to understand the current state of HIS in Portugal and its main challenges, as well as foresee future trends), an empirical and exploratory study was carried out, supported by a qualitative methodological approach, whose research design is presented below. So, the research protocol starts with a comprehensive literature review to define the boundaries of the research subject and the research questions, as well as to build the interview guidelines. Next, an intentional sample was selected from a population of HIS users in the private and public Healthcare organizations and Governmental entities involved in the definition of the HIS strategy in Portugal. In the third phase, the fieldwork that consisted of the execution of the interviews, as well as the selection of strategic documents, was conducted. The fourth stage comprised the processing and analysis of data, and finally, the analysis of findings and the consequent production of conclusions.

Figure 2 presents the research design followed in this study, where the main research question (Q1) was broken down into two sub-questions (Q1.1 and Q1.2), and these in other more specific questions. To collect data, different sources were used, which are broadly categorized as (i) analysis of strategic documents and legislation; and (ii) interviews. Strategic documents report a set of initiatives launched by government entities that can regulate and promote approaches for modernization in this sector. These documents included digital platforms, official documents, and legislation produced by the entities responsible for the definition of strategy and management of information on the health ecosystem at the national level, i.e., ENESIS-2020 [22] and ENESIS 20/22 [50]. The interviews were selected insofar, as they are considered one of the most proper methods to explore participants' experiences and/or reconstruct past events.

3.1. Data Collection Methods and Procedures

As mentioned, the qualitative approach was adopted in the data collection, combining: (i) the analysis of strategic documents and legislation; and (ii) semi-structured interviews conducted with different entities involved both in the definition of the HIS strategy in Portugal and as users of these HISs. The interviews (audio-recorded) were applied between August and December 2019, following previously developed scripts, oriented and adapted to the interviewees' profiles, and structured according to the specific goals mentioned in Figure 2. Each script included ten questions in addition to those that anonymously characterized the interviewee. Three types of interviewees were found, namely: (i) managers, which included health professionals with management or coordination positions; (ii) health professionals (physicians and nurses); and, also, (iii) users of HS. To conduct the interviews, the participant's consent was obtained, and confidentiality and anonymity were guaranteed. All participants were interviewed in person by the researchers.

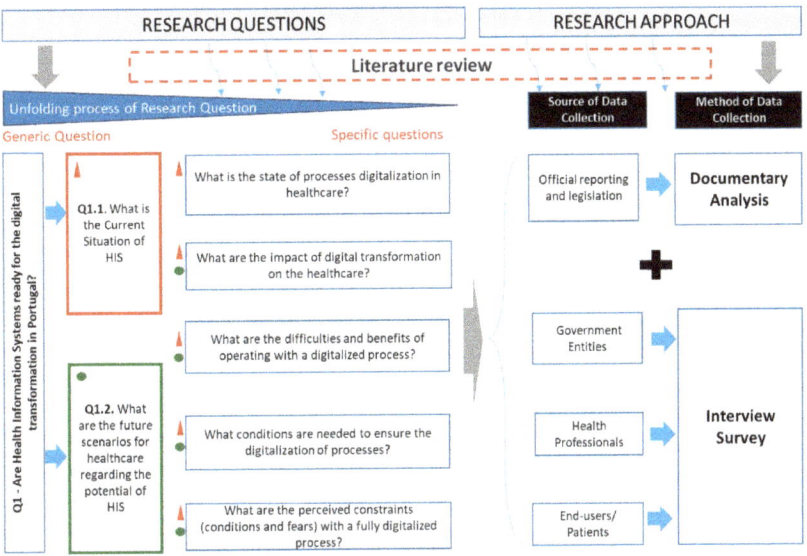

Figure 2. Research Design.

3.2. Data Analysis Methods

Due to the nature of the data obtained, qualitative analysis methods were used. Both the strategic reports and the transcribed interviews were subjected to a thematic-categorical content analysis, which represents a technique to capture the meaning of texts relating to a particular phenomenon under study [51]. For this purpose, the typical phases of content analysis were followed, which are based on: (i) the organization of the material and the definition of the procedures (pre-analysis); (ii) the identification of the categories that arise with the interpretation of the text (exploration); and finally, (iii) the treatment of the results, where we sought to interpret the data around the categories created. A manual coding procedure was used in this process.

3.3. The Sample Profile of Respondents

Given the exploratory nature of the study, intentional sampling was chosen. To minimize the limitations caused by the reduced sample size, particularly those related to the replicability and reliability of the study, we sought to diversify the demographic regions from where the participants originated. Table 1 presents the sample, composed of fourteen participants distributed as five health professionals (coded with the suffix P), six health professionals with management positions (suffix M), two users (suffix U), and one member of a governmental entity (suffix GE). Table 1 also presents other data that allow for a better characterization in terms of the region and organization to which they belong, the regime (public or private) in which they work, their profiles and positions, as well as their occupations and age groups. All interviewees mentioned being users of the HIS, although to different degrees.

4. Results

Based on the analysis of the interviews and some strategic documents [22,50,52,53], it was intended to answer the major research question Q1, see Figure 2. To achieve that goal, it is equally important to obtain answers to questions Q1.1 and Q1.2., i.e., to understand how the current state of the HIS in Portugal is as well as their trends in terms of future development.

Table 1. Characterization of the sample profile of respondents.

Interv.	Region	Organization	Regime	Profile (Position)	Profession	Age
E1-U	Azores	Praia and Vitoria Health Centre	Public	User	Computer Technician	–
E2-P	Azores	Praia and Vitoria Health Centre	Public	Professional	Doctor	41–51
E3-M	Azores	Praia and Vitória Health Centre	Public	Professional (Clinical Director)	Doctor	<40
E4-P	Azores	Praia and Vitória Health Centre	Public	Professional	Physician	<40
E5-M	Algarve	Family Health Unit Sol Nascente	Public	Professional (Hospital Coordinator)	Doctor	>51
E6-P	Algarve	Lusíadas Hospital	Private	Professional	Doctor	>51
E7-M	Aveiro	Finess Medical Clinic	Private	Professional (Clinical Director)	Doctor	>51
E8-P	Aveiro	Finess Medical Clinic	Private	Professional	Nurse	<40
E9-P	Aveiro	Tâmega e Sousa Hospital Centre	Public	Professional	Doctor	<40
E10-M	Aveiro	Ovar Hospital	Public	Professional (Chairman of the Board of Directors)	Manager	41–51
E11-M	Lisbon	Cascais Hospital	PPP	Professional (Chief Medical Information Officer)	Doctor	41–51
E12-M	Lisbon	Cascais Hospital	PPP	Professional (Chief Nursing Information Officer)	Nurse	41–51
E13-GE	Lisbon	Shared Services of the Ministry of Health	Public	Government Entity (President)	Manager/Doctor	41–51
E14-U	Faro	Retired	Public	User	Nurse	>51

4.1. Analysis and Reflection on the Current HIS

4.1.1. Current HIS Situation: Document Analysis

From document analysis, which includes online platforms, and legal documents that approved strategies for HIS, for the last two three-year-olds [22,50], it was perceived that police decision-makers consider that IS Healthcare can act in any organization that involves healthcare (public and private hospitals, clinics, clinics, pharmacies, nursing services, and primary healthcare, among others).

According to the Resolution of the Council of Ministers of 26 July 2017 [53], the HIS seems to include "all local and central information subsystems, in the entities of the National Health Service (NHS) and third parties integrated with it, to make available to several users all useful information to health literacy and health self-management (citizens), providing healthcare (health professionals), system management (local and central managers), health research and cross-cutting needs for public administration" [53].

The main strategy for HIS, entitled Health Information Ecosystem Strategy 2020 [20,40] adopted by Council of Ministers Resolution No. 62/2016, replaced by ENESIS 20/22 [50], was approved on 7 January 2020 after public consultation.

The ENESIS 2020 [22], as well as the following (ENESIS 20/22) [50], assume the objective of promoting the Digital Transformation of the health sector in Portugal and creating the conditions that allow the evolution of the Health Information Ecosystem (eSIS). They seek to respond to the priorities defined in terms of health policies, extending to the entire Health System and ensuring a common vision for the area of IS/IT. Analyzing the strategy still in place, ENESIS 20/22, the authors considered in general terms that it promotes a citizen-centered approach, ensuring simple and timely access to healthcare and improving his/her experience with the system [50]. The implementation of the strategy was structured in a set of six axes, namely: (i) access to healthcare throughout the life cycle of the citizen; (ii) training and empowerment of citizens; (iii) efficiency and sustainability of the health system; (iv) quality and safety of healthcare; (v) prevention, protection, and promotion of health; and (vi) training of professionals in organizations. Some of these axes (e.g., access to healthcare throughout the life cycle of the citizen; training and empowerment of citizens) seem in line with some literature which says that the healthcare sector begins to adopt a perspective less and is less based on hospital space and health professionals, and more focused on the citizen and their needs. Yet, in relation to the aspects such as efficiency and sustainability of the health system, quality and safety of healthcare, prevention, protection, and promotion of health, and training of professionals in organizations, they can find in other authors.

Regarding the Health Information Ecosystem in Portugal and considering the results obtained from documental analysis (placing references to the websites and the legislation/strategy), the authors could verify that there was a comprehensive evolution of the HIS in Portugal, which followed the evolution of the HIS in general [8,24] and in several countries of Europe, which is reflected in its strategy. Adopting a holistic and citizen-centered view, the HIS in Portugal increasingly tries to respond to the information needs around the life cycle of the person, from birth to death, being visible in the definition of the Health Information Ecosystem (eSIS), "a set of technologies, people and processes that intervene in the life cycle of information related to all dimensions of the health of citizens (...) regardless of the place of care and/or organizational barriers [52] (p. 3736)". It is precisely in this vision that the authors make the description of the HIS existing and/or under development in Portugal.

In Portugal, many HISs can be found in various areas (Administrative and patient management, Clinical, Financial, Management and planning, Informative and IT, and Communication), according to SPMS [54]. However, the authors focused their analysis primarily on HIS connected with the life cycle of citizens.

Thus, following a citizen-centric perspective, Table 2 presents the main HIS classified according to the different stages of people's life cycle, i.e., birth; (ii) health and well-being; (iii) disease, which may be acute and/or chronic; (iv) aging; and (v) Death. Some of the HISs are transversal to the various stages of the life cycle, serving the citizen from birth to death, such as the 'National Register of Users' (RNU) and the 'Electronic Health Record (RSE).

Table 2. Health Information Ecosystem in Portugal.

Categories of Ecosystem of His (ESIS)	Designation	Description
Transversal of all NHS (National health service)	RNU; SER; SNS24; SINAVE	These HIS have the role of centralizing and distribution of information for NHS users
Life Cycle—Birth	Birth News; to Birth Citizen, Health Child's Bulletin and Youth's Health Bulletin and, eBulletin of Vaccines.	To receive a new citizen in society and, to monitor him/she in terms of surveillance and/or monitoring of public health
Life Cycle—Health and Wellness	Daily of My Health, SISO, SIIMA; SClinic CSP, RENTEV and others	This cycle comprises the systems that accompany citizens in a perspective of prevention and promotion of health
Life Cycle—Acute or Chronic Disease	ICC, Sclinic Hospital, SClinic CSP; RCU2, SINAVE, PEM, SI VIDA; RNCCI; and, CNTS	These HIS serve to accompany the user in his/her disease process, allowing the recording, diagnosis, and treatments, in all clinical episodes
Life Cycle—Aging	RECM; RNCCI.	HIS is intended to support clinical practice in the adoption and maintenance of healthy life models by the elderly
Life Cycle—Death	SICO	The main objective is dematerializing the process of certification of deaths and better articulation between the entities involved in the process.

4.1.2. Findings from Interviews

The results obtained from these data are presented according to the following themes that have already emerged from the literature review with the unfolding of the research question: (i) the state of digitalization of clinical information processes; (ii) the impact of DT on the HS; (iii) difficulties and benefits of operating with digitalized processes; (iv) conditions for dematerialization; and (v) reduction of info exclusion.

The State of Processes' Digitalization

The results pointed to two distinct groups of respondents' thoughts. The first was satisfied when asked about the state-of-the-art of HIS in their organizations and claims to have achieved a significant digitalization through DT, as can be read in the following statements:

The process of digitalization is almost consolidated in the public sector, (...), we have problems with very large hospitals with very old and poorly computerized systems, both at the regional and national levels that delay the process as a whole. (E13-GE)

We are taking steps towards full integration in terms of system development (...). Portugal is far ahead, compared to many countries, in Europe. (E11-M)

We are even implementing the paperless hospital project which is a project that has had good and referenced results, our hospital can get 70% of patients to leave the hospital without paper. (E10-M)

The second was more cautious in their statements, being more skeptical, saying that although there has been a significant advance in recent years, there is still a lack of integration and communication between systems, as interpreted by the following comments:

(...) it has evolved into a strategic concept, but there is a long way to go, there is a lack of clinical information in the interaction between private and public institutions (...) (E6-P)

(...) there is still a lot of lack of communication between hospitals and health centers. A lot of time is spent transcribing the analyses (...) it is necessary to continuously improve the software (...) (E3-M)

Impact of Digital Transformation on the Healthcare

In general, respondents saw digital health as a way to promote health. One interviewee referred that DT supports another way of doing medicine and promoting citizens' health with an impact on the creation of new business models in clinical practice.

Digital health is another type of (...) health service, and another way of doing health" (...) to telehealth, but is connected for example with preventive medicine, with precision drugs. (E13-GE)

However, and in general, despite the potential benefits that the interviewees see in the eventual digitalization of processes, they were also unanimous about the challenges that DT brings. Thus, some benefits and challenges emerged, as can be seen in the following comments:

These platforms can complement and help diagnosis (...). (E1-U)

The potential to do good, to change health (...) the transformation of health, from one-on-one health practice to a population health practice. (E11-M)

There are already positive impacts, but the centrality of the patient in the system does not exist. (E6-P)

Difficulties and Benefits of Operating with Digitalized Processes

As regards difficulties in the operationalization and use of systems, some respondents with management responsibilities reported difficulties in the human resources area related to resistance to change. Additionally, difficulties associated with the functioning of the HIS, namely: slowness, redundancy, lack of response to clinical practice, and lack of interoperability (communication and integration) between public and private institutions, were also pointed out by some respondents:

(. . .) it has more to do with people than with technology (resistance to change, stability of teams, continuous training). (E11-M)

(. . .) lack of integration of systems between institutions. (E2-P)

(. . .) there is no standardization of the systems themselves, they are always different systems. (E4-P)

Regarding the expected benefits of digitalization, respondents (E3-M) (E4-P) (E5-M) (E10-M) perceived some benefits in the digitalization process, such as (i) greater agility in information flows, with access to information in almost real-time; (ii) more security, in terms of access to data only by authorized users; (iii) less propensity to human error; (iv) greater capacity to share information between services and organizations providing of healthcare; (v) easier of access to health services without geographical restrictions; (vi) reducing paper circulation and consumption, and (viii) cost savings often associated with the repetition of exams to better understand everything that surrounds the activity of healthcare.

Conditions for Digital Transformation

The results pointed to the need to work on the aspects related to legislation, namely with governmental entities of each country and even outside the country. For example, in the sharing and use of health data from wearable devices, the greatest trouble is with the issues of use and legitimacy. With the technologies currently available, citizens can generate data to complement their health records through smart devices. To support the use of these technologies for health, legislation and some joint work with health regulators is needed. The results of the interviews also suggested the need for standards to achieve the interoperability of systems at the international level.

(. . .) it is necessary to create legitimacy to define the use of digital technology [wearables] because it is different if I use it to record my health data, or if the data generated by these technologies can be used to make diagnoses or suggest therapy. This is too important to be seen at an international level to define interoperability standards and rules. (E13-EG)

The ENESIS-2020 [22] and ENESIS-20/22 [50] assume the objective of promoting the DT of the health sector in Portugal and creating the conditions that allow responding to the priorities defined in terms of health policies, extending to the entire HS and also privileges a citizen-centered approach, ensuring simple and timely access to healthcare and improving his experience with the system.

Conditions Required for the Reduction of Info Exclusion

When it comes to ensuring the digitalization of processes, the question arises about citizens' digital literacy, which may represent an enabling factor or an obstacle in the functioning of processes in that ecosystem. The reduction of info exclusion was referred to by 12 (86%) of interviewees as an important challenge to overcome.

On this subject, the interviewees reported that the reduction of illiteracy depends on several factors, namely: (i) the opening of the user to this type of knowledge and innovation; (ii) the development of accessible systems; and, also, (iii) the monitoring and training of older people and/or those with greater difficulty to ensure health literacy and digital literacy. The responsibility for this training and monitoring would be on the government, health organizations, educational institutions, and community entities, such as the town halls and parish councils. Supporting these findings, we have the comments presented below:

(. . .) the current population is very ageing (. . .) it does not easily adapt to IT. The state should ensure minimal training, monitoring, and simplify the development of these technologies (. . .). (E1-U)

(. . .) we have pioneering projects such as Citizen HOSP that, through our social workers, support users to take advantage of the use of IT in access to health services (. . .). (E10-M)

The implementation of ENESIS-2020 [22] and ENESIS20/22 [50], developed in recent years, is still in the consolidation phase, presenting, however, a significant advance in terms of the number of new digital services, namely digital platforms such as the Health Web Portal and other applications that can be accessed from the Citizen Web Portal.

These systems, in turn, tend to respond to certain stages of a citizen's life, ranging from the simple registration of citizen follow-up to systems that accommodate the entire clinical history when in situations of illness (severe or chronic). This concept appears implicitly in ENESIS-2020 [22], defining it as a set of technologies, people, and processes that intervene in the life cycle of information related to all dimensions of citizen health and related, regardless of the place of care and/or organizational barriers.

4.2. Prospects for HIS in Portugal and Scenarios

Looking at future scenarios and based on the analysis of the results, three categories were shown that allowed to outline some scenarios for the future of the HIS in Portugal (i) Medicine Practices; (ii) Technologies; and (iii) Fears and Challenges.

4.2.1. Medicine Practices

The way the interviewees saw the evolution of the practice of medicine associated with technological evolution was dual. If, on the one hand, they understand that technologies combined with the greater use of AI will make medicine richer, more dematerialized, advanced, and effective (namely by speeding and accuracy in diagnosis, self-learning, and knowledge because it is based on predictability and allows personalized treatments), with an impact on the increase in quality average life expectancy, on the other hand, they consider that these benefits cannot imply the loss of the doctor–patient relationship nor can the data overlap to the psychological reality of the patient in the interpretation of his disease.

(. . .) Reduction of doctor-patient contact because computer solutions will compare certain standards by AI and allow diagnostics, without the patient presence. (. . .) a great combination of general medical knowledge with computer knowledge. (E1-U)

The practice of medicine in the future will be more dematerialized, remote[telemedicine], a preventive and precision medicine (. . .) the citizen will be more involved in his/her health/disease and the decisions about it, he/she will now his/her test results, and he will already bring the data stored in digital media. (E13-GE)

The last comment highlighted the core of current health strategies and policies, which focus on the citizen/patient, and the centrality of the patient with his/her information would enhance preventive models in health and accuracy in personalized diagnostics and treatment.

However, it should be noted that the results also identified a gap between the progression of technology and the way healthcare is organized, the latter being associated with strategy, management, and legislation:

(. . .) technology is evolving and the way we organize ourselves to supply healthcare is not advancing at the same pace. (E13-GE)

4.2.2. Technologies

On this subject, some health professionals were skeptical about the adoption of technologies in medical practices, believing in a worsening of the social part, with a negative impact on the patient–doctor relationship, contrary to other professionals who had an optimistic view of the future of HIS.

(. . .) there has been a decrease in doctor-patient confidence and (. . .), this system, although useful, can aggravate even more this situation. (E9-P)

(. . .) we need robust systems to treat this information, such as Business Intelligence or Data Mining, which are being implemented in our hospital (. . .) technologies allow us to innovate health, better manage resources, know patients (. . .). (E10-M)

In ten years, I think it will be possible to computerize almost total medical information. (E9-P)

Some interviewees, such as (E13-GE) and (E1-U), went further in what they think is the future of HIS, referring to the decrease in interfaces between technology and users, the existence of speech recognition software to support health professionals filling EHRs, the increasing existence of intra-devices, and the storage of data by patients themselves for reasons of cybersecurity, privacy, and data sharing.

4.2.3. Fears and Challenges

Faced with the idea of a fully digitalized reality, most respondents mentioned concerns about the confidentiality of information (who accesses the data and with what intention), as well as other types of threats (e.g., cyberattacks). The dehumanization in the provision of care also arises as an apprehension that stems from the reduction of physical contact between doctors and patients (for example, health professionals can make decisions based on a set of data without the need for the physical presence of the patient), thus losing the emotional aspects characteristic of human interaction, typical of traditional medicine. The comments presented below reflect these fears:

(...) lose the patients' data (...). (E2-P)

(...) who has access to this information and what are you going to do with it? (E11-M)

Some of the pointed-out fears and negative opinions about ICT evolution can be seen as challenges by HIS developers. The following comments can be illustrative:

(...) doctors must rediscover themselves, as coaches, people who guide reading. (E13-GE)

(...) the data still cannot sustain the psychological reality of the patient in the interpretation of his disease (...) we must not forget that we have biological complexity and that the Human being is not purely data. We do not treat data, we treat people. (E6-P)

4.2.4. Possible Scenarios

Based on the previous categories and following the three dimensions identified and described above, some relevant concepts were found, which allowed the design of future scenarios for the HIS.

(i) Medical Practices—the following concepts were found:
- Precision Medicine/Individualized—a medicine whose treatment is specific to a particular patient.
- Preventive Medicine—in which the focus is to keep healthy instead of curing the disease.
- Point-of-Care (Telemedicine)—allowing citizens to be physically distant from medical centers to have access to expert diagnoses.
- Assisted Medical Practices—in which machines (robots) with AI embedded start to help or even replace health professionals in various medical acts.

(ii) Technologies—the concepts found are:
- Interoperability—integration of HIS with intra- and inter-organizational information exchange between public and private and national and international entities.
- Digital Health Transformation—health processes are aided by technologies.
- Technology to Assist Medical Practices—such as robots and AI, among others, helping health professionals.
- Use of electronic devices (wearables), robotics, and intra-devices—used by citizens to monitor their health, with the possibility of collecting data and sending these data to health entities/health professionals.

(iii) Challenges and Risks—the concepts are:
- Resistance to change (people, users, and health professionals)
- Information exclusion (Training/Monitoring)
- Information (privacy, quality, and security—access and loss)

Given these perspectives, three HIS scenarios were developed for the future, i.e.,: (i) realistic; (ii) pessimistic; and (iii) optimistic scenarios.

Table 3 describes the scenarios for Medical Practices.

Table 3. Scenarios for Medical Practices.

Medical Practices	Pessimist	Realist	Optimistic
Precision Medicine/Individualized	The costs (financial and adaptation) are enormous, and for this reason, it will not be the usual practice.	It will be used to solve serious and critical diseases, where the cost/benefit justifies it.	Medicine will be fully focused on the citizen, with better accuracy in the personalized diagnoses and treatments.
Preventive medicine	Those responsible for the health area still have difficulties adapting to a reality focused on prevention.	Health officials will try to make health digital, with a focus on health rather than a disease, optimizing the entire HS.	The practice of medicine is focused on prevention and health promotion.
Point-of-Care (Telemedicine)	It is already a current practice when distance obliges. One should bet on its development.	Telemedicine will be used regularly, regardless of distance, and more focused on solving the problems of the citizen.	Telemedicine will be used frequently, facilitating the sharing of information between professionals for cases of complex diagnosis, and the citizen will have privileged consultations with healthcare professionals through Telemedicine and Telehealth.
Assisted Medical Practices	Healthcare professionals will have digital assistants who will help make diagnoses, but the presence of the health professional will be required.	Healthcare professionals, in some diagnoses, will be replaced by machines. The use of machines (robots) to help some medical practices (e.g., surgeries) will be more common.	The diagnoses will be made by machines, and these machines (robots) will replace healthcare professionals in clinical practices, such as surgeries.

Table 4 shows the concepts related to Technologies.

Table 4. Scenarios for Technologies.

Technologies	Pessimist	Realist	Optimistic
Interoperability (integration)	Health organizations (public and private), due to the existence of legacy systems or heterogeneous HISs, do not allow interoperability of the systems. Thus, the sharing of a citizen's health data between several entities will be a distant reality.	Health organizations (public and private) collaborate in defining a set of shared services that allows the integration and access of a citizen's EHRs.	HIS providers adopt international, European, and national recommendations, enabling interoperability between existing HISs and facilitating the sharing of EHR between different health organizations, respecting existing (legislation) standards.
Digital Health Transformation	It will occur when health organizations/entities (public and private) can change/innovate their processes, improve their leadership, and reduce resistance.	There are health organizations/entities (public and private) that innovate their processes, achieving significant efficiency gains. These cases will be examples to follow by other entities.	Health organizations/entities (public and private) present advanced dematerialization with significant gains in process performance. Success stories are shared and replicated.
Technology to Assist Medical Practices	Gradually technology that incorporates intelligence will be applied in the HIS; there is a need to create legitimacy for this to happen. Health professionals will resist but will eventually adopt the technologies.	The technology is currently able to assist professionals in medical practices. However, there are still obstacles to overcome: legitimation (legislation) and acceptance by all involved (health professionals and citizens) of the existed possibilities and limitations.	Health organizations and professionals perceive the positive side of incorporating intelligent technology and force legitimation (legislation) to occur. Intelligent technology, being incorporated into all medical processes and practices, leads to a huge efficiency gain and cost reduction.
Use of wearables	There are more and more devices able to collect data on citizens' health. However, these data will not be used without regulation to process it. On the other hand, the existent healthcare services do not have the capacity to treat such data.	The collection of device-generated data is already a reality, and it does not raise technical issues; it is a matter of work, standards, and interoperability. The legitimacy of these systems and devices will occur, and healthcare models will adapt to this reality.	In the short term, legislation will be created to enable the collection and use of health data from electronic devices. Clinical practices will already use these data to promote models of healthy living for citizens.

Table 5 outlines the scenarios related to Challenges and Risks.

Table 5. Scenarios for Challenges and Risks.

Challenges and Risks	Pessimist	Realist	Optimistic
Changing the existing culture (resistance to change among users and health professionals)	It will only occur when all stakeholders can understand the benefits to be obtained and realize that they will have to change/innovate their processes, and this will be a time-consuming process.	Health and care processes need to be innovated. There is little research and literature in this area. It is necessary to study the way care is organized and identify advantages and benefits causing changes in culture. Technology is a means and not the solution.	Healthcare delivery models will be studied and changed by accommodating emerging technologies with a positive and high impact on citizens' health.
Info Exclusion (Training)	There is a need to simplify and disseminate the HIS, mainly those that are in place and those that will appear in the future, and the advantage of their use (to health professionals and citizens). It will be necessary to reduce digital illiteracy, mainly among older people.	Copying good practices successfully implemented by some organizations (training, monitoring, and involving all stakeholders), showing the advantages/benefits of using it.	All entities realize the advantages/benefits of using technological solutions and increasingly seek technological solutions to solve their problems.
Information (privacy, quality, security)	Legislation is needed to regulate the collection, access, treatment, and security of health information. This will be one of the biggest challenges of the next years.	The question of legitimacy (legislation) will be resolved quickly (by national or European directives). The next step will be to ensure the quality and security of this information so that all stakeholders maintain confidence in it.	The collection, access, and sharing of health data will already be sufficiently regulated to maintain high standards of data security and privacy. All stakeholders (within their legitimacy) can add and share health information with confidence with other entities.

5. Discussion and Reflection

The health sector has, from early on, incorporated the benefits of the use of information technologies, with the first era of HIS that supported caregivers in their practice arising in the 1960s. The evolution has been remarkable, and the EHR, initially held by healthcare providers, is now run by citizens, true owners, and interested parties of their health data. However, this change requires an effort from all agents in terms of the design of "future" HIS, where aspects such as interoperability, standardization, privacy, security, and actions to address info exclusion appear as striking challenges. These factors and challenges are shared by the respondents, as shown in Table 6. However, a greater or lesser sharing should not be understood as a greater or lesser importance of each of the challenges but rather demonstrates the degree of awareness of these factors and challenges by the group of respondents.

Table 6. Important factors in HIS.

Key Factors to Address	Number of Interviewers Who Referenced It
Interoperability	4 (29%)
Standardization	3 (21%)
Privacy	6 (43%)
Security	7 (50%)
Actions to address info exclusion	12 (86%)

This change has been driven by social factors, such as increased life expectancy and mobility, but also by a set of ambitious responses and strategies created by healthcare decision-makers and managers. The reinforcement and commitment to HIS are expected to provide an adequate response to the health of citizens and guarantee the overall sustainability of the system. Thus, the technological advances brought by the 4IR have had a significant impact and acceptance in this sector, and the pandemic has further accelerated its adoption. The Health 4.0 concept incorporates innovative technologies which promote substantial improvements in health services and facilitates a more citizen-focused model concerning health.

The implementation of HIS in Portugal takes place at different paces, depending on the areas of activity, the type of sectors involved, and the legal regime of the organization, among other factors. It should also be noted that existing strategic documents specifically cover the public health sector and therefore do not consider private health entities; although the proposed ENESIS-20/22 strategy refers to a wider health ecosystem, in practice, the approaches presented are very much public-sector-focused.

In line with Ciasullo et al. [29], to make the HS sustainable, it is necessary for citizens to actively take part in their health process, and to do so, they need to have adequate means and knowledge, as well as the ability to interpret their health data, which requires an investment in their digital and health literacy. Likewise, health organizations, public or private, together with industry regulators and those responsible for health strategy, need to work together to enable (digital) communication and integrated sharing of health data, particularly in the context of healthcare.

In addition, emerging I4.0 technologies have enhanced the creation of better conditions for data collection and information sharing, as reported by some studies [25,26]. Nowadays, there are already several types of equipment that can be used or applications that can be installed on personal mobile devices, which allow monitoring the parameters of health and quality of life [28]. It is, therefore, a pressing thing to evaluate and classify these types of devices and applications in terms of quality and reliability and to legitimize them for this purpose so that citizens and health professionals can see their usefulness, have confidence in their use, and can use them by increasing preventive and predictive health models.

Given the current technological context, the citizen, in addition to standing as a fundamental actor in data creation, can also play a relevant role as a consumer of information. As such, it is also important to provide citizens with access to health-related information, for example, through the EHR, thereby enhancing better decisions about their health while promoting a preventive health model, concerns already seen in other studies that seek citizen centricity [31,35].

Another important aspect that was highlighted in this study is the need to raise awareness and training of the citizen so that they are the main promoters of their health which confirms the results presented by Rahi et al. [55]. Thus, it is necessary to ensure that the citizen has conditions for this, such as (i) empowering the citizen to use the systems; (ii) raising citizens' awareness to manage their health data, as well as sharing them with the professionals responsible for monitoring them, in a digital, holistic, and integrated ecosystem; and (iii) empowering citizens on the correct use of digital health solutions.

The existence of health data, part of them collected by the citizens themselves, as well as the later exchange of this data between patients and health professionals, would certainly allow an increase in the value of the services, enhancing benefits for both parties. It is essential to ensure reliable and quality data sources, as well as their protection, privacy, and security.

In a more social and human aspect, and in line with what has been proposed in several other studies [1], it is important to highlight the importance of health professionals with adequate skills to implement DT in the health sector and, consequently, develop new processes in terms of healthcare and/or restructure existing ones, as well as the training of citizen so that they are the main responsible and promoters of their health.

To conclude, several new trends, described around three scenarios, which may emerge in the future with the adoption of emerging I 4.0 technologies, should be noted, although these require a new strategic approach, with appropriate action plans to promote accessibility for all citizens and professionals.

5.1. Theoretical Implications

This study reinforced the literature's long-held view about the importance of future health promotion strategies through I4.0 emerging technologies. The findings of this study confirmed that the digitalization of processes in healthcare can bring benefits to stakeholders while also bringing some challenges that should be properly addressed beforehand to maximize positive results. Furthermore, these findings could be used by researchers in Business Management and Information Systems areas to advance novel solutions to e-health-related sectors.

5.2. Managerial and Societal Implications

This study has several implications that are useful not only for health providers and receivers but also for society in general. Our findings showed that there is a need for greater commitment from managers and decision makers to invest in solutions that allow a more equal DT approach between different health institutions/sectors. In addition, decision makers should promote processes that not only support interoperability, respecting data privacy and security but also increase the digital literacy of all healthcare providers and the health literacy of system users.

6. Conclusions and Recommendations

This study assessed the current state of HIS in Portugal, verifying that there is a strategic alignment with our European partner countries in terms of legislation and definitions of health policies. Although there has been continuous redesign and the emergence of several applications/solutions for different processes, Portuguese health systems remain incomplete, with gaps to be filled in legislation and the adoption of innovative healthcare processes by organizations. Moreover, struggles with integration and interoperability between solutions from the public and other sectors (private and social), or even from the same institution, not only lead to substantial costs in terms of redundancy and consistency of information but also reduce the interaction with users (healthcare providers, managers, and citizens). There were some other issues found, namely the need to improve digital literacy in all actors involved, as well as the urgency to increase citizens' health literacy, both tasks requiring significant educational effort. Additionally, there remains some resistance to change. Nevertheless, the benefits expected (some already verified) by the different parties involved, such as the dematerialization, digitalization, and incorporation of emerging technologies, showed that there is an effective process of health DT in Portugal.

When it comes to the future of HIS, three possibilities which include a pessimistic, optimistic as well as a more realistic scenario, were outlined based on the Portuguese case. These fell into three main categories: (i) Medical Practices; (ii) Technologies; and (iii) Fears and Challenges.

Limitations and Future Work

Firstly, a limited number of interviews were included. Even though an attempt was made to have individuals from different professional backgrounds and geographical areas, the results stood for a fraction of each sector's professionals, and carefulness was warranted when generalizing the results.

Moreover, important, the interviews were performed before the COVID-19 pandemic, so the consequent increase in health's DT was not considered, meaning that the results might not mirror the post-pandemic reality. Another limitation assumed by the authors is the fact that the results achieved and reported here came from qualitative research only, and there was no data to quantify the results.

As such, the authors would like, in the future, to re-evaluate the Portuguese HISs, checking the impact of the pandemic on the users' health literacy and HISs' development, using a more comprehensive method of data collection, with emphasis on quantitative approaches to data collection and analysis. In addition, it would be interesting to understand if the coronavirus outbreak forced a greater articulation of HISs between the public and private sectors. Finally, it should be noted that considering the utmost importance of issues related to data protection, privacy, and security of health data, it would be interesting to extend this study to evaluate questions related to these themes.

Author Contributions: Conceptualization, L.T., I.C., J.O.e.S. and F.M.; methodology, L.T., I.C., J.O.e.S. and F.M.; formal analysis, L.T., I.C., J.O.e.S. and F.M.; investigation, L.T., I.C., J.O.e.S. and F.M.; resources, L.T., I.C., J.O.e.S. and F.M.; data curation, L.T., I.C., J.O.e.S. and F.M.; writing—original draft preparation, L.T., I.C., J.O.e.S. and F.M.; writing—review and editing, L.T., I.C., J.O.e.S. and F.M. All authors have read and agreed to the published version of the manuscript.

Funding: This research was funded by Foundation for Science and Technology, in the context of the project UIDB/00127/2020.

Institutional Review Board Statement: Not applicable.

Informed Consent Statement: Informed consent was obtained from all subjects involved in the study.

Data Availability Statement: Data sharing not applicable.

Conflicts of Interest: The authors declare no conflict of interest.

References

1. Gökalp, E.; Kayabay, K.; Gökalp, M.O. Leveraging Digital Transformation Technologies to Tackle COVID-19: Proposing a Privacy-First Holistic Framework. In *Emerging Technologies During the Era of COVID-19 Pandemic*; Arpaci, I., Al-Emran, M., Al-Sharafi, M.A., Marques, G., Eds.; Studies in Systems, Decision and Control; Springer: Cham, Switzerland, 2021; Volume 348, pp. 149–166. [CrossRef]
2. Loeza-Mejía, C.-I.; Sánchez-DelaCruz, E.; Pozos-Parra, P.; Landero-Hernández, L.-A. The potential and challenges of Health 4.0 to face COVID-19 pandemic: A rapid review. *Health Technol.* **2021**, *11*, 1321–1330. [CrossRef]
3. Nomura, S.; Siesjö, V.; Tomson, G.; Mohr, W.; Fukuchi, E.; Shibuya, K.; Tangcharoensathien, V.; Miyata, H. Contributions of information and communications technology to future health systems and Universal Health Coverage: Application of Japan's experiences. *Health Res. Policy Syst.* **2020**, *18*, 73. [CrossRef]
4. WHO. Global Observatory for eHealth, World Health Organization. 2019. Available online: http://www.who.int/goe/en/ (accessed on 22 March 2019).
5. El-Sherif, D.M.; Abouzid, M.; Elzarif, M.T.; Ahmed, A.A.; Albakri, A.; Alshehri, M.M. Telehealth and Artificial Intelligence Insights into Healthcare during the COVID-19 Pandemic. *Healthcare* **2022**, *10*, 385. [CrossRef]
6. Epizitone, A.; Moyane, S.P.; Agbehadji, I.E. Health Information System and Health Care Applications Performance in the Healthcare Arena: A Bibliometric Analysis. *Healthcare* **2022**, *10*, 2273. [CrossRef]
7. Massaro, M. Digital transformation in the healthcare sector through blockchain technology. Insights from academic research and business developments. *Technovation* **2023**, *120*, 102386. [CrossRef]
8. Haux, R. Health Information Systems—From Present to Future? The German Medical Informatics Initiative. *Methods Inf. Med.* **2018**, *57*, e43–e45. [CrossRef]
9. Can, O.; Sezer, E.; Bursa, O.; Unalir, M.O. Comparing Relational and Ontological Triple Stores in Healthcare Domain. *Entropy* **2017**, *19*, 30. [CrossRef]
10. Haarbrandt, B.; Schreiweis, B.; Rey, S.; Sax, U.; Scheithauer, S.; Rienhoff, O.; Knaup-Gregori, P.; Bavendiek, U.; Dieterich, C.; Brors, B.; et al. HiGHmed—An Open Platform Approach to Enhance Care and Research across Institutional Boundaries. *Methods Inf. Med.* **2018**, *57*, e66–e81. [CrossRef]
11. Al-Jaroodi, J.; Mohamed, N.; Abukhousa, E. Health 4.0: On the Way to Realizing the Healthcare of the Future. *IEEE Access* **2020**, *8*, 211189–211210. [CrossRef]
12. Marinelli, S.; Basile, G.; Zaami, S. Telemedicine, Telepsychiatry and COVID-19 Pandemic: Future Prospects for Global Health. *Healthcare* **2022**, *10*, 2085. [CrossRef]
13. Haddara, M.; Staaby, A. RFID Applications and Adoptions in Healthcare: A Review on Patient Safety. *Procedia Comput. Sci.* **2018**, *138*, 80–88. [CrossRef]
14. Prasser, F.; Kohlbacher, O.; Mansmann, U.; Bauer, B.; Kuhn, K.A. Data Integration for Future Medicine (DIFUTURE). *Methods Inf. Med.* **2018**, *57*, e57–e65. [CrossRef]
15. Cavallone, M.; Palumbo, R. Debunking the myth of industry 4.0 in health care: Insights from a systematic literature review. *TQM J.* **2020**, *32*, 849–868. [CrossRef]
16. Nayak, S.; Patgiri, R. A Vision on Intelligent Medical Service for Emergency on 5G and 6G Communication Era. *EAI Endorsed Trans. Internet Things* **2020**, *6*. [CrossRef]

17. Nayak, S.; Patgiri, R. 6G Communication Technology: A Vision on Intelligent Healthcare. In *Health Informatics: A Computational Perspective in Healthcare*; Patgiri, R., Biswas, A., Roy, P., Eds.; Studies in Computational Intelligence; Springer: Singapore, 2021; Volume 932, pp. 1–18. [CrossRef]
18. Gehring, S.; Eulenfeld, R. German Medical Informatics Initiative: Unlocking Data for Research and Health Care. *Methods Inf. Med.* **2018**, *57*, e46–e49. [CrossRef]
19. Tummers, J.; Tekinerdogan, B.; Tobi, H.; Catal, C.; Schalk, B. Obstacles and features of health information systems: A systematic literature review. *Comput. Biol. Med.* **2021**, *137*, 104785. [CrossRef]
20. Khubone, T.; Tlou, B.; Mashamba-Thompson, T.P. Electronic Health Information Systems to Improve Disease Diagnosis and Management at Point-of-Care in Low and Middle Income Countries: A Narrative Review. *Diagnostics* **2020**, *10*, 327. [CrossRef]
21. Tabaeeian, R.A.; Hajrahimi, B.; Khoshfetrat, A. A systematic review of telemedicine systems use barriers: Primary health care providers' perspective. *J. Sci. Technol. Policy Manag.* **2022**. [CrossRef]
22. SPMS. ENESIS-2020. 2017. Available online: https://enesis.spms.min-saude.pt (accessed on 29 December 2019).
23. Marutha, N.S. Landscaping health-care system using functional records management activities. *Collect. Curation* **2020**, *40*, 9–14. [CrossRef]
24. Hawthorne, K.H.; Richards, L. Personal health records: A new type of electronic medical record. *Rec. Manag. J.* **2017**, *27*, 286–301. [CrossRef]
25. Carolina, A.; Monteiro, B.; França, R.P.; Estrela, V.V.; Iano, Y.; Khelassi, A.; Razmjooy, N. Health 4.0: Applications, Management Technologies and Review. *Med. Technol. J.* **2019**, *2*, 262–276. [CrossRef]
26. Alcácer, V.; Cruz-Machado, V. Scanning the Industry 4.0: A Literature Review on Technologies for Manufacturing Systems. *Eng. Sci. Technol. Int. J.* **2019**, *22*, 899–919. [CrossRef]
27. Karatas, M.; Eriskin, L.; Deveci, M.; Pamucar, D.; Garg, H. Big Data for Healthcare Industry 4.0: Applications, challenges and future perspectives. *Expert Syst. Appl.* **2022**, *200*, 116912. [CrossRef]
28. Javaid, M.; Haleem, A. Industry 4.0 applications in medical field: A brief review. *Curr. Med. Res. Pract.* **2019**, *9*, 102–109. [CrossRef]
29. Ciasullo, M.V.; Orciuoli, F.; Douglas, A.; Palumbo, R. Putting Health 4.0 at the service of Society 5.0: Exploratory insights from a pilot study. *Socio-Econ. Plan. Sci.* **2022**, *80*, 101163. [CrossRef]
30. Cáceres, C.; Rosário, J.M.; Amaya, D. Towards Health 4.0: e-Hospital Proposal Based Industry 4.0 and Artificial Intelligence Concepts. In *Artificial Intelligence in Medicine*; Riaño, D., Wilk, S., ten Teije, A., Eds.; Lecture Notes in Computer Science; Springer International Publishing: Cham, Switzerland, 2019; Volume 11526. [CrossRef]
31. Liu, Z.; Ren, L.; Xiao, C.; Zhang, K.; Demian, P. Virtual Reality Aided Therapy towards Health 4.0: A Two-Decade Bibliometric Analysis. *Int. J. Environ. Res. Public Health* **2022**, *19*, 1525. [CrossRef]
32. Wehde, M. Healthcare 4.0. *IEEE Eng. Manag. Rev.* **2019**, *47*, 24–28. [CrossRef]
33. Schnurr, H.-P.; Aronsky, D.; Wenke, D. MEDICINE 4.0—Interplay of Intelligent Systems and Medical Experts. In *Knowledge Management in Digital Change*; North, K., Maier, R., Haas, O., Eds.; Springer: Cham, Switzerland, 2018; pp. 51–63. [CrossRef]
34. Chute, C.; French, T. Introducing Care 4.0: An Integrated Care Paradigm Built on Industry 4.0 Capabilities. *Int. J. Environ. Res. Public Health* **2019**, *16*, 2247. [CrossRef]
35. Afferni, P.; Merone, M.; Soda, P. Hospital 4.0 and Its Innovation in Methodologies and Technologies. In Proceedings of the 2018 IEEE 31st International Symposium on Computer-Based Medical Systems (CBMS), Karlstad, Sweden, 18–21 June 2018; pp. 333–338. [CrossRef]
36. Unterhofer, M.; Rauch, E.; Matt, D.T. Hospital 4.0 roadmap: An agile implementation guideline for hospital manager. *Int. J. Agil. Syst. Manag.* **2021**, *14*, 635. [CrossRef]
37. Feussner, H.; Ostler, D.; Kranzfelde, M.; Kohn, N.; Koller, S.; Wilhelm, D.; Thuemmler, C.; Schneider, A. Surgery 4.0. In *Health 4.0: How Virtualization and Big Data Are Revolutionizing Healthcare*; Thuemmler, C., Bai, C., Eds.; Springer International Publishing: Cham, Switzerland, 2017; pp. 91–107. [CrossRef]
38. Kickbusch, I. Health promotion 4.0. *Health Promot. Int.* **2019**, *34*, 179–181. [CrossRef]
39. Hou, Y.; Wang, W.; Bartolo, P. Application of additively manufactured 3D scaffolds for bone cancer treatment: A review. *Bio-Des. Manuf.* **2022**; in press. [CrossRef]
40. Siuly, S.; Aickelin, U.; Kabir, E.; Huang, Z.; Zhang, Y. Guest Editorial: Special issue on "Artificial Intelligence in Health Informatics". *Health Inf. Sci. Syst.* **2021**, *9*, 23. [CrossRef] [PubMed]
41. Lohr, S. What Ever Happened to IBM's Watson? *The New York Times*. 17 July 2021. Available online: https://www.nytimes.com/2021/07/16/technology/what-happened-ibm-watson.html (accessed on 18 February 2023).
42. Zou, F.W.; Tang, Y.F.; Liu, C.Y.; Ma, J.A.; Hu, C.H. Concordance Study between IBM Watson for On-cology and Real Clinical Practice for Cervical Cancer Patients in China: A Retrospective Analysis. *Front. Genet.* **2020**, *11*, 200. [CrossRef] [PubMed]
43. Wong, A.N.N.; He, Z.; Leung, K.L.; To, C.C.K.; Wong, C.Y.; Wong, S.C.C.; Yoo, J.S.; Chan, C.K.R.; Chan, A.Z.; Lacambra, M.D.; et al. Current Developments of Artificial Intelligence in Digital Pathology and Its Future Clinical Applications in Gastrointestinal Cancers. *Cancers* **2022**, *14*, 3780. [CrossRef]
44. Koh, D.M.; Papanikolaou, N.; Bick, U.; Illing, R.; Kahn, C.E.; Kalpathi-Cramer, J.; Matos, C.; Martí-Bonmatí, L.; Miles, A.; Mun, S.K.; et al. Artificial intelligence and machine learning in cancer imaging. *Commun. Med.* **2022**, *2*, 133. [CrossRef]

45. Oren, O.; Gersh, B.J.; Bhatt, D.L. Artificial intelligence in medical imaging: Switching from radiographic pathological data to clinically meaningful endpoints. *Lancet Digit. Health* **2020**, *2*, e486–e488. [CrossRef]
46. Kung, T.H.; Cheatham, M.; Medenilla, A.; Sillos, C.; De Leon, L.; Elepaño, C.; Madriaga, M.; Aggabao, R.; Diaz-Candido, G.; Maningo, J.; et al. Performance of ChatGPT on USMLE: Potential for AI-Assisted Medical Education Using Large Language Models. *medRxiv* **2022**. [CrossRef]
47. Clayton, B.J. Microsoft Unveils New Bing with ChatGPT Powers. *BBC News*. 7 February 2023. Available online: https://www.bbc.com/news/business-64562672?at_medium=RSS (accessed on 18 February 2023).
48. Eurostat. One in Two EU Citizens Look for Health Information Online. 6 April 2021. Available online: https://ec.europa.eu/eurostat/web/products-eurostat-news/-/edn-20210406-1 (accessed on 18 February 2023).
49. Sam Altman on. (11 December 2022). Twitter. Available online: https://mobile.twitter.com/sama/status/1601731295792414720?cxt=HHwWgICq9ZGIv7osAAAA (accessed on 18 February 2023).
50. SPMS. ENESIS-20/22. 2019. Available online: https://www.spms.min-saude.pt/wp-content/uploads2019/10/ENESIS2022_VersaoParaConsultaPublicaOut2019.pdf (accessed on 29 December 2022).
51. Krippendorf, K. *Content Analysis: An Introduction to Its Methodology*, 4th ed.; Sage Publication: Thousand Oaks, CA, USA, 2018.
52. PCM. Resolução do Conselho de Ministros 62/2016, 2016-10-17—DRE. 2016. Available online: https://dre.pt/dre/detalhe/resolucao-conselho-ministros/62-2016-7554212413/2/2023 (accessed on 18 February 2023).
53. PCM. Resolução do Conselho de Ministros 108/2017, 2017-07-26—DRE. 2017. Available online: https://dre.pt/dre/detalhe/resolucao-conselho-ministros/108-2017-107757007 (accessed on 18 February 2023).
54. SPMS. Sobre os Sistemas de Informação. 2023. Available online: https://www.spms.min-saude.pt/sobre-os-sistemas-de-informacao/ (accessed on 18 February 2023).
55. Rahi, S.; Khan, M.M.; Alghizzawi, M. Factors influencing the adoption of telemedicine health services during COVID-19 pandemic crisis: An integrative research model. *Enterp. Inf. Syst.* **2021**, *15*, 769–793. [CrossRef]

Disclaimer/Publisher's Note: The statements, opinions and data contained in all publications are solely those of the individual author(s) and contributor(s) and not of MDPI and/or the editor(s). MDPI and/or the editor(s) disclaim responsibility for any injury to people or property resulting from any ideas, methods, instructions or products referred to in the content.

Article

Detection of Depression-Related Tweets in Mexico Using Crosslingual Schemes and Knowledge Distillation

Jorge Pool-Cen [1,†], Hugo Carlos-Martínez [1,2,3,*,†], Gandhi Hernández-Chan [1,2,3,*,†] and Oscar Sánchez-Siordia [1,3]

1. Geospatial Information Sciences Research Center, Mexico City 14240, Mexico
2. IxM CONACyT, Mexico City 14240, Mexico
3. Laboratorio Nacional de Geointeligencia (GeoInt), Mexico City 14240, Mexico
* Correspondence: hcarlos@centrogeo.edu.mx or hugo.martinez@conacyt.mx (H.C.-M.); ghernandezc@centrogeo.edu.mx or ghernandezc@conacyt.mx (G.H.-C.); Tel.: +52-1-473-756-2152 (H.C.-M.); +52-9992-68-45-79 (G.H.-C.)
† These authors contributed equally to this work.

Abstract: Mental health problems are one of the various ills that afflict the world's population. Early diagnosis and medical care are public health problems addressed from various perspectives. Among the mental illnesses that most afflict the population is depression; its early diagnosis is vitally important, as it can trigger more severe illnesses, such as suicidal ideation. Due to the lack of homogeneity in current diagnostic tools, the community has focused on using AI tools for opportune diagnosis. Unfortunately, there is a lack of data that allows the use of IA tools for the Spanish language. Our work has a cross-lingual scheme to address this issue, allowing us to identify Spanish and English texts. The experiments demonstrated the methodology's effectiveness with an F1-score of 0.95. With this methodology, we propose a method to solve a classification problem for depression tweets (or short texts) by reusing English language databases with insufficient data to generate a classification model, such as in the Spanish language. We also validated the information obtained with public data to analyze the behavior of depression in Mexico during the COVID-19 pandemic. Our results show that the use of these methodologies can serve as support, not only in the diagnosis of depression, but also in the construction of different language databases that allow the creation of more efficient diagnostic tools.

Keywords: depression; text classification; knowledge distillation; dimensionality reduction; Twitter; COVID-19

1. Introduction

Mental health problems are an area of medical and social sciences that have become very important in recent decades, because the number of people who have suffered, or are suffering, a mental illness is increasing. Some studies estimate that almost one billion people worldwide have a mental disorder. Due to this, even on a global scale, multiple initiatives are trying to address mental health problems in a comprehensive way [1].

Due to the COVID-19 pandemic, many mental health problems have increased in recent years. Only a few years after the COVID-19 pandemic, it is possible to explore the effects of the pandemic on mental health. Recent studies suggest there has been a rise in mental health problems in people who were mentally healthy before the pandemic. On the other hand, people who had some previous condition prior to the pandemic have seen the effects of their mental illnesses increase [2,3]. In particular, the mental health of young people has drastically reduced [4].

Some mental illnesses have become so widespread among the population that they have become a subject of public health policy. In particular, depression is one of the leading causes of disability and can increase the risk of suicidal ideation and suicide attempts [5].

The latter has led to the creation of public policies that promote the treatment of depression in its early stages and the receipt of psychological and psychiatric care [6,7]. Like other diseases, mental health problems harm people's well-being and directly impact activities of other natures, such as economic ones. For example, lost productivity due to two of the most common mental disorders, anxiety and depression, costs the global economy one trillion dollars annually [8].

In the case of Latin America, some studies suggest that 50% of people with depression do not receive adequate treatment, one of the leading causes being lack of diagnosis [9]. Some studies even suggest that a possible way to address the problem is by using the Internet to facilitate detection and treatment mechanisms [10]. Along the same lines, some studies suggest the use of apps to treat depression in Latino and Hispanic populations [11]. However, much work remains to be done. As we cite later, the literature promotes the creation of multilingual care schemes for Latino populations, with a particular emphasis on immigrants.

Depression is typically diagnosed based on individual self-reporting or specific questionnaires designed to detect characteristic patterns of feelings or social interactions [12]. However, these tools generally have some subjective components or are not applied homogeneously, which complicates the diagnosis process [13]. Due to the above, the opportune detection and diagnosis of mental illnesses have become very active research topics. The idea is to have more robust tools for early diagnosis that allow diseases to be treated promptly. From this idea, the use of computational tools for diagnosing and detecting mental illnesses has spread [14].

Machine Learning (ML), particularly Deep Learning (DL) algorithms, has successfully detected mental diseases and characterized behavior patterns. For depression, for example, there exist multidisciplinary solutions that use demographic and genetic information to improve antidepressant treatments [15], or applications based on Natural Language Processing (NLP) that successfully detect depression [16–18]. Since DL algorithms generally require a considerable volume of data, social networks have become an indispensable source of information [19,20]. In particular, Twitter has become a primary data source for feeding these algorithms [21,22]. However, one of the main problems is that the data sets are usually not public or homogenized, which often prevents reproduction of the results. In the case of NLP, the most used language processing models are those that are based on schemes such as Bidirectional Encoder Representations from Transformers (BERT) [23]. BERT-type models often lead to specific models for different languages. In Candida et al. [24] we find a general summary of the application of these models to mental health problems.

One of the under-researched areas in the detection and diagnosis of depression is the use of multilingual methodologies within the framework of NLP, which is the main idea of this work. We develop a methodology that allows the use of existing data sets of tweets in the English language to detect depressive tweets in Spanish. From a technical point of view, the detection of depression from Twitter posts requires the following two steps: the detection of tweets depression-related or that manifest depression; and the incorporation of a temporal component, which requires that users must publish tweets associated with depression with some frequency. The use of a temporal component is due to the fact that depression is a complex disease, the severity of which tends to vary over time. Therefore, it is necessary to consider the frequency of publications since it is impossible to determine the state of depression based on only a small group of publications. This is one of the weak points of using social networks to identify depression, as users must post texts with a certain regularity that allow the identification of a depressive state. In this work, we focused on the first element. Although our work did not focus on detecting depression, it is valuable as a first step in complete methodologies. We limited the scope of the current research because a complete methodology requires a database of user profiles diagnosed with depression by experts in the field and, unfortunately, this is a lack of such research in Spanish. In future investigations, we will address this issue.

The organization of the text is as follows. Section 2 presents related works that were taken as a reference for the present investigation. Section 3 introduces the theoretical framework used to develop the methodology. Section 4 details the methodology, based on the framework presented in section three. After that, in Sections 5 and 6, we present the materials and experimentation schemes employed. In Section 7, we present the results obtained by our methodology and compare translations. Finally, Section 8 presents the results of applying our methodology to geo-referenced Twitter data in Mexico for the years 2018–2021.

2. Related Works

The use of NLP models in health problems has been a very popular topic of study. There are applications in the field of medicine in general [25,26]. In the literature, we find works that refer to the importance of creating multilingual schemes to address mental health problems. For example, in Brisset et al. [27], the authors describe the problems of providing primary mental health care to immigrants in Montreal due to language barriers. In Límon et al. [28], the authors highlight the problems in regard to early detection of depression in Spanish-speaking immigrants; in this research, the authors emphasized the problems of translating depression instruments from English to Spanish. In Garcia et al. [29], the authors mention that people with limited English proficiency are the ones who most frequently suffer from depression, mainly Latin American immigrants (see Figure 1).

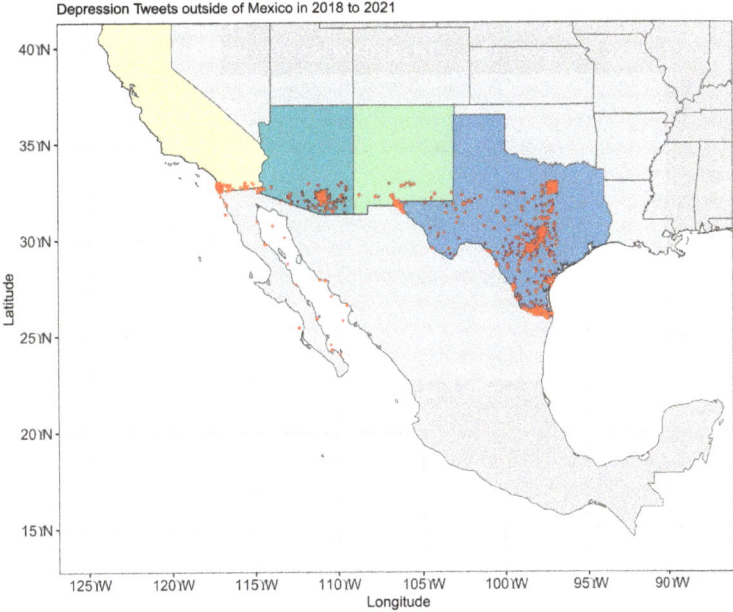

Figure 1. Tweets depression-related in Spanish in the United States border States from 2018–2021. Latin American immigrants are a population that frequently suffers from depression.

In Figure 1, we can see the distribution of tweets related to depression in Spanish in the US border states, where much of the immigrant population is concentrated. The Figure illustrates the feasibility of using Twitter to monitor mental health issues in immigrant populations.

The detection of depression using social networks and learning algorithms is not new. Many works address the problem using different strategies. In particular, they can be distinguished by considering social networks that serve as sources of information, NLP models used to represent text, and classification algorithms used to distinguish factors. A complete description of existing works in the English language can be found in [30].

For the analysis of depressive tweets in the Spanish language, there are few works. Most focus on constructing dictionaries (or translated phrases) that include words related to depression, and then use these dictionaries to select depression-related tweets to generate statistical descriptors, or to train classification algorithms. For example, in [31], the authors introduce a comprehensive collection of Spanish words commonly used by depressive patients and gave insight into the relevance of these words in identifying posts on social media related to depression. One of the central affirmations of this work is that using dictionaries to identify post-depressive patients is inadequate, because the words are frequently used in different contexts. In Leis et al. [32], the authors present a methodology to identify signs of depression based on the linguistic characteristics of the tweets in Spanish. The authors selected Twitter users who indicated potential signs of depressive symptoms based on the 20 most common Spanish words expressed by patients with depression in clinical settings. Once users were selected, the authors used statistical descriptions of language and behavior to identify a sign of depression. In Valeriano et al. [33], the authors use a dictionary of English phrases translated into Spanish to identify tweets related to suicide. Once phrases were identified, a manual selection was made to differentiate tweets that could correspond to expressions of sarcasm, song lyrics, etc., and, then, a machine learning algorithm was trained to classify depressive tweets. In Shekerbekova et al. [34], the authors compare different machine learning algorithms to identify posts related to depression. As in our work, the authors selected a set of posts related to depression and general posts.

If the literature on identifying depression in the Spanish language is insufficient, it is almost null in the case of multilingual models. Moreover, most studies have a comparative approach, rather than considering it as a multi-language problem. For example, in Ramirez et al. [35], the authors use computational methods to compare expressions of depression in English and Spanish. It is a comparative study of variations in expressions of depression in both languages. There is research that, although formulated for the English language, implicitly used NLP models allow working with text in other languages. For example, in Basco et al. [12], the authors incorporate multilingual NLP models to detect depression and gambling disorders. The authors argue that many users generate posts in languages other than their native ones (e.g., English).

Some works intend to detect signs of depression regardless of the language used. For this, data from conversations in different languages and algorithms for extracting speech features are employed. For example, Kiss et al. in [36], evaluate the possibility of extracting speech characteristics as descriptors to identify depression. This work suggests that the descriptors found are similar regardless of language. On the other hand, in Demiroglu et al. [37], the authors use a combination of sound and text descriptors. For this, the extracted speech features are merged with sentiment analysis expressions obtained through text. Finally, in Kiss, G. [38], the author discusses and evaluates the possibility of generating models to identify depression using speech in different languages and assesses the ability to identify depression regardless of the language used.

To our knowledge, the works closest to ours are the ones presented in [39,40]. These papers present a methodology for detecting depression based on the construction of linear transformations that are capable of aligning words in different languages. For a set of equivalent words in both languages, it is possible to find a linear transformation W (viewed as an embedding space) that maps between languages. This transformation makes it possible to train a classification algorithm in a language (e.g., English) and use this classifier for texts in Spanish. Among the main differences from our work are the type of transformation used and the inclusion of attention mechanisms to maintain semantic properties. While in [39,40], the mapping is only between words, our methodology used knowledge distillation to find more complex mapping functions, while incorporating semantic properties.

It is important to note that using knowledge distillation to manage multilingual schemes is not the only viable option to identify or classify depression. There are models in the literature designed to handle multi-language sentences, such as that presented in

Feng et al. [41]. Some works apply these models, for example, in sentiment analysis [42]. However, the results presented in Reimers et al. [43] showed a better vectorial representation of sentences in different languages. Due to this, in this work, we focused on applying knowledge distillation to detect tweets that were depression-related.

Derived from the literature review, we detected the following limitations. The models based on dictionaries or associated phrases restrict the detection capability to the quality of the dictionaries; furthermore, these schemes are not practical for multilanguage problems, since the dictionaries could vary significantly between different languages. On the other hand, the explicit use of translators only partially solves the problem, since the texts found on social networks are usually concise, and, in the translation, there may be a loss of context. Finally, the lack of data in other languages (besides English) complicates reproduction of the results. With these limitations in mind, the question arises as to whether it is possible to build a model trained with a limited amount of data and easily generalized to other languages without re-training or building new databases.

Our proposal arose as a response to these problems. The main idea was to build an embedded space containing phrases with similar semantic and syntactic content so that dictionaries or translations are not explicitly needed. This space could be used to train classification models in a specific language (e.g., English) to be used to detect similar phrases in other languages. One way to generate such a space is through knowledge distillation and dimensionality reduction schemes. The following section presents the necessary concepts to build this space.

3. Framework

As previously mentioned, we used knowledge distillation to obtain the vector representation of the tweets. Unfortunately, since they were usually concise texts, it was convenient to use a dimensionality reduction scheme; in particular, we used the proposal presented in [44] (known as IVIS). In the following sections, we describe, in a general manner, the mathematical foundations of both methodologies.

3.1. Knowledge Distillation

The general concept of knowledge distillation refers to the process of knowledge transfer from large models (i.e., a large number of parameters) to simpler models designed to perform specific tasks. These models are formulated in terms of teacher and student models. The idea is that the student model can be trained on specific tasks from the master model. These methodologies are trendy in NLP tasks, where large models have been trained with a large amount of data. In our work, we used the knowledge distillation methodology presented in [43]. This model proposes mapping a translated sentence of a language to the same vector space as the original language's sentence to mimic the language's properties, i.e., this knowledge distillation aims to extend one language's characteristics or properties to another. In other words, the original and translated vector representations of semantically similar declarations must be neighbors.

As we mentioned earlier, the idea starts with a teacher model, denoted by M for a language s, and a parallel set of translated sentences, denoted $((s_1, t_1), \ldots (s_n, t_n))$; with t_i being the translation of s_i. Then a model called Student, denoted by $\hat{M}$, is trained as $\hat{M}(s_i) \approx M(s_i)$ and $\hat{M}(t_i) \approx M(s_i)$, through the loss function:

$$\frac{1}{|\beta|} \sum_{j \in \beta} [(M(s_j) - \hat{M}(s_j))^2 + (M(s_j) - \hat{M}(t_j))^2], \quad (1)$$

where β represents a batch of sentences. In the first part of the equation, $(M(s_j) - \hat{M}(s_j))^2$, the student model learns to project the sentence onto the same vector space as the teacher model. The second part of the cost function, named $(M(s_j) - \hat{M}(t_j))^2$, aims to teach the student model how to project the translated sentences to the exact location in the vector space as the original sentences. That is, sentences with similar semantic content are close (in Euclidean distance) regardless of language.

In practice, we used the model distiluse-base-multilingual-cased-v1 (DBM) formulated in [45] and implemented in [46]. This model is a sentence-transformers model; that is, it maps sentences and paragraphs to a 512-dimensional dense vector space and can be used for tasks like clustering or semantic search. The model was trained in 15 languages, including English and Spanish.

3.2. Dimensionality Reduction with Ivis

Text is an unstructured data type that can be encountered with various lengths, so extracting features in a corpus can generate high-dimensionality sparse vector representations. Using dimensionality reduction algorithms reduces the vector dimension while maintaining the quality of the vector representation of the original data. Several dimensionality reduction algorithms exist in the literature, such as PCA, LDA, t-SNE, IVIS, and ISOMAP [47]. However, IVIS has shown performance comparable to, or superior to, the algorithms mentioned above. IVIS was conceived as a Siamese neural network with a triple loss function. The results reported by the authors emphasize that IVIS preserves global data structures in a low-dimensional space for real and simulated data sets.

The IVIS algorithm is a non-linear dimensionality reduction method, based on a neural network model with three training schemes: supervised, unsupervised, and semi-supervised. The cost function used in training the neural network is a variant of the standard triple loss function.

$$L_{tri}(\theta) = [\Sigma_{a,p,n} D_{a,p} - min(D_{a,n}, D_{p,n}) + m], \qquad (2)$$

where a, p, and n correspond to a sample of interest, a positive sample, and a negative sample, respectively. D is a distance function, and m is the margin parameter. The distance function D corresponds to the Euclidean distance and measures the similarity between the points a and b in the embedded space.

$$D_{a,b} = \sqrt{\sum_{i=1}^{n}(a_i - b_i)^2} \qquad (3)$$

The loss function minimizes the distance between the point of interest and the positive sample, while maximizing the distance to the negative sample. At each point of interest in the dataset, positive and negative samples are received according to the k-nearest neighbor algorithm.

We used IVIS for the conversion of vector representations with dimension 512 to two-dimensional representation vectors. This number of dimensions was selected because the results did not improve significantly in the experiments carried out when considering larger dimensions.

In the following sections, we present the proposed methodology. We first introduce the set of data used and then describe the characteristics of each stage.

4. Methodology

Identifying depression-related tweets was carried out in four stages: (1) pre-processing, (2) feature extraction, using knowledge distillation methodology, (3) dimensionality reduction, using IVIS and (4) tweet classification. In the following sections, we describe each of the stages and give a hint of the importance attached to its application.

4.1. Pre-Processing and Feature Extraction

In this phase, we processed each tweet to normalize the text to lowercase and removed the blank spaces found at the beginning and end. Next, we removed null records, duplicate records, emojis, hyperlinks, mentions, punctuation signs, and words that contained the symbol @ or #. During the cleaning process we removed records with single-word phrases that did not carry any meaning, for example, abbreviations such as thx and thd. It is essential to mention that after the pre-processing phase, the size of the datasets did not change significantly. Finally, we used the DMB model to obtain its vector representation.

4.2. Dimensionality Reduction

Extracting text features generates high-dimensional vector representations; however, these representations can be sparse vectors due to text features, such as the length of each text. The dimensionality reduction method helps to compress the information and maintain the qualities of the original data. As mentioned above, we used existing IVIS training schemes to evaluate our methodology, considering the problems encountered in practice. For example, a semi-supervised strategy can be used when one of the datasets contains a few unclassified tweets. On the contrary, an unsupervised strategy is usually used when there are no labels, but the text refers to fewer topics.

4.3. Tweet Identification

The ultimate goal of our methodology was to correctly identify tweets related to depression. In principle, after obtaining the 2D vector representation, it would be possible to apply a simple classification algorithm. In particular, we compared the results obtained by the following algorithms: Logistic regression (LR), Support Vector Machines (SVM), Gaussian Process (GP), and Quadratic Discriminant Analysis (QDA). The idea was to evaluate whether our methodology was robust, regardless of the classification algorithm. For all classifiers, the hyperparameters were determined using a grid search cross-validation strategy, and experiments were performed using the Scikit Learn library [48].

5. Materials And Methods

In this work, we considered three possible classes. The first class, C_D, labeled as 1, corresponds to tweets related to depression. The second class, C_N, labeled as 0, corresponds to tweets that are not related to depression. Finally, the third class, C_U, corresponds to tweets with unknown content. We used the C_U class to evaluate semi-supervised dimensionality reduction methods.

We created four data sets, all of which contained phrases related to the topic of depression: D_1, D_2, D_3, and D_4. Some texts might contain news or reports on depression, while others were posted by users who expressed depressive emotional feelings. The data set D_1 was obtained from Kaggle (https://www.kaggle.com/general/234873, accessed on 22 December 2022) and consisted of 4493 tweets in English, of which 2385 were tagged with class C_D and 2263 with class C_N. The data set D_2 contained 2000 Spanish tweets published in 2019 extracted from the AGEI platform (http://agei.geoint.mx/, accessed on 9 November 2022), with 50% of the data corresponding to tweets related to depression; all tweets were labeled by experts. The data set D_3 contained 5093 tweets and was made up of a mixture of D_1 and 600 tweets randomly obtained from D_2. The data set D_4 was a subset of D_2 and contained 1400 tweets distributed in 50% for the depression class and the other 50% for the non-depression class. This data set was used as a test for the semi-supervised dimensionality reduction experiment, explained in Section 4.2. Table 1 shows the results of the exploratory analysis of the texts with respect to the length and number of words.

Table 1. Statistics for characters and words for datasets D_1–D_4.

Statistics	D_1	D_2	D_3	D_4
		Characters		
Maximum	874	278	460	277
Minimum	9	7	7	8
Average	117.5	98.64	97.79	99.66
		Words		
Maximum	73	57	73	57
Minimum	2	2	2	1
Average	17.22	18.38	17.3	18.64

6. Experiments

We divided the experiments according to the dimensionality reduction methodology employed. Specifically, we designed the experiments according to supervised, semi-supervised, and unsupervised methodologies. This was because each methodology represents a different approach to the problem encountered in practice when finding depression-related tweets.

6.1. Supervised Dimensionality Reduction

Once the vector representations of the data sets D_1 and D_2 were obtained, we trained IVIS using the supervised scheme. In this series of experiments, we used only the data set D_1 for the training phase, that is, IVIS and the classification algorithm were trained only on English data. The idea was to evaluate whether it was possible to use depression-related tweets written in English to detect tweets with similar content in Spanish.

6.2. Unsupervised Dimensionality Reduction

We trained IVIS and the classification algorithms for these tests using the data set D_1 without including labels. The idea was to assess whether the methodology was robust when there were no labeled data, but one of the topics (in this case depression) was predominant. This experiment could be understood if we assumed that the syntax of depression-related tweets has a semantic structure that makes it possible to differentiate them from other topics (i.e., not depressive).

6.3. Semi-Supervised Dimensionality Reduction

This experiment's training and test data corresponded to the data sets D_3 and D_4, respectively. For these experiments, we evaluated the ability of our methodology to assign labels to data that could be mislabeled. On many occasions, when evaluating whether a tweet is depression-related, there may be discrepancies between experts when labeling it. One way to address this problem is to leave these tweets unlabeled, letting the methodology assign the corresponding class from its vector representation. On the other hand, in some cases, if the dataset of the language of interest contains little data, it might be convenient to use the semi-supervised methodology.

The experimentation phase was carried out in the months of October and November of 2022 on a computer with i5 at 4.10 GHz and 16 GB RAM on OS Debian.

6.4. Experiments with Translations

Although it is a naive idea, the use of translations to identify tweets in different languages has been frequently used in other problems, such as sentiment analysis. We compared this strategy using translations obtained through the Google Translate platform using English phrases from the data set D_1 as a source of information. Once we obtained the translations, we used the BETO model to obtain the vector representation [49]. In other words, in this set of experiments, the only language used was Spanish. To do this, we built data sets in Spanish from the original sets D_1 and D_3 and used the BETO model to build the vector representation of all tweets (including translations). Throughout the document, we distinguish between tweets written in Spanish (i.e., native Spanish) and the translations obtained with Google. Table 2 summarizes the contents for each data set used, with translations and knowledge distillation.

The experimentation phase was carried out in October and November of 2022 on a computer with an i5 processor at 4.10 GHz and 16 GB RAM on OS Debian. This work was part of the Self-inflicted Death Study Seminar (SIEMAI) (http://siemai.geoint.mx/, accessed on 22 December 2022), with the voluntary collaboration of mental health experts.

Table 2. Datasets used for the training and testing of the dimensionality reduction and classification models.

Datasets	Language	Samples	C_D	C_N	C_U
		Knowledge distillation			
D_1	English	4648	2385	2263	0
D_2	Spanish	2000	1000	1000	0
D_3	English & Spanish	5093	2685	2263	600
D_4	Spanish	1400	700	700	0
		Translations			
T_1	Translations	4648	2385	2263	0
T_2	Spanish	2000	1000	1000	0
T_3	Translations & Spanish	5248	2685	2563	600
T_4	Spanish	1400	700	700	0

7. Results

In this section, we present the results obtained from the experiments. To compare the performance of the classification algorithms, we used the following metrics: accuracy, precision, recall, and F1 metrics.

7.1. Evaluation of Experiments with Translations

This section presents the performance measures of the classification models using translations, the extraction of text features using the BETO model, and the various dimensionality reduction schemes. Table 3 shows the results obtained for this strategy. In these experiments, the best score was obtained with unsupervised dimensionality. The best models were Logistic Regression and Linear SVM with 0.85 on the F1-Score. In general, there did not seem to be any significant difference between the different dimensionality reduction schemes, which was understandable if we consider that, during translation, there were changes in the syntax that made classification difficult.

Table 3. Results using translations.

IVIS	Model	Accuracy	Precision	Recall	F1
Supervised	LR	0.82	0.84	0.81	0.82
	SVM	0.82	0.84	0.81	0.82
	GP	0.82	0.81	0.81	0.82
	QDA	0.84	0.84	0.83	**0.83**
Unsupervised	LR	0.84	0.84	0.85	**0.85**
	SVM	0.84	0.84	0.85	0.85
	GP	0.84	0.84	0.85	0.84
	QDA	0.84	0.83	0.84	0.84
Semi-supervised	LR	0.82	0.82	0.84	0.84
	SVM	0.83	0.83	0.83	0.83
	GP	0.82	0.82	0.83	0.83
	QDA	0.83	0.82	0.86	**0.84**

Figure 2 shows the classification results obtained using QDA. Note that the unsupervised scheme presented considerable dispersion, although, in general, it had the highest classification percentages. On the contrary, in the supervised and semi-supervised schemes, the data were in more compact regions but overlapped, which explains the classification percentages obtained. In the same sense, we must emphasize that adding a priori information about the classes did not seem to provide any significant advantage when using translations.

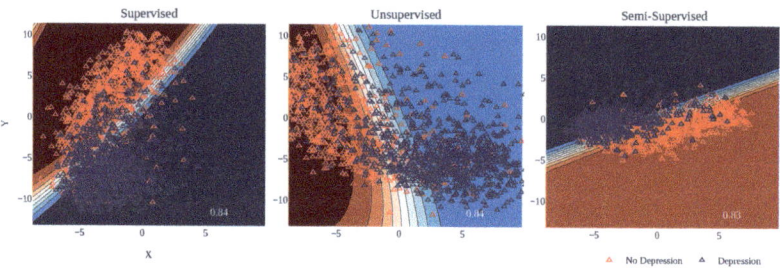

Figure 2. Results using translated sentences. Note that there was a substantial overlap between the different classes, which made the classification process difficult.

7.2. Evaluation of Experiments with Knowledge Distillation

Table 4 shows the results of classifying depression-related tweets using knowledge distillation. The best results for the supervised and unsupervised schemes were obtained by GP, with an accuracy and an F1 score of 0.93. Depression-related tweets could be classified using this model with reasonable accuracy.

Finally, the semi-supervised scheme obtained very high accuracy percentages for all the classifiers, which could be explained if we consider that few tweets in Spanish were included during the training. Concerning the F1 score, the best results were obtained by Logistic Regression; however, the differences between the classifiers were insignificant.

Table 4. Results for Knowledge Distillation.

IVIS	Model	Accuracy	Precision	Recall	F1
Supervised	LR	0.91	0.98	0.84	0.90
	SVM	0.90	0.99	0.80	0.89
	GP	0.90	0.98	0.82	0.89
	QDA	0.93	0.97	0.89	**0.93**
Unsupervised	LR	0.92	0.93	0.90	0.91
	SVM	0.92	0.93	0.90	0.92
	GP	0.91	0.94	0.88	0.91
	QDA	0.93	0.92	0.94	**0.93**
Semi-supervised	LR	0.95	0.96	0.94	**0.95**
	SVM	0.95	0.96	0.93	0.95
	GP	0.95	0.96	0.93	0.94
	QDA	0.95	0.96	0.93	0.94

In Figure 3 we show the results obtained using QDA. In the figures, the boundary surfaces were constructed using the training data (i.e., tweets in English). Note that the data was much more concentrated for the semi-supervised scheme, while the supervised and unsupervised schemes were much more dispersed.

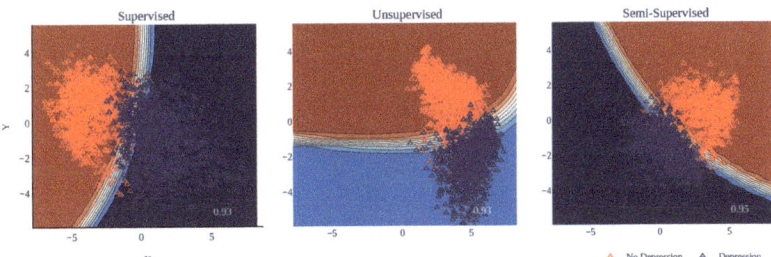

Figure 3. Results using knowledge distillation. Note that, regardless of the dimensionality reduction scheme employed, there was little overlap between the different classes.

8. Geospatial Analysis of Depressive Tweets in Mexico

We employed our methodology to perform a space–time analysis of the depressión-related Tweets obtained through the AGEI platform. For this analysis, we only used public tweets containing geo-referenced information because we wanted to identify the State and date of publication. The objective was to analyze the tweets' content, State of publication, and dates to compare the information with official data published in the same period. We built two descriptors based on geo-reference and user IDs. The first descriptor corresponded to the rate of tweets per State, and we built it using our methodology with the semi-supervised IVIS scheme and QDF as a classifier. Once we identified tweets with depression-related content, we used the user ID to generate a rate of user accounts that posted these tweets. We defined the State to which each user belonged depending on the State wherefrom the user posted most frequently; this was because some users posted in different States over time. Both rates described the rate per 100,000 inhabitants and were estimated using the information corresponding to the INEGI Population and Housing Census for 2020.

8.1. Analysis during the COVID-19 Pandemic Period

As we previously mentioned, the COVID-19 pandemic caused changes in the population's behavior patterns. Although there are studies on the effects of the COVID pandemic in Mexico, we used our methodology to capture the variations in the publication of depression-related Tweets on a time scale that we divided into two periods. The first period corresponded to 2018–2019, which practically enclosed the interval before the pandemic. The second period corresponded to 2020–2021, when the pandemic had its most significant peak.

8.1.1. Tweet Distributions

One of the most significant aspects to study during the pandemic was the change in behavior due to long periods of confinement. As a first analysis, we used tweets with content related to depression to analyze behavioral changes, especially in the periods of the highest contagion. To do this, we identified the date and place of publication and calculated the monthly distribution for each period.

Figure 4 shows the distribution of tweets related to depression for the different periods. The distributions illustrate the change in behavior in the publication of tweets. In the 2018–2019 period, the publications seemed to be more evenly distributed throughout the year, while, for the 2020–2021 period, the distribution shifted to the left, which corresponded to the second quarter of the year, months in which the highest COVID infection rates occurred.

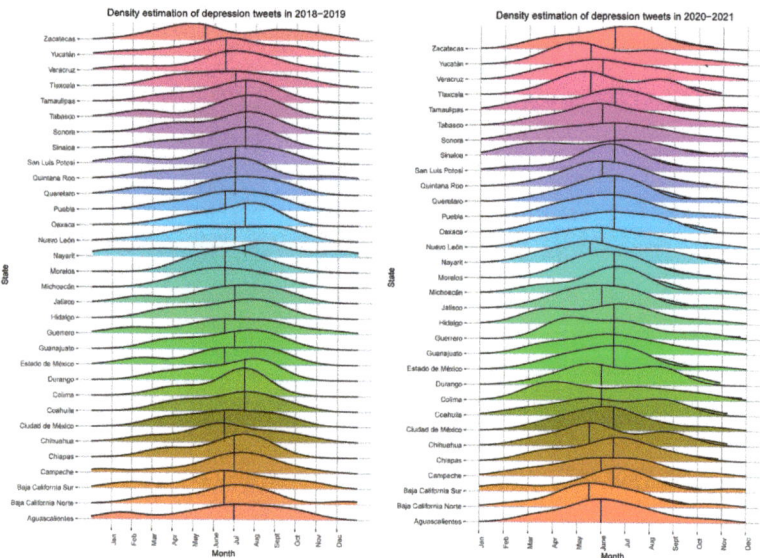

Figure 4. The density of tweets related to depression per month for the periods 2018–2019 and 2020–2021. It is possible to observe an increase in tweets for the second quarter of 2020–2021, consistent with the dates of the highest COVID-19 contagion in Mexico.

8.1.2. Content Analysis

To describe the content of the tweets related to depression in the evaluated period, we used the importance scores of each word obtained through TD–IDF. The main idea was to identify which words were commonly used each year.

The results can be seen in Figure 5. The results showed a change in the most relevant words in the years evaluated. For example, in 2018, the most relevant words referred to concepts related to the family, parents, etc. On the other hand, 2019 showed that words related to security and violence were gaining more importance. For the 2020–2021 period, the terms associated with the pandemic became essential. Note that the words that referred to family and parents remained relevant for all years.

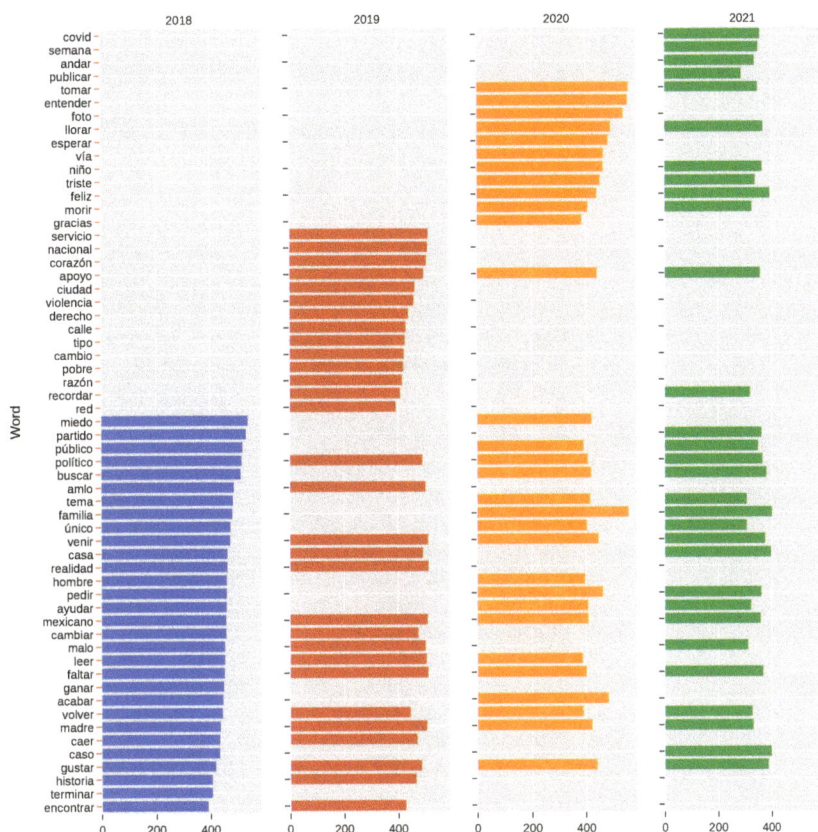

Figure 5. Importance of words for the period 2018–2021 using TI–IDF. For each year, the tweets related to depression refer to different topics. Note that, for the period 2020–2021, words associated with the pandemic became more relevant.

8.2. State of Mood of Twitter Users in Mexico

Among the indicators published by INEGI is one associated with the state of mind of Twitter users in Mexico. INEGI calls this indicator the positivity rate and defines it as the number of positive tweets divided by the number of negative tweets for a given geographical area for a given period. Using depression-related tweets, we calculated an equivalent ratio by dividing the number of non-depression-related tweets by the number of depression-related tweets. We illustrated the behavior of both curves for the States with the highest suicide rate in Mexico.

The results can be seen in Figure 6. Note that, for the four States, both curves maintained the same trend. In the cases of Mexico City and Aguascalientes, the curves had very similar measurements. On the contrary, in Yucatán, although the trend was the same, the ratio between Tweets was generally above the curve. On the other hand, Coahuila was the opposite of Yucatan; in this State, the positivity rate was generally above the depression rate.

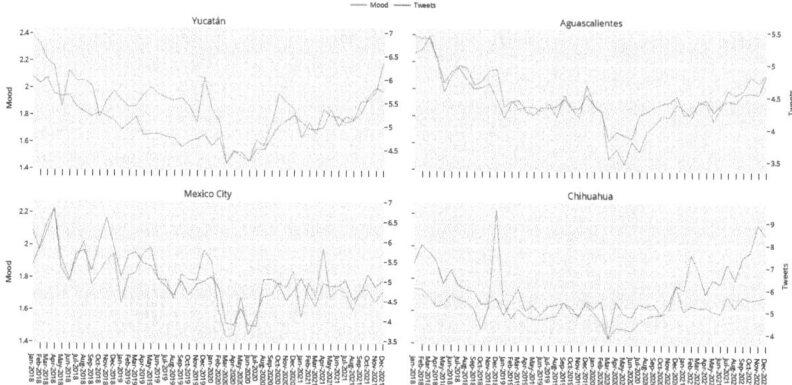

Figure 6. Comparison between the positivity rate, estimated by INEGI, and the depression ratio, estimated using the information obtained from Twitter. Note that the curves generally had the same trend.

8.3. Depression in MéXico

To assess the ability to use the information from Twitter as an indicator of depression levels in Mexico, we compared the rates of Tweets and User Accounts against the official data provided by INEGI. Unfortunately, there were few official data on depression in Mexico; the existing data corresponds to depression rates per 100,000 inhabitants published by INEGI in 2021. However, there are official rates related to suicide, which we included in our analysis to make it more complete. First, we estimated the correlation between these official data against the user's accounts and tweet rates for the different States.

The correlation analysis can be seen in Table 5. The results show a weak negative correlation between suicide and depression rates published by INEGI; this is relevant because, in many studies, this behavior suggests an under-reporting of depression cases at the national level. However, there was a negative correlation between depression rates and Tweets and User Accounts. A significant positive correlation could also be observed between the descriptors used and the published suicide rates. The results show that it is possible to use the rates of Tweets and User Accounts as an auxiliary estimator in the construction of national measures of suicide.

Table 5. Correlations between the rates of depression published by the INEGI and the descriptors obtained with Twitter.

Rate		Correlation	p-Value
Suicide	Depression	−0.3448455	7.23×10^{-2}
Suicide	Tweets	0.5186857	4.68×10^{-3}
Suicide	User accounts	0.4935746	7.60×10^{-3}
Depression	Tweets	−0.4858599	8.76×10^{-3}
Depression	User accounts	−0.4750278	1.06×10^{-2}

9. Discussion

During this study, we highlighted the importance of the fact that depression, as a mental health problem, can lead to other more serious problems, such as suicide, which is considered one of the leading causes of death in young people around the world. With this in mind, and from the existing limitations of the models and methodologies reported in state-of-the-art, our study proposes a method to generate classification models using the knowledge distillation technique.

Our results showed that the explicit use of translations in short texts reduces the accuracy of text classification, because there is a loss of context when translating short texts. On the other hand, the results using knowledge distillation showed better performance than translations, even with unlabeled data. The dimensionality reduction schemes used in this research generated similar vector representations, regardless of the use of labels. This allows models to be trained even with unlabeled or few data. Our results also illustrated the robustness of the methodology; regardless of the classification method used, the values in the f1-score were comparable.

We applied the model with the highest f1-scores to analyze tweets published in the period 2018–2021. We selected these years in order to determine the effect of the COVID-19 pandemic on the mood of the population in Mexico. The findings showed that the pandemic affected the mood of the Mexican population and that this was reflected in an increase in depression levels, which coincided with the results of the National Self-Reported Well-being Survey (ENBIARE). This scenario allows us to assume that, based on geospatial analysis, it is possible to have an approximation of the state of mental health in Mexico, state by state, to detect risk zones, and even to detect other factors that could be associated, almost in real-time.

10. Conclusions

The existing depression-related tweet classification models in the literature have three fundamental disadvantages: the use of dictionaries, possible loss of context, and the lack of generalization to multilanguage schemes. This work presents a methodology that responds to the existing state-of-the-art limitations. In particular, this work presented a methodology to classify depression-related tweets in Spanish. The methodology uses knowledge distillation and dimensionality reduction to train classification algorithms that allow the distinguishing of tweets related to depression in English and Spanish. The results obtained by the unsupervised schemes show that it is feasible to apply our proposal even in the absence of labeled data.

One of the drawbacks of this work is that it requires two stages. The first is in control of building the embedded space where the sentences in different languages are represented, and the second one uses dimensionality reduction algorithms. This has a computational cost that could be reduced; for example, designing a model that performs both tasks simultaneously. Furthermore, we tested this methodology only for short texts. Applications with long texts from other sources, like forums or news, require different models.

With this classification model, we aimed to generate a database of depression-related tweets to detect depression in its early stages and other related mental illnesses. Future work will apply this same methodology to other problems, such as suicide, misogyny, or bullying topics in Twitter. Since it is difficult to determine the state of mental health of a person with only some of their publications, as future work, we intend to expand this study to analyze the timeline of some Twitter accounts in order to generate models that allow us to identify individual cases of risk, but also, to be able to take advantage of the geospatial component (location of the person), to define prevention strategies.

Author Contributions: For this research the authors contributes as follows: Conceptualization, H.C.-M. and G.H.-C.; methodology, H.C.-M.; software, J.P.-C.; validation, O.S.-S. and G.H.-C.; formal analysis, H.C.-M., G.H.-C. and J.P.-C.; investigation, H.C.-M. and O.S.-S.; resources, O.S.-S. and J.P.-C.; data curation, J.P.-C. and H.C.-M.; writing—original draft preparation, H.C.-M. and G.H.-C.; writing—review and editing, O.S.-S. and G.H.-C.; visualization, J.P.-C. and H.C.-M.; supervision, H.C.-M. and G.H.-C.; project administration, J.P.-C. and H.C.-M. All authors have read and agreed to the published version of the manuscript.

Funding: This research received no external funding.

Data Availability Statement: The source files and datasets used during this research are available in: https://github.com/jpoolcen/classification-tweets-depresive, accessed on 22 December 2022. The repository includes the datasets and the codes for data processing. Even when we use data from twitter accounts, they are public data and the personal data was not used, so it is not part of this study.

Acknowledgments: The authors would like to thank the Geospatial Information Sciences Research Center (CentroGeo) and, especially, the National GeoIntelligence Laboratory (GeoInt) for the facilities provided to carry out this study, specifically for the use of AGEI in the data extraction process from twitter accounts.

Conflicts of Interest: The authors declare no conflict of interest.

References

1. Patel, V. Why mental health matters to global health. *Transcult. Psychiatry* **2014**, *51*, 777–789. [CrossRef] [PubMed]
2. Moreno, C.; Wykes, T.; Galderisi, S.; Nordentoft, M.; Crossley, N.; Jones, N.; Cannon, M.; Correll, C.U.; Byrne, L.; Carr, S.; et al. How mental health care should change as a consequence of the COVID-19 pandemic. *Lancet Psychiatry* **2020**, *7*, 813–824. [CrossRef]
3. Kumar, A.; Nayar, K.R. COVID 19 and its mental health consequences. *J. Ment. Health* **2021**, *30*, 1–2. [CrossRef] [PubMed]
4. Hertz, M.F.; Barrios, L.C. Adolescent mental health, COVID-19, and the value of school-community partnerships. *Inj. Prev.* **2021**, *27*, 85–86. [CrossRef] [PubMed]
5. Killgore, W.D.; Cloonan, S.A.; Taylor, E.C.; Dailey, N.S. Loneliness: A signature mental health concern in the era of COVID-19. *Psychiatry Res.* **2020**, *290*, 113117. [CrossRef]
6. Walker, C. *Depression and Globalization: The Politics of Mental Health in the 21st Century*; Springer Science & Business Media: Berlin, Germany, 2007.
7. Ebert, D.D.; Cuijpers, P. It is time to invest in the prevention of depression. *JAMA Netw. Open* **2018**, *1*, e180335. [CrossRef] [PubMed]
8. Health, T.L.G. Mental health matters. *Lancet. Glob. Health* **2020**, *8*, e1352. [CrossRef]
9. Escobar-Viera, C.G.; Cernuzzi, L.C.; Miller, R.S.; Rodríguez-Marín, H.J.; Vieta, E.; Gonzalez Tonanez, M.; Marsch, L.A.; Hidalgo-Mazzei, D. Feasibility of mHealth interventions for depressive symptoms in Latin America: A systematic review. *Int. Rev. Psychiatry* **2021**, *33*, 300–311. [CrossRef]
10. Jiménez-Molina, Á.; Franco, P.; Martínez, V.; Martínez, P.; Rojas, G.; Araya, R. Internet-based interventions for the prevention and treatment of mental disorders in Latin America: A scoping review. *Front. Psychiatry* **2019**, *10*, 664. [CrossRef]
11. Pratap, A.; Renn, B.N.; Volponi, J.; Mooney, S.D.; Gazzaley, A.; Arean, P.A.; Anguera, J.A.; et al. Using mobile apps to assess and treat depression in Hispanic and Latino populations: Fully remote randomized clinical trial. *J. Med Internet Res.* **2018**, *20*, e10130. [CrossRef]
12. Basco, M.R.; Bostic, J.Q.; Davies, D.; Rush, A.J.; Witte, B.; Hendrickse, W.; Barnett, V. Methods to improve diagnostic accuracy in a community mental health setting. *Am. J. Psychiatry* **2000**, *157*, 1599–1605. [CrossRef] [PubMed]
13. First, M.B.; Rebello, T.J.; Keeley, J.W.; Bhargava, R.; Dai, Y.; Kulygina, M.; Matsumoto, C.; Robles, R.; Stona, A.C.; Reed, G.M. Do mental health professionals use diagnostic classifications the way we think they do? A global survey. *World Psychiatry* **2018**, *17*, 187–195. [CrossRef] [PubMed]
14. Shatte, A.B.; Hutchinson, D.M.; Teague, S.J. Machine learning in mental health: A scoping review of methods and applications. *Psychol. Med.* **2019**, *49*, 1426–1448. [CrossRef] [PubMed]
15. Taliaz, D.; Spinrad, A.; Barzilay, R.; Barnett-Itzhaki, Z.; Averbuch, D.; Teltsh, O.; Schurr, R.; Darki-Morag, S.; Lerer, B. Optimizing prediction of response to antidepressant medications using machine learning and integrated genetic, clinical, and demographic data. *Transl. Psychiatry* **2021**, *11*, 381. [CrossRef] [PubMed]
16. Le Glaz, A.; Haralambous, Y.; Kim-Dufor, D.H.; Lenca, P.; Billot, R.; Ryan, T.C.; Marsh, J.; Devylder, J.; Walter, M.; Berrouiguet, S.; et al. Machine learning and natural language processing in mental health: systematic review. *J. Med Internet Res.* **2021**, *23*, e15708. [CrossRef] [PubMed]
17. Squarcina, L.; Villa, F.M.; Nobile, M.; Grisan, E.; Brambilla, P. Deep learning for the prediction of treatment response in depression. *J. Affect. Disord.* **2021**, *281*, 618–622. [CrossRef] [PubMed]
18. Calvo, R.A.; Milne, D.N.; Hussain, M.S.; Christensen, H. Natural language processing in mental health applications using non-clinical texts. *Nat. Lang. Eng.* **2017**, *23*, 649–685. [CrossRef]
19. Kim, J.; Uddin, Z.A.; Lee, Y.; Nasri, F.; Gill, H.; Subramanieapillai, M.; Lee, R.; Udovica, A.; Phan, L.; Lui, L.; et al. A systematic review of the validity of screening depression through Facebook, Twitter, Instagram, and Snapchat. *J. Affect. Disord.* **2021**, *286*, 360–369. [CrossRef] [PubMed]
20. Guntuku, S.C.; Yaden, D.B.; Kern, M.L.; Ungar, L.H.; Eichstaedt, J.C. Detecting depression and mental illness on social media: An integrative review. *Curr. Opin. Behav. Sci.* **2017**, *18*, 43–49. [CrossRef]
21. Shrestha, K. Machine learning for depression diagnosis using twitter data. *Int. J. Comput. Eng. Res. Trends* **2018**, *5*, 57–61.

22. Zhang, Y.; Lyu, H.; Liu, Y.; Zhang, X.; Wang, Y.; Luo, J. Monitoring depression trend on Twitter during the COVID-19 pandemic. *arXiv* **2020**, arXiv:2007.00228.
23. Devlin, J.; Chang, M.W.; Lee, K.; Toutanova, K. Bert: Pre-training of deep bidirectional transformers for language understanding. *arXiv* **2018**, arXiv:1810.04805.
24. Greco, C.M.; Simeri, A.; Tagarelli, A.; Zumpano, E. Transformer-based Language Models for Mental Health issues: A Survey. *Pattern Recognit. Lett.* **2023**, *167*, 204–211. [CrossRef]
25. Tamine, L.; Goeuriot, L. Semantic information retrieval on medical texts: Research challenges, survey, and open issues. *ACM Comput. Surv. (CSUR)* **2021**, *54*, 1–38. [CrossRef]
26. Arnaud, É.; Elbattah, M.; Gignon, M.; Dequen, G. Learning Embeddings from Free-text Triage Notes using Pretrained Transformer Models. In Proceedings of the 15th International Joint Conference on Biomedical Engineering Systems and Technologies (BIOSTEC 2022), Vienna, Austria, 9–11 February 2022; pp. 835–841.
27. Brisset, C.; Leanza, Y.; Rosenberg, E.; Vissandjée, B.; Kirmayer, L.J.; Muckle, G.; Xenocostas, S.; Laforce, H. Language barriers in mental health care: A survey of primary care practitioners. *J. Immigr. Minor. Health* **2014**, *16*, 1238–1246. [CrossRef]
28. Limon, F.J.; Lamson, A.L.; Hodgson, J.; Bowler, M.; Saeed, S. Screening for depression in Latino immigrants: A systematic review of depression screening instruments translated into Spanish. *J. Immigr. Minor. Health* **2016**, *18*, 787–798. [CrossRef] [PubMed]
29. Garcia, M.E.; Ochoa-Frongia, L.; Moise, N.; Aguilera, A.; Fernandez, A. Collaborative care for depression among patients with limited English proficiency: A systematic review. *J. Gen. Intern. Med.* **2018**, *33*, 347–357. [CrossRef] [PubMed]
30. Salas-Zárate, R.; Alor-Hernández, G.; Salas-Zárate, M.d.P.; Paredes-Valverde, M.A.; Bustos-López, M.; Sánchez-Cervantes, J.L. Detecting depression signs on social media: A systematic literature review. *Healthcare* **2022**, *10*, 291. [CrossRef]
31. Leis, A.; Mayer, M.A.; Ronzano, F.; Torrens, M.; Castillo, C.; Furlong, L.I.; Sanz, F. Clinical-Based and Expert Selection of Terms Related to Depression for Twitter Streaming and Language Analysis. In *Digital Personalized Health and Medicine*; IOS Press: Amsterdam, The Netherlands, 2020; pp. 921–925.
32. Leis, A.; Ronzano, F.; Mayer, M.A.; Furlong, L.I.; Sanz, F. Detecting signs of depression in tweets in Spanish: Behavioral and linguistic analysis. *J. Med. Internet Res.* **2019**, *21*, e14199. [CrossRef] [PubMed]
33. Valeriano, K.; Condori-Larico, A.; Sulla-Torres, J. Detection of suicidal intent in Spanish language social networks using machine learning. *Int. J. Adv. Comput. Sci. Appl.* **2020**, *11*, 689–690. [CrossRef]
34. Shekerbekova, S.; Yerekesheva, M.; Tukenova, L.; Turganbay, K.; Kozhamkulova, Z.; Omarov, B. Applying Machine Learning to Detect Depression-Related Texts on Social Networks. In *Advanced Informatics for Computing Research. ICAICR 2020. Communications in Computer and Information Science*; Springer: Singapore, 2020; Volume 1393, pp. 161–169.
35. Ramirez-Esparza, N.; Chung, C.; Kacewic, E.; Pennebaker, J. The psychology of word use in depression forums in English and in Spanish: Testing two text analytic approaches. In Proceedings of the International AAAI Conference on Web and Social Media, Seattle, WA, USA, 30 March–2 April 2008; Volume 2, pp. 102–108.
36. Kiss, G.; Tulics, M.G.; Sztahó, D.; Esposito, A.; Vicsi, K., Language Independent Detection Possibilities of Depression by Speech. In *Recent Advances in Nonlinear Speech Processing*; Esposito, A., Faundez-Zanuy, M., Esposito, A.M., Cordasco, G., Drugman, T., Solé-Casals, J., Morabito, F.C., Eds.; Springer International Publishing: Cham, Switzerland, 2016; pp. 103–114. [CrossRef]
37. Demiroglu, C.; Beşirli, A.; Ozkanca, Y.; Çelik, S. Depression-level assessment from multi-lingual conversational speech data using acoustic and text features. *EURASIP J. Audio Speech Music Process.* **2020**, *2020*, 1–17. [CrossRef]
38. Kiss, G. Investigation of speech-based language-independent possibilities of depression recognition. In Proceedings of the 2022 45th International Conference on Telecommunications and Signal Processing (TSP), Virtual, 13–15 July 2022; pp. 226–229.
39. Coello-Guilarte, L.; Ortega-Mendoza, R.M.; Villaseñor-Pineda, L.; Montes-y Gómez, M. Crosslingual depression detection in twitter using bilingual word alignments. In Proceedings of the International Conference of the Cross-Language Evaluation Forum for European Languages, Lugano, Switzerland, 9–12 September 2019.
40. Villasenor-Pineda, L.; Montes-y Gómez, M. Crosslingual Depression Detection in Twitter Using Bilingual Word Alignments. In Proceedings of the Experimental IR Meets Multilinguality, Multimodality, and Interaction: 10th International Conference of the CLEF Association, CLEF 2019, Lugano, Switzerland, 9–12 September 2019; p. 49.
41. Feng, F.; Yang, Y.; Cer, D.; Arivazhagan, N.; Wang, W. Language-agnostic bert sentence embedding. *arXiv* **2020**, arXiv:2007.01852.
42. Robnik-Šikonja, M.; Reba, K.; Mozetic, I. Cross-lingual transfer of twitter sentiment models using a common vector space. *arXiv* **2020**, arXiv:2005.07456.
43. Reimers, N.; Gurevych, I. Making Monolingual Sentence Embeddings Multilingual using Knowledge Distillation. In Proceedings of the 2020 Conference on Empirical Methods in Natural Language Processing, Online, 16–20 November 2020.
44. Szubert, B.; Drozdov, I. ivis: Dimensionality reduction in very large datasets using Siamese Networks. *J. Open Source Softw.* **2019**, *4*, 1596. [CrossRef]
45. Wolf, T.; Debut, L.; Sanh, V.; Chaumond, J.; Delangue, C.; Moi, A.; Cistac, P.; Rault, T.; Louf, R.; Funtowicz, M.; et al. Transformers: State-of-the-art natural language processing. In Proceedings of the 2020 Conference on Empirical Methods in Natural Language Processing: System Demonstrations, Online, 16–20 November 2020; pp. 38–45.
46. Sanh, V.; Debut, L.; Chaumond, J.; Wolf, T. DistilBERT, a distilled version of BERT: Smaller, faster, cheaper and lighter. *arXiv* **2019**, arXiv:1910.01108.
47. Anowar, F.; Sadaoui, S.; Selim, B. Conceptual and empirical comparison of dimensionality reduction algorithms (pca, kpca, lda, mds, svd, lle, isomap, le, ica, t-sne). *Comput. Sci. Rev.* **2021**, *40*, 100378. [CrossRef]

48. Pedregosa, F.; Varoquaux, G.; Gramfort, A.; Michel, V.; Thirion, B.; Grisel, O.; Blondel, M.; Prettenhofer, P.; Weiss, R.; Dubourg, V.; et al. Scikit-learn: Machine Learning in Python. *J. Mach. Learn. Res.* **2011**, *12*, 2825–2830.
49. Cañete, J.; Chaperon, G.; Fuentes, R.; Ho, J.H.; Kang, H.; Pérez, J. Spanish Pre-Trained BERT Model and Evaluation Data. In Proceedings of the PML4DC at ICLR 2020, Ethiopia, Online, 26 April–1 May 2020.

Disclaimer/Publisher's Note: The statements, opinions and data contained in all publications are solely those of the individual author(s) and contributor(s) and not of MDPI and/or the editor(s). MDPI and/or the editor(s) disclaim responsibility for any injury to people or property resulting from any ideas, methods, instructions or products referred to in the content.

Article

A Robust Design-Based Expert System for Feature Selection and COVID-19 Pandemic Prediction in Japan

Chien-Ta Ho and Cheng-Yi Wang *

Graduate Institute of Technology Management, National Chung Hsing University, 145 Xingda Rd., Taichung City 402, Taiwan
* Correspondence: rockytallen@gmail.com

Abstract: Expert systems are frequently used to make predictions in various areas. However, the practical robustness of expert systems is not as good as expected, mainly due to the fact that finding an ideal system configuration from a specific dataset is a challenging task. Therefore, how to optimize an expert system has become an important issue of research. In this paper, a new method called the robust design-based expert system is proposed to bridge this gap. The technical process of this system consists of data initialization, configuration generation, a genetic algorithm (GA) framework for feature selection, and a robust mechanism that helps the system find a configuration with the highest robustness. The system will finally obtain a set of features, which can be used to predict a pandemic based on given data. The robust mechanism can increase the efficiency of the system. The configuration for training is optimized by means of a genetic algorithm (GA) and the Taguchi method. The effectiveness of the proposed system in predicting epidemic trends is examined using a real COVID-19 dataset from Japan. For this dataset, the average prediction accuracy was 60%. Additionally, 10 representative features were also selected, resulting in a selection rate of 67% with a reduction rate of 33%. The critical features for predicting the epidemic trend of COVID-19 were also obtained, including new confirmed cases, ICU patients, people vaccinated, population, population density, hospital beds per thousand, middle age, aged 70 or older, and GDP per capital. The main contribution of this paper is two-fold: Firstly, this paper has bridged the gap between the pandemic research and expert systems with robust predictive performance. Secondly, this paper proposes a feature selection method for extracting representative variables and predicting the epidemic trend of a pandemic disease. The prediction results indicate that the system is valuable to healthcare authorities and can help governments get hold of the epidemic trend and strategize their use of healthcare resources.

Keywords: artificial intelligence; expert system; robust design; feature selection; COVID-19; disease prediction; genetic algorithm; healthcare

Citation: Ho, C.-T.; Wang, C.-Y. A Robust Design-Based Expert System for Feature Selection and COVID-19 Pandemic Prediction in Japan. *Healthcare* **2022**, *10*, 1759. https://doi.org/10.3390/healthcare10091759

Academic Editors:
Giner Alor-Hernández,
Jezreel Mejía-Miranda, José
Luis Sánchez-Cervantes and
Alejandro Rodríguez-González

Received: 11 July 2022
Accepted: 8 September 2022
Published: 13 September 2022

Publisher's Note: MDPI stays neutral with regard to jurisdictional claims in published maps and institutional affiliations.

Copyright: © 2022 by the authors. Licensee MDPI, Basel, Switzerland. This article is an open access article distributed under the terms and conditions of the Creative Commons Attribution (CC BY) license (https://creativecommons.org/licenses/by/4.0/).

1. Introduction

Infectious diseases, if not effectively monitored and controlled, often result in mass human infections and pose risks of mass mortality, economic recession, and depletion of medical resources. For instance, coronavirus disease 19 (COVID-19) first broke out in late 2019 in Wuhan, Hubei Province, China [1,2], and then spread rapidly around the world. It was soon recognized as a global pandemic by the World Health Organization (WHO). In the next year, COVID-19 began to cause enormous impacts across the world. As of 2022, COVID-19 continues to spread aggressively from country to country, causing not only over 500 million sick people but also 6 million deaths. The unpredictable nature of COVID-19 has placed a great deal of pressure on governments to set up policies to curb the spread of the epidemic, and it is likely to cause medical resource depletion [3]. Moreover, COVID-19 has also had many negative effects on the global economic environment [4].

Since the Omicron variant began to sweep the world, the COVID-19 pandemic has intensified in many countries due to the high transmissibility of this variant. Despite the emergence of vaccines approved by emergency-use authorization (EUA), because human knowledge of this virus is still insufficient, many people are dying from it every day. In 2022, WHO once again warned the world that the spread of COVID-19 should be closely monitored to prevent a simultaneous increase in the number of moderate and severe cases and the number of deaths as the virus continues to mutate. WHO also suggested that all governments should adjust management policies and quarantine measures for severe cases in a timely manner. Therefore, health authorities' ability to get hold of the development trends of the pandemic is particularly important.

Nowadays, the main tasks of healthcare management authorities in epidemic prevention include infection prevention, spread prediction, infection control, treatment of confirmed cases, and mortality reduction [5]. Until a highly effective method of eliminating this virus is found (including the improvement of vaccines), monitoring the spread trends and reducing the mortality of COVID-19 is a priority in epidemic management for governments seeking to maintain economic activity during the pandemic [6,7]. The prediction of the epidemic trend is linked to the government's epidemic prevention policies [8–10]. How to effectively get hold of the changes in the number of deaths is an imperative task for health authorities because it is conducive to the deployment of medical resources and the improvement of healthcare policies. Therefore, developing a support system that can capture the trends of COVID-19 has become an emerging area of research.

The recent years have seen a substantial development of expert systems and publication of outstanding research findings about such systems in all fields. A human expert can quickly find feasible solutions to a new problem based on his or her experiences that accumulate over time. The introduction of the LISP programming language by John McCarthy in 1960 ushered in the development of research on expert systems [11]. Early expert systems can be represented by the problem-solving model proposed by Feigenbaum et al. in 1970 to determine the structures of chemical molecules [12]. This type of expert system is a rule-based reasoning system that can be applied to disease diagnosis. Due to certain limitations, this type of expert system soon hit a bottleneck in its development. For example, rule-based reasoning systems require the establishment of very complicated conditional formulas, where the cause–effect relationships are highly restrictive, so they are less flexible, and the cross-references between cases may be easily ignored. Hence, there is still a gap in decision-making behavior between expert systems and real human experts. Fortunately, with the advancement of computer science and the extensive use of personal computers, scholars have begun to promote new techniques of expert systems. For instance, Aliev et al. [13] proposed an if–then rules-based fuzzy technique for reasoning with imperfect information and applied it to evaluate job satisfaction and students' educational achievement related to psychological and perceptual issues. Tang and Pedrycz [14] demonstrated the stability of an expert system. They investigated the oscillation-bound estimation of perturbations for Bandler–Kohout subproduct (BKS) and constructed upper and lower bounds of BKS output deviation derived from the simple perturbation of the input set.

Among the numerous techniques of expert systems, artificial intelligence systems (AIs) have received particular attention from researchers [15–17]. AIs are algorithms developed to mimic the operation and behavioral patterns of living organisms. They process information based on past experiences. Through improvements, AIs can more efficiently enhance the learning performance of expert systems and expand the scope of problem search, gradually pushing the decision-making ability of expert systems closer to the level of human experts. As expert systems can be integrated with various types of AIs, many cross-disciplinary applications have been attempted in areas such as clinical, healthcare, environmental, and industrial [18–22]. These applications have also contributed to the flourishing development of expert systems. AIs use datasets to train a model and create an input–output mapping. This type of operation makes AIs highly appropriate for application in data mining, knowledge engineering, medical assessment, diseases prediction, etc. [15,16,18,23,24].

In the last decade, big data analysis was a new line of research. Advancements in this area of research have contributed to the growth of expert systems. With the development of data mining, many big data-driven expert systems have been proposed [25–27]. This suggests that big data can widen the search scope of expert systems and also improve their training performances.

Due to the outstanding contribution of AIs to various research fields, it has also received attention in the field of healthcare. For instance, Malki et al. [28] developed a supervised decision tree model to predict the spread of COVID-19 infection in many countries. Khalilpourazari and Hashemi Doulabi [29] designed a hybrid reinforcement training-based framework to predict the COVID-19 pandemic and help policy makers to optimize the use of healthcare system capacity and resource allocation. Alam et al. [30] developed a disease diagnosis using the Internet of Things (IoT) integrated with a fuzzy inference system to diagnose various diseases.

Recent research has shown that feature selection is the most representative technique of expert systems. Expert systems usually rely on supervised learning. They need to be given a set of training data to learn the relationship between "input features" and "outcomes" [31,32].

Feature selection is a technique for extracting relevant features in data. The basic concept of feature selection is to find distinctive case features in order to enhance the learning efficiency of the expert system [33,34].

If the system can omit certain unnecessary features, it can reduce the data comparison time and achieve a higher accuracy. The main advantage of feature selection lies in its ability to adopt supervised or hierarchical feature extraction algorithms to replace manual ways [31]. The growth of feature selection (including feature extraction) is manifested in the fruitful results of recent research. The outcomes derived from a large amount of data, in particular, have drawn the attention of experts across all fields [27,32]. However, the goal of feature selection for expert systems is to learn autonomously from a large amount of data to create a better model with better training results. In addition, the benefit of using supervised feature selection in an expert system is that the system model can autonomously extract appropriate features and define the recognition or prediction result for each instance.

In an early application of feature selection, Siedlecki and Sklansky [35] used genetic algorithms (GAs) to deal with large-scale feature selection. They attempted to design a set of automatic pattern classifiers. Feature selection could help the system extract the features of patterns suitable for recognition and then deliver the selection results to the system for prediction.

As to the applications of feature selection in recent years, Lin et al. [36] proposed a technique using layered genetic programming for feature extraction to deal with the problem of optimizing the classification of data into two groups. Here, the concept of "feature extraction" refers to transferring the good features obtained in each evolution to the GA processing of the next layer, in order to achieve hierarchical optimization. Their experimental results confirmed that this technique can enhance the problem-solving performance of GAs. Quan and Ren [37] proposed a method of product feature extraction for feature-oriented opinion determination. The feature extraction technique was applied to deal with opinion mining and perform sentiment analysis for product improvement. They showed a high applicability of feature selection using comparative domain corpora.

Zhang et al. [38] proposed an ant colony algorithm-based feature selection method for intelligent fault diagnosis of rotating machinery using a support vector machine. Some scholars have applied GAs as a data mining technique to extract informative and significant features in breast cancer diagnosis [39,40]. In addition, GA-based feature selection methods can deliver a better performance [39]. Gokulnath and Shantharajah [41] employed a GA as an optimization function based on a support vector machine (SVM) for heart disease diagnosis. Khan et al. [42] proposed a hybrid feature selection and reduction scheme for selecting the high discriminative characteristics in hypertension features. Kwon et al. [43] employed feature selection methods to support the prediction of osteoporosis. They conducted a comparison of machine learning with different models and found that features selected by "survey+checkup" led to a better prediction accuracy than survey or checkup only. Moreover, machine learning could achieve good performance in disease prediction. More recent studies of expert systems have empirically demonstrated the effectiveness of applying feature selection in disease prediction, equipment examination, and mental state prediction [43–45].

A feature-selection-based expert system is a smart technique that can be used for describing structured data. It can converge in its own database by inputs of specific structures, so it can accept a wide range of real cases. With this characteristic, feature selection is most appropriate for experiments that involve observations with an expert system [28,29]. As the development trend of a pandemic disease is the result of a features-and-effect phenomenon that evolves over time, feature selection is a very suitable solution for the prediction of epidemic trends. A comparison of previous research on feature selection for prediction is presented in Table 1.

Table 1. Comparison of previous research on feature selection.

Method	Issue	Year	Reference
GA-based algorithm	Large set feature extraction	1989	[35]
GA-based algorithm	Diagnostic classification	2008	[36]
Comparative domain corpora	Product improvement	2014	[37]
Ant-colony-based algorithm	Fault diagnosis	2015	[38]
GA-based algorithm	Breast cancer diagnosis	2016	[39]
GA-based algorithm	Breast cancer diagnosis	2017	[40]
GA-based algorithm	Heart disease diagnosis	2019	[41]
Machine learning	Hypertension Detection	2021	[42]
Machine learning	Comparison of different classifier ensemble methods	2021	[43]
Machine learning	Prediction of osteoporosis	2022	[45]

Previous studies have shown that it is not easy to get impressive learning results from expert systems [46–49]. Oftentimes, it is necessary to repetitively adjust and test the parameters of the system. This procedure is very time-consuming and will increase the cost of system modeling. In addition, the obtained parameter values cannot always guarantee good prediction performance in the future. The abovementioned situation reduces the robustness of the expert system. Moreover, when building a pandemic prediction system, it is necessary to adjust the system parameters whenever needed. In other words, system adaptability and reliability are also of high importance. Therefore, an effective and stable system building method must be developed so as to exploit the excellent performance of expert systems. This study aims to fill two major gaps in the literature: Firstly, the extant research of pandemics lacks studies on expert systems with a robust predictive performance. Secondly, little research has attempted to investigate feature selection methods for predicting the epidemic trend of a pandemic disease.

In this paper, a modified expert system, called the robust design-based expert system, is proposed to address the above issues. In addition, a genetic algorithm (GA) framework and the Taguchi method are integrated into the system to optimize the performance of the system. A good system configuration can not only increase the system's prediction accuracy but also ensure the stable quality of the system. Features selected by the system can support inferences of epidemic trends. Finally, the feasibility and efficiency of the proposed system is verified using COVID-19 as an example.

2. The Expert System with Robust Design

2.1. System Architecture

In this study, a robust design-based expert system is proposed for pandemic prediction. The system architecture is shown in Figure 1. The operational steps of the expert system are as follows: First, the dataset is imported from the case database and normalized. Later, systematic training with the selected parameter levels is conducted. The system will learn the best pattern of features from the dataset, meaning that the system will obtain a feasible solution from the genetic algorithm (GA) framework (see Figure 2). However, this solution does not represent a robust solution of the system under different configurations. The system will repeatedly execute the procedure under different configurations through the robustness mechanism until all the runs have been completed. After all the runs have been completed, a robust result can be obtained.

2.2. Optimization of the System

To enhance the predictive performance of the proposed system, this study applies a genetic algorithm (GA) to feature selection. GAs have been widely used as a means to optimize expert systems [34–36]. It is an optimizing technique that mimics the evolutionary process of biological chromosomes. Based on the concept of genetic evolution, it repeatedly searches for feasible solutions in order to find an optimal solution to the given problem. The operating process of the GA is briefly explained as follows:

First of all, the GA stochastically generates an initial set of feasible solutions (called the initial population), in which each feasible solution is called chromosome and coded by a value of 0 or 1 (see Figure 3). Later, the fitness of each feasible solution is computed. The fitness function can be customized by users. A higher fitness value usually indicates a better solution. In the optimization of an expert system, the fitness function is usually defined as the accuracy of the inferential result.

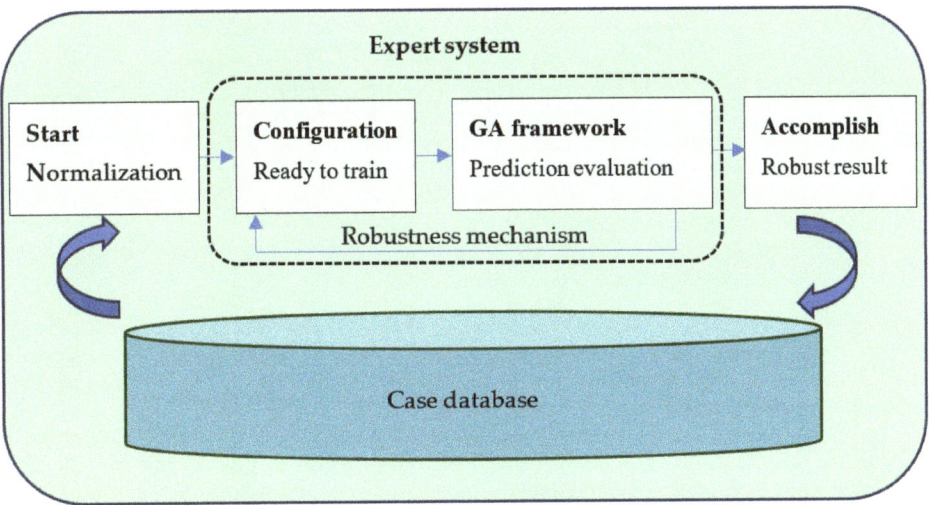

Figure 1. Architecture of the robust design-based expert system.

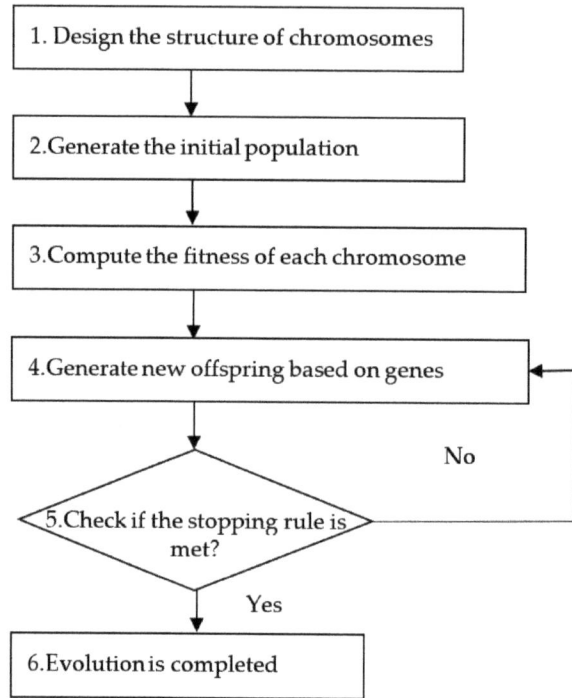

Figure 2. GA framework of the system.

| Population | Feature selection | | | | |
(A generation)	F_1	F_2	F_3	...	F_x
Chromosome 1	1	0	1	...	0
Chromosome 2	1	1	0	...	1
⋮			⋮		
Chromosome n	0	0	1	...	0

Figure 3. Structure of chromosomes.

Next, the GA uses the genes in the chromosomes to compute the next generation. The proposed system adopts the GA as a predictor because of its evolution mechanism including selection, crossover, and mutation. These mechanisms can help the system achieve a high prediction accuracy. Selection decides which chromosomes can survive or should be eliminated; crossover is used to exchange partial sections of the chromosomes among parents to create the chromosomes of the next generation. Finally, mutation selects one gene from chromosomes for mutation. The probability of mutation is usually very low. Through repetitive executions of the above genetic operation, offspring with better fitness can be generated, and this operation stops when the stopping rule is met.

The GA framework designed for this system is as illustrated in Figure 2. The proposed system applies the GA to select features in the dataset. This training procedure consists of six steps as explained below:

Step 1. Design the structure of chromosomes

In order to obtain an optimal combination of features, we encode feature selection in chromosomes using 0 or 1, as shown in Figure 3. For instance, "0" denotes the corresponding item is unselected, while "1" denotes the item is selected.

Step 2. Generate the initial population

Before execution of the genetic algorithm, the system has to generate an initial population comprising n chromosomes, each containing randomly generated parameter values. Each chromosome represents a possible solution (the initial feature selection). Given a total of x features, each chromosome is represented by x genes, and each generation has n chromosomes. Through evolution from one generation to the next of each generation, a better solution can be progressively obtained.

Step 3. Compute the fitness of each chromosome

To compute the fitness of each chromosome, we divide the case dataset into two subsets, including a training dataset and a test dataset. The training dataset is the main dataset for training the expert system, whereas the test dataset provides the subject to be tested by the expert system. The training dataset is larger than the test dataset.

For any chromosome i $(i = 1, 2, \ldots, n)$, some features in both the training dataset and the test dataset need to be removed, weakened, or reinforced according to the set of features stored in the chromosome. Assume that $Train_i$ and $Test_i$, respectively, denote the modified training dataset and the modified test dataset. The fitness of chromosome i can be computed through the following steps:

(1) Compute the predicted level (PL) for each case in the test dataset. For each case t in the test dataset, we apply the nearest neighbor method to find the most similar case in the training dataset $Train_i$ to predict the level of this case (the level is represented by PL_t). The similarity between cases is measured using Euclidean distance.

(2) Compute the fitness of chromosome i $(i = 1, 2, \ldots, n)$. The fitness of chromosome i can be expressed using the following function:

$$fitness(chromosome\ i) = \sum_{t=1}^{D} \frac{match_t(Train_i, Test_i)}{D} \quad (1)$$

where D denotes the number of cases in $Test_i$; $match_t$ indicates whether the predicted level (PL_t) matches the actual level (AL_t). If $PL_t = AL_t$, $match_t(Train_i, Test_i) = 1$; otherwise, $match_t(Train_i, Test_i) = 0$. The fitness of a chromosome represents the prediction accuracy obtained based on the corresponding feature selection. This value is continuously updated as the evolution progresses. Moreover, it is also used as an indicator to assess the quality of each chromosome. It provides a reference for subsequent genetic evolution. A better chromosome is more likely to be chosen for crossover.

Step 4. Apply genetic operators to derive new offspring

After a new generation is generated, the max fitness value searched for in the previous generation may be changed. As mentioned above, these genetic operators, including chromosome selection, crossover, and mutation, are intended to help generate new chromosomes. The selection operator determines whether a chromosome should be kept or eliminated depending on its fitness value. Chromosomes with a higher fitness value are more likely to survive. For crossover and mutation, the probabilities should be defined in advance.

Step 5. Repeat Step 3 and Step 4 until the stopping rule is met

Step 3 and Step 4 are iteratively executed until the stopping rule is satisfied. By the time that the expert system terminates the evolution based on the stopping rule, an optimal solution will be generated. This solution contains the finally selected features, which are most useful for the prediction of new cases and optimization of the weighting of features in the system.

Step 6. Evolution is completed

After genetic evolutions, the system outputs selected features.

However, system configuration affects the solution performance of the GA framework and further reduce the robustness of the system.

To enhance the robustness of the proposed system, a GA and the Taguchi method are integrated into the expert system.

The Taguchi method [50] is utilized to optimize the system. It uses an orthogonal array and a signal-to-noise ratio (SN ratio) to help expert systems find an optimal system configuration. The advantage of using an orthogonal array is that it can significantly reduce the total number of runs of the experiment to slash the time cost, whereas the advantage of using an SN ratio is that the quality of the system can be measured. The Taguchi method designed for the system consists of three processes: firstly, set up the parameters of system; secondly, define the levels of each parameter; finally, generate the orthogonal array for the system. An example is given as follows. Assume that there are three parameters, and each parameter has three levels. For a full factorial experiment, the system needs to perform 27 experiments, which is quite time-consuming. Using the Taguchi method, this system generates an orthogonal array and needs to perform only nine sets (i.e., $L_9(3^3)$) of experiments to obtain a reliable solution. In this way, while the system execution time is being drastically reduced, the system quality can also be ensured.

After the configuration training is completed, the system will measure the mean-square error (MSE) of the expected results based on the data from each run. The MSE value has a smaller-the-better characteristic. It is expressed as follows:

$$MSE = \frac{1}{n} \times \sum_{i=1}^{n}(A_i - P_i)^2 \qquad (2)$$

where n is the number of observations in the test data, P_i is the predicted value for the ith observation, and A_i is the actual value of the ith observation.

After measuring the MSE value for each run, the system will estimate the SN ratio for each configuration. The SN ratio has a larger-the-better characteristic. It is expressed as follows:

$$SN = -10 \times \log\left(\frac{1}{m}\sum_{j=1}^{m}\frac{1}{MSE_j^2}\right) \qquad (3)$$

where m is the number of repetitions for each configuration, and MSE_j denotes the result of the jth run.

Finally, the system will obtain the robust configuration P_q ($q = 1, 2, \ldots, Q$) with the highest total SN ratio from all the runs. It can be expressed as follows:

$$R(configuration\ P_q) = \text{Max}\ SN_{P_{q_k(k=1,2,\ldots,K)}} \qquad (4)$$

where K is the number of levels for each parameter.

Details on data collection and performance of the system are provided in Section 3.

3. Results

3.1. Data Collection

Microsoft Excel 2016 (https://www.microsoft.com (accessed on 30 June 2022)) was installed as the runtime environment to implement the program. The proposed system was built using Evolver version 8.2 (https://www.palisade.com (accessed on 30 June 2022)) to process the genetic operations in the training. VBA (Visual Basic for Application) programming was also integrated to build the proposed expert system.

The data used in this study are real COVID-19 data reported from around the world. The data comprise the statistics of the pandemic provided on WHO's official website (https://covid19.who.int (accessed on 30 June 2022). According to the statistics, the number of daily confirmed cases remains very high and COVID-19 is still severely spreading across the world (see Figure 4). A large number of infections have been caused since the outbreak of COVID-19. Daily infections of the virus peaked in the first quarter of 2022. Despite devotion to pandemic control, all governments around the world have been unable to resist the repeated growth of infections due to the virus' continuous mutation and evolution. At present, hundreds of thousands of confirmed COVID-19 cases are being reported in many countries every day, indicating that the pandemic is still raging [3,4].

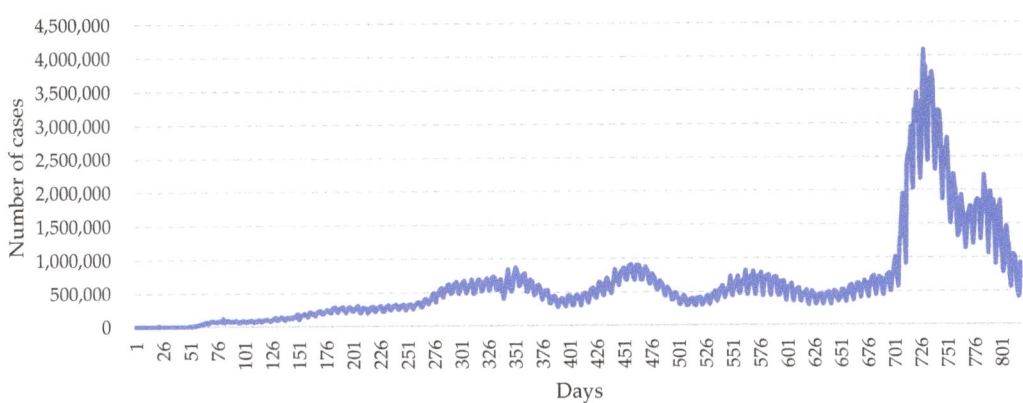

Figure 4. Daily new confirmed COVID-19 cases worldwide.

Japan has been one of the hardest-hit areas for COVID-19 in Asia since 2020. More importantly, Japan decided to postpone the 2020 World Olympic Games in 2021. As shown in Figure 5, the number of confirmed cases in Japan peaked two times within one year. The case of Japan shows that it has experienced several ups and downs of the epidemic. In particular, under the invasion of the Omicron variant virus, Japan also reached the peak of the epidemic in March 2022. Thus, the spread of the epidemic in Japan has virtually become the focus of global attention. In this study, we chose Japan as the subject in the hope of performing an effective prediction of the epidemic trend in Japan. Real data from Japan were collected for data compilation and subsequent analysis. The data include daily statistics of confirmed COVID-19 cases in Japan from the outbreak of the pandemic.

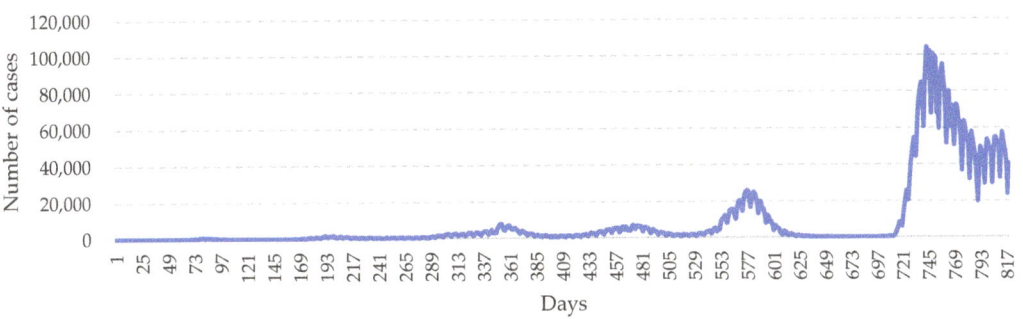

Figure 5. Daily new confirmed COVID-19 cases in Japan. (January 2020–April 2022).

This period spans from January 2020 to April 2022. In this period, the highest number of daily confirmed cases was 104,345, and the average number of daily confirmed cases was approximately 9080 with a standard deviation of 19,819. In addition, the death rate of COVID-19 in Japan was usually higher than 0.003. The highest number of daily death cases was 322, and the average number of daily death rate was approximately 0.018 with a standard deviation of 0.012 (see Figure 6). The historical peak of COVID-19 mortality in Japan occurred in the first year. Although the death rate has gradually decreased, the death toll remains high because the variant virus increases its susceptibility. It shows that reducing the mortality rate has become an important challenge for the government in Japan.

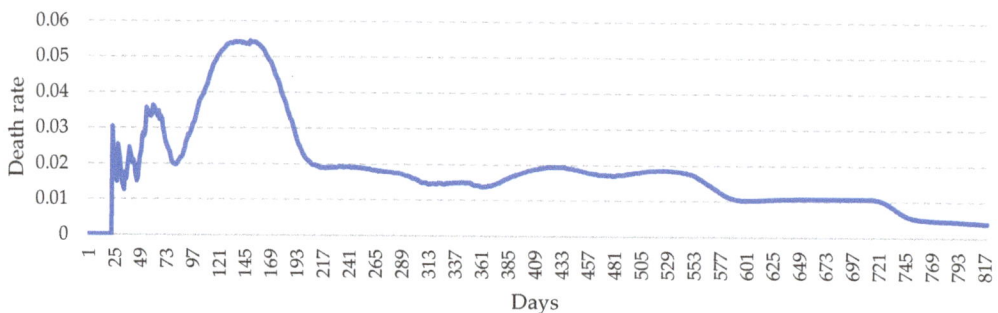

Figure 6. Daily death rate in Japan. (January 2020–April 2022).

When a pandemic begins to spread in large numbers (such as COVID-19), governments are most concerned about the rapid rise in confirmed cases. In the post-epidemic era, governments would pay particular attention to issues relating to the reduction of the mortality rate, including the treatment of severe cases and how to provide effective treatments. Therefore, this study uses the death rate as an important evaluation index and classifies the epidemic situation into different alert states as a reference for healthcare policy makers.

The dataset comprises 819 days of COVID-19 data in Japan. We defined death rates for a daily number equivalent to or below 0.005 as low death rates (I), death rates between 0.005 and 0.01 as mid death rates (II), and death rates equivalent to or above 0.01 as high death rates (III). In the dataset, 83 cases were classified into the low death rate level (I), 34 cases into the mid death rate level (II), and 702 cases into the high death rate level (III). Death rate is defined as follows:

$$\text{Death rate} = \frac{\text{total deaths}}{\text{total confirmed cases}} \quad (5)$$

In addition, this study uses demographic variables including population, population density, middle-aged population (middle age), population aged 65 or older (aged 65 older), population aged 70 or order (aged 70 older), cardiovascular disease death rate (CVD death rate), diabetes prevalence, hospital beds per thousand, and GDP per capital. The descriptive variables of COVID-19 collected in this study include new confirmed cases, hospital patients, ICU patients, people vaccinated, people fully vaccinated, and stringency index. These variables are used as the features of the COVID-19 dataset (as shown in Table 2). All the above variables are continuous variables.

Table 2. Levels of key parameters for system.

Parameter	Level 1	Level 2	Level 3
PS × GN	200 × 100	400 × 50	100 × 200
CR	0.5	0.75	1.0
MR	0.05	0.075	0.1

3.2. Performance of the System

After data collection, we imported all the data into the system to train features through the GA framework. After training, the selected features could be used to predict the number of death cases over the next 120 days. The training dataset consists of data from 22 January 2020 to 31 December 2021. The validation dataset spans from 1 January 2022 to 21 April 2022.

To optimize training and system performance, using an appropriate system configuration is very important. Different configurations of parameters affect the performance of the system in finding a solution. Based on the Taguchi method [30], we adopted an orthogonal array and *SN* ratio to find the optimal set of parameters, including (1) population × generation (PS × GN), (2) crossover rate (CR), and (3) mutation rate (MR).

As shown in Table 2, for each parameter, three different levels were adopted in the system. Considering the solution seeking time, the number of PS × GN was set as 20,000. Therefore, the system was designed to execute an $L_9(3^3)$ experiment, where 9 at the bottom indicates the number of experimental runs, 3 at the center stands for the number of levels, and 3 at the top represents the number of parameters.

The experimental runs were based on the three levels of each parameter. The *MSE* value in each configuration of parameter levels was measured three times ($m = 3$) to obtain MSE_1, MSE_2, and MSE_3. The *MSE* function has a smaller-the-better characteristic.

After measuring *MSE*s, the system estimated the *SN* ratio for each configuration. Then, the system computed the results from all the runs of the experiment respectively.

Table 3 shows the *MSE*s and *SN* ratios from all of the runs of the experiment. As mentioned above, the *SN* ratio has a larger-the-better characteristic. The system summed the *SN* ratios from all of the runs of the experiment (as shown in Table 4) and selected the levels with the max sum. Consequently, the system obtained the configuration with the highest robustness of parameters as follows: PS = 400, GN = 50, CR = 0.75, and MR = 0.05. The average accuracy for the death rate level inference of the proposed system was 60%.

Table 3. The $L_9(3^3)$ orthogonal array for system.

Experiment	PS × GN	CR	MR	MSE_1	MSE_2	MSE_3	SN
1	1	1	1	0.0006010890	0.0006444215	0.0006394203	−64.0492353808
2	1	2	2	0.0004591344	0.0013799845	0.0004216040	−65.6009508154
3	1	3	3	0.0004056524	0.0004205735	0.0004056524	−67.7348319371
4	2	1	2	0.0004388614	0.0005938607	0.0006411313	−65.4242838171
5	2	2	3	0.0006394203	0.0006298676	0.0005938010	−64.1511242272
6	2	3	1	0.0011535812	0.0006116809	0.0004205265	−64.8091223273
7	3	1	3	0.0004344587	0.0006110492	0.0004362107	−66.4448913140
8	3	2	1	0.0005938607	0.0005989994	0.0004195332	−65.7611583377
9	3	3	2	0.0006485800	0.0004341845	0.0004500635	−66.2389340652

Table 4. Sum of *SN* ratios.

	PS × GN	CR	MR
Level 1	−197.3850181333	−195.9184105119	−194.6195160458 *
Level 2	−194.3845303716 *	−195.5132333803 *	−197.2641686977
Level 3	−198.4449837169	−198.7828883296	−198.3308474783

* Selected level with the max sum of *SN*.

After a robust configuration was obtained (including selected features), Japan's epidemic data from 22 January 2020 to 31 December 2021 was imported into the system to predict the epidemic trend over the next 120 days. The results are shown in Table 4 and Figure 7. In addition, from the 15 features, 10 more representative ones were also selected, resulting in a selection rate of 67%.

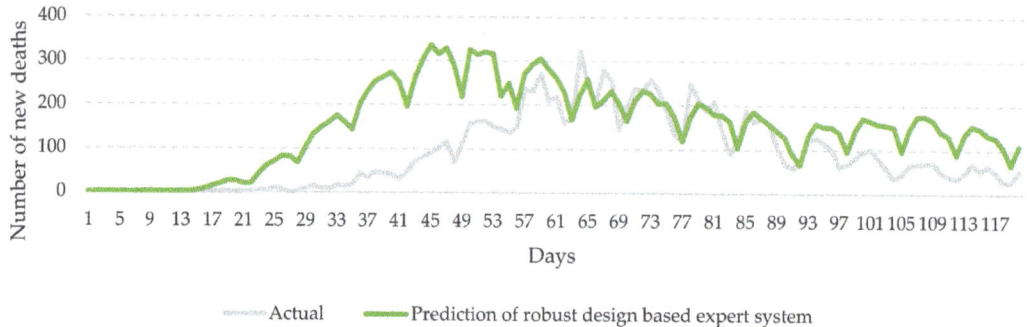

Figure 7. Prediction of robust design-based expert system. (January 2022–April 2022, in Japan).

It can be easily discovered from Table 5 that these features are mainly descriptive variables and demographic variables relating to COVID-19. As to the features associated with CVD death rate and diabetes prevalence, no significant difference was found between different epidemic levels. The gray line shows the actual data, while the green one represents the epidemic trend predicted by the proposed system (see Figure 7). The results indicate that the mean number of daily death cases of the proposed system was 151 persons with a standard deviation of 94 persons in this span of prediction. In the actual data, the mean number of daily death cases was 89 persons with a standard deviation of 84 persons. The time point of the increase in the death rate estimated by this system is slightly earlier than the actual data. Overall, it is quite close to the actual situation. A further comparison of the system's prediction with the actual data was conducted.

Table 5. Selected and unselected informative features of COVID-19 in Japan.

Feature	Selection
Descriptive variables of COVID-19	
New confirmed cases	Selected
Hospital patients	Unselected
ICU patients	Selected
People vaccinated	Selected
People fully vaccinated	Selected
Stringency_index	Unselected
Demographic variables	
Population	Selected
Population density	Selected
Cardiovasc death rate	Unselected
Diabetes prevalence	Unselected
Hospital beds per thousand	Selected
median_age	Selected
Aged 65 older	Unselected
Aged 70 older	Selected
GDP per capital	Selected

It can be discovered that the prediction of the proposed system was close to the real situations. As shown in Figure 7, the system has captured the potential trend pattern of the outbreak. For example, there was an obvious increase in the number of death cases in Japan during the prediction period (i.e., 18 January 2022–22 February 2022). In the actual data, the number of daily death cases grew rapidly from 10 to 332 within four weeks. This abnormal surge was captured by the system effectively. In addition, the following trend in the epidemic situation was also predicted by the system. The results indicate that the performance of the proposed system is good.

Furthermore, the robust design-based expert system can give a warning when the number of death cases is about to rise (as shown in Figure 8). For management authorities, this represents a practical and managerial implication concerning the formulation of measures.

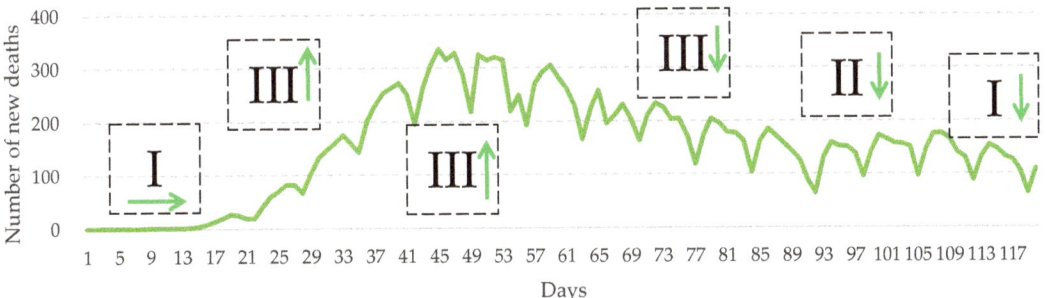

Figure 8. Monitoring levels for COVID-19 suggested by the system. (January 2022–April 2022, in Japan).

For instance, when the epidemic is mild, the system suggests a monitoring level of I, meaning that healthcare authorities can maintain the current management policy. When the death rate grows higher than 0.01, the system will suggest a monitoring level of III. Healthcare authorities should: impose restrictions on the economic activity of certain industries and adopt crowd diversion measures for confirmed cases; require patients with mild illnesses to implement home quarantine, so as to reserve the medical capacity for patients in moderate or severe conditions; provide special care to patients in high-risk groups (including elderly and young children); and set up easily accessible PCR testing sites to instantly identify confirmed cases and provide medication. After the epidemic reaches its peak and begins to slow down through levels III and II with a decreasing death rate, the system suggests a monitoring level of I. Healthcare authorities are advised to progressively relax social management measures, lift the limitations on the economic activity of all industries, and also encourage people to conduct self-health management (as shown in Figure 8).

The monitoring levels suggested by the system based on dynamic changes in the pandemic cases can assist the authorities concerned to plan and deploy in advance. Therefore, the proposed system can be very helpful in practical applications where it is used as a policy support system.

To explore whether different compositions of the training dataset would affect the system's predictive performance, we created three training datasets respectively consisting of 100%, 95%, and 90% of the training data. For each training dataset, the training was carried out five times, and the *MSE* was measured. Finally, analysis of variance (ANOVA) was applied to test if there was any significant difference in prediction accuracy between datasets.

Table 6 shows the ANOVA results, which suggest that under the significance level of 0.05, there was no significant difference in prediction accuracy between the training results with 5% and 10% reduced data.

Table 6. ANOVA result for training datasets.

	Sum of Squares	Degree of Freedom	Mean Sum of Square	F-Test	p Value
Between groups	1.1043×10^{-5}	2	5.52149×10^{-6}	2.776	0.102
Within groups	2.38722×10^{-5}	12	1.98935×10^{-6}		
Total	3.49152×10^{-5}	14			

4. Discussion

This study applied a robust design-based expert system with a feature selection learning mechanism to predict the possible trends of a pandemic disease. This system was designed to combine the strengths of the GA framework and the Taguchi method. In the case of Japan, the experimental results indicate that the proposed system can successfully capture the possible trends of COVID-19. The predicted results are close to the actual situations, meaning that the system's capability is assured. With a unique design, the system can deliver good performance and stable quality. The prediction results of the system may be helpful for the government when estimating the death rate of the virus. Healthcare authorities can also employ the system to support formulation of epidemic management policies and allocation of medical resources.

Building an effective expert system is not easy [41,43]. This type of system usually requires repetitive tests, which are very time-consuming but do not guarantee the robustness of its results [44]. From the experimental results, we can find that the robust design has demonstrated its outstanding efficacy in improving the expert system. The orthogonal array design has significantly reduced the total number of runs of the experiment. The *SN* ratio can correctly reflect the differences between system configurations to further ensure the prediction quality of the system.

We also found that although the GA architecture is suitable for learning feature selection when performing an expert system experiment, the amount of the training dataset must be sufficient to obtain good results. In addition, due to the selection, crossover, and mutation mechanisms designed in the GA framework, the evolution of each feature in the chromosome can be evaluated accurately. This is key to the success of the feature selection expert system. Hence, the GA framework can achieve a good performance.

The proposed system can signal an alert when there is a rise in the level of death cases, which means that the number of severe cases has risen too. This alert has an important management implication for healthcare management authorities. If we can capture an imminent increase in the domestic death of the epidemic, the government can make necessary deployments in advance [7,8]. This way, the government's use of medical resources can be more effective, and its responses to the pandemic can also be more timely. Thus, the proposed system is effective in supporting epidemic management.

Policy implications: In fact, the government's healthcare resources are limited. If changes in the number of severe cases or death cases in the country cannot be correctly assessed, the government's deployment of medical resources will be affected first. For example, when the epidemic is slowing down, health authorities can allow more patients with mild symptoms to stay in the hospital for treatment and observation and also delay their discharge time. Conversely, if the epidemic becomes more serious (i.e., the number of death cases grows rapidly), health authorities need to deploy ahead of time and change the management policies for mild and severe patients. For instance, when medical resources are insufficient, some countries may advise patients with a mild illness to stay at home to avoid medical collapse. Take Japan as an example. The repeated changes in the epidemic situation have put tremendous pressure on the government's patient care policy. This highlights the importance of situational judgment. Therefore, the government's primary task in COVID-19 management is to get hold of changes in the death rate in its territory and then make immediate adjustments of health care policies according to the epidemic trend. This kind of timely and situation-based response is conducive to the deployment of medical resources and patient management.

In the present, many countries have adopted policies that favor living with the virus. However, the spread and mutation of COVID-19 still pose a high threat. For example, despite the release of EUA-approved vaccines, many countries still experience large numbers of breakthrough infections after widespread vaccination. This phenomenon shows that epidemic control cannot rely solely on EUA-approved vaccines. All countries should be cautious in monitoring, isolation, and social management of confirmed cases. In the meantime, how to effectively get hold of the daily number of death cases becomes crucial and is an important task for governments when building an epidemic management information system. It determines whether governments can monitor the epidemic development in real time and adjust the degree of relaxation or contraction of medical resources at any time, so as to control the rate of virus transmission within the limits of national resources. Therefore, effective epidemic predictions with an appropriate support system can contribute greatly to epidemic control.

Management implications: In this study, the features considered in the prediction of COVID-19 epidemic trends consisted of descriptive variables and demographic variables. From the dataset of Japan, the average prediction accuracy was 60%. The system shows better accuracy given three levels of alert state were considered (e.g., there are three levels which means the probability is 33.33%). In addition, 10 representative features were also selected, resulting in a selection rate of 67% with a reduced rate of 33%. According to the experimental results, among the descriptive variables, new confirmed cases, ICU patients, and people vaccinated are important factors affecting the mortality of COVID-19 in Japan. In comparison, hospital patients and stringency index are not critical to the prediction result.

Among the demographic variables, population, population density, hospital beds per thousand, middle age, aged 70 older, and GDP per capita are all critical to the mortality of COVID-19 in Japan. However, CVD death rate, diabetes prevalence, and aged 65 older are not significantly related to mortality.

From the above findings, we can infer that, in addition to the descriptive variables that healthcare units are more concerned about (e.g., people vaccinated), the demographic structure of a country also significantly affects the pandemic mortality. This suggests that governments should pay attention to the correlation between the country's demographic structure and the important descriptive variables of COVID-19, because the mortality of the pandemic may increase significantly with increases in certain descriptive variables such as new confirmed cases and ICU patients.

Therefore, the interactions between the abovementioned features and the differences in their importance should be a research topic worthy of further research.

This study was subject to two main limitations. First, the descriptive variables of COVID-19 are aggregated statistical data, and the database is not classified by virus variants determined through gene sequencing. Second, the subject of this study was Japan but the spread of the epidemic varies from country to country. With a good predictive performance, the proposed system can be applied to predict the spread of new COVID-19 variants and explore the differences in death rate between different variants. Moreover, the differences in importance between features is also an interesting research issue. The system architecture of this study can be modified for the research of the above issues in the future.

5. Conclusions

Expert systems are increasingly used in important applications. Healthcare management organizations across the world are seeking more accurate methods to predict the spread of epidemics and support policy adjustment. Considering the tremendous impacts that COVID-19 has brought to the world and the need for healthcare authorities to build a highly adaptive prediction system, this study proposes an enhanced artificial intelligence prediction technique called the robust design-based expert system. The GA framework and Taguchi method are integrated into this system to optimize the performance of the system.

The epidemic data in Japan are employed to develop a prediction system for COVID-19. The prediction accuracy of proposed system was 60%. In addition, the feature selection rate of the system was 67% with a reduction of 33%. The experimental results indicate that the proposed system is effective in predicting the epidemic trend over the next four months (about 120 days). The proposed system can be utilized to support the prediction of epidemic trends as well as the deployment of resources. Future researchers can apply this system to analyze the epidemic trends in countries with more serious outbreaks and further explore the differences between countries and what can be improved with regard to the application of this expert system.

Author Contributions: Conceptualization, C.-Y.W. and C.-T.H.; methodology, C.-Y.W.; software, C.-Y.W.; formal analysis, C.-Y.W.; supervision, C.-T.H. All authors have read and agreed to the published version of the manuscript.

Funding: This research received no external funding.

Institutional Review Board Statement: Not applicable.

Informed Consent Statement: Not applicable.

Data Availability Statement: The raw data are available from the corresponding author on reasonable request.

Conflicts of Interest: The authors declare no conflict of interest.

References

1. Li, Q.; Guan, X.; Wu, P.; Wang, X.; Zhou, L.; Tong, Y.; Ren, R.; Leung, K.S.M.; Lau, E.H.Y.; Wong, J.Y.; et al. Early Transmission Dynamics in Wuhan, China, of Novel Coronavirus–Infected Pneumonia. *N. Engl. J. Med.* **2020**, *382*, 1199–1207. [CrossRef] [PubMed]
2. Wang, C.; Horby, P.W.; Hayden, F.G.; Gao, G.F. A novel coronavirus outbreak of global health concern. *Lancet* **2020**, *395*, 470–473. [CrossRef]
3. World Health Organization. *Looking Back at a Year That Changed the World: WHO's Response to COVID-19*; WHO: Geneva, Switzerland, 2021.
4. Rousset, S.; Level, A.; François, F.; Muller, L. Behavioral and Emotional Changes One Year after the First Lockdown Induced by COVID-19 in a French Adult Population. *Healthcare* **2022**, *10*, 1042. [CrossRef]
5. Karako, K.; Song, P.; Chen, Y.; Tang, W. Analysis of COVID-19 infection spread in Japan based on stochastic transition model. *BioSci. Trends* **2020**, *14*, 134–138. [CrossRef]
6. Shimizu, K.; Negita, M. Lessons Learned from Japan's Response to the First Wave of Covid-19: A Content Analysis. *Healthcare* **2020**, *8*, 426. [CrossRef] [PubMed]
7. Sumikawa, Y.; Honda, C.; Yoshioka-Maeda, K.; Yamamoto-Mitani, N. Characteristics of COVID-19-Related Free Telephone Consultations by Public Health Nurses in Japan: A Retrospective Study. *Healthcare* **2021**, *9*, 1022. [CrossRef]
8. Tomar, A.; Gupta, N. Prediction for the spread of COVID-19 in India and effectiveness of preventive measures. *Sci. Total Environ.* **2020**, *728*, 138762. [CrossRef]
9. Tuli, S.; Tuli, S.; Tuli, R.; Gill, S.S. Predicting the growth and trend of COVID-19 pandemic using machine learning and cloud computing. *Internet Things* **2020**, *11*, 100222. [CrossRef]
10. Araja, D.; Berkis, U.; Murovska, M. COVID-19 Pandemic-Revealed Consistencies and Inconsistencies in Healthcare: A Medical and Organizational View. *Healthcare* **2022**, *10*, 1018. [CrossRef]
11. McCarthy, J. Recursive functions of symbolic expressions and their computation by machine, part I. *Commun. ACM* **1960**, *3*, 184–195. [CrossRef]
12. Feigenbaum, E.A.; Buchanan, B.G.; Lederberg, J. On generality and problem solving: A case study using the DENDRAL program. *Mach. Intell.* **1971**, *6*, 165–190.
13. Aliev, R.A.; Pedrycz, W.; Huseynov, O.H.; Eyupoglu, S.Z. Approximate reasoning on a basis of Z-number-valued if–then rules. *IEEE Trans. Fuzzy Syst.* **2017**, *25*, 1589–1600. [CrossRef]
14. Tang, Y.; Pedrycz, W. Oscillation-bound estimation of perturbations under Bandler-Kohout subproduct. *IEEE Trans. Cybern.* **2022**, *52*, 6269–6282. [CrossRef]
15. Michael, N. *Artificial Intelligence*, 2nd ed.; Addison-Wesley: Boston, MA, USA, 2005.
16. Pandit, M. Expert system–A review article. *Int. J. Eng. Sci. Res. Technol.* **2013**, *2*, 1583–1585.
17. Alma, Z.; Mansiya, K.; Torgyn, M.; Marzhan, K.; Kanat, N. The methodology of expert systems. *Int. J. Comput. Sci. Netw. Secur.* **2014**, *14*, 62–63.
18. Ahmed, I.M.; Alfonse, M.; Aref, M.; Salem, A.-B.M. Reasoning Techniques for Diabetics Expert Systems. *Procedia Comput. Sci.* **2015**, *65*, 813–820. [CrossRef]

19. Bennett, C.C.; Doub, T.W. Chapter 2—Expert Systems in Mental Health Care: AI Applications in Decision-Making and Consultation. In *Artificial Intelligence in Behavioral and Mental Health Care*; Academic Press: Cambridge, MA, USA, 2016; pp. 27–51.
20. Bullon, J.; González Arrieta, A.; Hernández Encinas, A.; Queiruga Dios, A. Manufacturing processes in the textile industry. Expert Systems for fabrics production. *Adv. Distrib. Comput. Artif. Intell. J.* **2017**, *6*, 41–50.
21. Ahmed, K.; Shahid, S.; Haroon, S.B.; Wang, X.J. Multilayer perceptron neural network for downscaling rainfall in arid region: A case study of Baluchistan, Pakistan. *J. Earth Syst. Sci.* **2015**, *124*, 1325–1341. [CrossRef]
22. Deng, Y.; Zhou, X.; Shen, J.; Xiao, G.; Hong, H.; Lin, H.; Wu, F.; Liao, B.Q. New methods based on back propagation (BP) and radial basis function (RBF) artificial neural networks (ANNs) for predicting the occurrence of haloketones in tap water. *Sci. Total Environ.* **2021**, *772*, 145534. [CrossRef]
23. Gholipour, C.; Rahim, F.; Fakhree, A.; Ziapour, B. Using an Artificial Neural Networks (ANNs) Model for Prediction of Intensive Care Unit (ICU) Outcome and Length of Stay at Hospital in Traumatic Patients. *J. Clin. Diagn. Res.* **2015**, *9*, OC19–OC23. [CrossRef]
24. Kumari, R.; Kumar, S.; Paonia, R.C.; Singh, V.; Raja, L.; Bhatnagar, V.; Agarwal, P. Analysis and predictions of spread, recovery, and death caused by COVID-19 in India. *Big Data Min. Anal.* **2021**, *4*, 65–75. [CrossRef]
25. Duan, Y.; Edwards, J.S.; Dwivedi, Y.K. Artificial intelligence for decision making in the era of Big Data—evolution, challenges and research agenda. *Int. J. Inf. Manag.* **2019**, *48*, 63–71. [CrossRef]
26. Tkatek, S.; Belmzoukia, A.; Nafai, S.; Abouchabaka, J.; Ibnou-Ratib, Y. Putting the world back to work: An expert system using big data and artificial intelligence in combating the spread of COVID-19 and similar contagious diseases. *Work* **2020**, *67*, 557–572. [CrossRef] [PubMed]
27. Choi, D.; Lee, H.; Bok, K.; Yoo, J. Design and implementation of an academic expert system through big data analysis. *J. Supercomput.* **2021**, *77*, 7854–7878. [CrossRef]
28. Malki, Z.; Atlam, E.S.; Ewis, A.; Dagnew, G.; Ghoneim, O.A.; Mohamed, A.A.; Abdel-Daim, M.M.; Gad, I. The COVID-19 pandemic: Prediction study based on machine learning models. *Environ. Sci. Pollut. Res.* **2021**, *28*, 40496–40506. [CrossRef]
29. Khalilpourazari, S.; Hashemi Doulabi, H. Designing a hybrid reinforcement learning based algorithm with application in prediction of the COVID-19 pandemic in Quebec. *Ann. Oper. Res.* **2022**, *312*, 1261–1305. [CrossRef]
30. Alam, T.M.; Shaukat, K.; Khelifi, A.; Khan, W.A.; Raza, H.M.E.; Idrees, M.; Luo, S.; Hameed, I.A. Disease diagnosis system using IoT empowered with fuzzy inference system. *Comput. Mater. Contin.* **2022**, *7*, 5305–5319.
31. Kumar, V.; Minz, S. Feature Selection: A literature Review. *Smart Comput. Rev.* **2014**, *4*, 211–229. [CrossRef]
32. Jović, A.; Brkić, K.; Bogunović, N. A review of feature selection methods with applications. In Proceedings of the 2015 38th International Convention on Information and Communication Technology, Electronics and Microelectronics (MIPRO), Opatija, Croatia, 25–29 May 2015; IEEE: Piscataway, NJ, USA, 2015; pp. 1200–1205.
33. Khalid, S.; Khalil, T.; Nasreen, S. A Survey of Feature Selection and Feature Extraction Techniques in Machine Learning. In Proceedings of the Science and Information Conference 2014, London, UK, 27–29 August 2014.
34. Chen, Y.K.; Wang, C.Y.; Feng, Y.Y. Application of a 3NN+1 based CBR system to segmentation of the notebook computers market. *Expert Syst. Appl.* **2010**, *37*, 276–281. [CrossRef]
35. Siedlecki, W.; Sklanski, J. A note on genetic algorithms for large-scale feature selection. *Pattern Recognit. Lett.* **1989**, *10*, 335–347. [CrossRef]
36. Lin, J.Y.; Ke, H.R.; Chien, B.C.; Yang, W.P. Classifier design with feature selection and feature extraction using layered genetic programming. *Expert Syst. Appl.* **2008**, *34*, 1284–1293. [CrossRef]
37. Quan, C.; Ren, F. Unsupervised product feature extraction for feature-oriented opinion determination. *Inf. Sci.* **2014**, *272*, 16–28. [CrossRef]
38. Zhang, X.L.; Chen, W.; Wang, B.J.; Chen, X.F. Intelligent fault diagnosis of rotating machinery using support vector machine with ant colony algorithm for synchronous feature selection and parameter optimization. *Neurocomputing* **2015**, *167*, 260–279. [CrossRef]
39. Aalaei, S.; Shahraki, H.; Rowhanimanesh, A.; Eslami, S. Feature selection using genetic algorithm for breast cancer diagnosis: Experiment on three different datasets. *Iran. J. Basic Med. Sci.* **2016**, *19*, 476.
40. Aličković, E.; Subasi, A. Breast cancer diagnosis using GA feature selection and Rotation Forest. *Neural Comput. Appl.* **2017**, *28*, 753–763. [CrossRef]
41. Gokulnath, C.B.; Shantharajah, S.P. An optimized feature selection based on genetic approach and support vector machine for heart disease. *Clust. Comput.* **2019**, *22*, 14777–14787. [CrossRef]
42. Khan, M.U.; Aziz, S.; Akram, T.; Amjad, F.; Iqtidar, K.; Nam, Y.; Khan, M.A. Expert Hypertension Detection System Featuring Pulse Plethysmograph Signals and Hybrid Feature Selection and Reduction Scheme. *Sensors* **2021**, *21*, 247. [CrossRef]
43. Kwon, Y.; Lee, J.; Park, J.H.; Kim, Y.M.; Kim, S.H.; Won, Y.J.; Kim, H.-Y. Osteoporosis Pre-Screening Using Ensemble Machine Learning in Postmenopausal Korean Women. *Healthcare* **2022**, *10*, 1107. [CrossRef]
44. Khaire, U.M.; Dhanalakshmi, R. Stability of feature selection algorithm: A review. *J. King Saud Univ.-Comput. Inf. Sci.* **2022**, *34*, 1060–1073. [CrossRef]
45. Kiziloz, H.E. Classifier ensemble methods in feature selection. *Neurocomputing* **2021**, *419*, 97–107. [CrossRef]
46. Yusup, N.; Zain, A.M.; Hashim, S.T.M. Evolutionary techniques in optimizing machining parameters: Review and recent applications (2007–2011). *Expert Syst. Appl.* **2012**, *39*, 9909–9927. [CrossRef]

47. Zahraee, S.M.; Khalaji Assadi, M.; Saidur, R. Application of Artificial Intelligence Methods for Hybrid Energy System Optimization. *Renew. Sustain. Energy Rev.* **2016**, *66*, 617–630. [CrossRef]
48. Li, S.; Fang, H.; Liu, X. Parameter optimization of support vector regression based on sine cosine algorithm. *Expert Syst. Appl.* **2018**, *91*, 63–77. [CrossRef]
49. Krishnamoorthy, C.S.; Rajeev, S. *Artificial Intelligence and Expert Systems for Artificial Intelligence Engineers*; CRC Press: Boca Raton, FL, USA, 2018.
50. Taguchi, G. *Taguchi Methods/Design of Experiments*; English edition; Dearborn MI/ASI: Tokyo, Japan, 1949.

Article

Accurate Prediction of Anxiety Levels in Asian Countries Using a Fuzzy Expert System

Mouz Ramzan [1], Muhammad Hamid [2], Amel Ali Alhussan [3], Hussah Nasser AlEisa [3] and Hanaa A. Abdallah [4,*]

1. Department of Computer Science, National College of Business Administration and Economics (NCBA&E), Lahore 54000, Pakistan
2. Department of Statistics and Computer Science, University of Veterinary and Animal Sciences, Lahore 54000, Pakistan
3. Department of Computer Sciences, College of Computer and Information Sciences, Princess Nourah bint Abdulrahman University, P.O. Box 84428, Riyadh 11671, Saudi Arabia; haleisa@pnu.edu.sa (H.N.A.)
4. Department of Information Technology, College of Computer and Information Sciences, Princess Nourah bint Abdulrahman University, Riyadh 84428, Saudi Arabia
* Correspondence: haabdullah@pnu.edu.sa

Abstract: Anxiety is a common mental health issue that affects a significant portion of the global population and can lead to severe physical and psychological consequences. The proposed system aims to provide an objective and reliable method for the early detection of anxiety levels by using patients' physical symptoms as input variables. This paper introduces an expert system utilizing a fuzzy inference system (FIS) to predict anxiety levels. The system is designed to address anxiety's complex and uncertain nature by utilizing a comprehensive set of input variables and fuzzy logic techniques. It is based on a set of rules that represent medical knowledge of anxiety disorders, making it a valuable tool for clinicians in diagnosing and treating these disorders. The system was tested on real datasets, demonstrating high accuracy in the prediction of anxiety levels. The FIS-based expert system offers a powerful approach to cope with imprecision and uncertainty and can potentially assist in addressing the lack of effective remedies for anxiety disorders. The research primarily focused on Asian countries, such as Pakistan, and the system achieved an accuracy of 87%, which is noteworthy.

Keywords: anxiety; anxiety prediction; fuzzy logic; fuzzy inference system

1. Introduction

Anxiety, sociopathy, emotional stability, and desire are just a few psychological factors affecting human behavior [1]. Human behavior can also be described in terms of different patterns, including the thoughts and emotions through which each person adapts to the circumstances of their existence. Another factor that affects behavior is anxiety, described as an unpleasant unease of the mind related to an imminent or expected illness. It symbolizes an internal rather than an exterior threat or danger [2]. Anxiety is a disorder that can be brought on by ongoing stress and if it is, sufferers may face significant hazards. While some disorders only manifest in adulthood, anxiety can start as early as childhood and threaten the individual and the larger community. According to reports [3,4], anxiety can negatively affect up to one-third of the population. In light of this observation, the WHO estimates that 450 million individuals globally experience stress and anxiety [5]. Moreover, the characteristics of emotional integrity that are highly complex and associated with an individual's well-being are directly linked to their quality of life. Stress is a frequently cited factor, and extensive research has been conducted on its diagnosis and treatment [6]. However, the diagnosis of an anxiety disorder is a challenging and intricate process [7]. Anxiety is highly neglected in Asian countries, such as Pakistan [8,9], and it is often not taken seriously. However, this can lead to serious health hazards because diseases such as heart attacks and brain hemorrhages are related to anxiety. People must take anxiety symptoms seriously and observe their surroundings to identify whether someone has anxiety symptoms. The early treatment of anxiety can save a person's life.

Various techniques can be used to predict anxiety at early stages [10]. One of these techniques is the development of a fuzzy-rule-based system, which can predict the disorder based on the created knowledge base. The development of a machine-learning model is another approach that can be utilized to predict anxiety. Different machine-learning techniques are available, such as SVM, decision trees, and ANN. However, problems can arise due to a lack of inputs or low accuracy in the system.

The importance of the fuzzy expert system lies in its ability to predict anxiety at an early stage. It is a useful tool for healthcare professionals to identify the risk of anxiety disorder and to provide early intervention to patients. The system utilizes a knowledge base, which makes it more reliable and accurate in predicting anxiety. On the other hand, the use of machine-learning techniques also provides a promising approach to the prediction of anxiety, but the accuracy of the model is dependent on the quality and quantity of the input data. The questions that arose after learning more about the issue were:

RQ1. Which symptoms can lead to anxiety in a patient?

RQ2. How much accuracy can be obtained by predicting anxiety at certain levels?

To answer these research questions, our research objectives and key research contributions are outlined below.

A fuzzy inference system (FIS) is proposed, taking the symptoms as inputs using them to predict outcomes. Patients suffering from anxiety report their basic symptoms to the system. The system uses these symptoms as inputs and predicts whether the patients are suffering from anxiety. The study offers a workable method for the precise diagnosis and prediction of anxiety while advancing our understanding of anxiety and its effects on people and society. The early diagnosis and treatment of anxiety can have major advantages because they can improve treatment outcomes and lessen the toll the condition takes on sufferers, their families, and society. In its early stages, when symptoms may be subtle or ambiguous, anxiety is a complicated disorder that can present in various ways and may not always be obvious. As a result, the creation of a fuzzy expert system that can reliably predict anxiety at an early stage may be extremely helpful in assisting medical practitioners in making prompt and knowledgeable therapeutic decisions.

The rest of this paper is organized as follows. In Section 2, a review of the literature is discussed. Next, the detailed methodology is described, in Section 3. In Section 4, demonstrations and evaluations are presented, while Section 5 presents the answers to the research question. Finally, in Section 6, concludes the research and provides directions for future work.

2. Literature Review

In recent years, systems based on artificial intelligence (AI) have been increasingly utilized to improve the quality, sensitivity, and timeliness of the diagnosis of psychiatric disorders, including anxiety disorders. These systems employ AI techniques such as fuzzy logic, neural networks (NNs), support vector machines (SVMs), and decision trees (DTs) [11–13]. Altintaş et al. [14] conducted a review of research on machine-learning techniques for diagnosing anxiety disorders between 2015 and 2021. The study examined databases for information on different categories of anxiety disorder and identified thirty different ML techniques used in their research. They compared these techniques in terms of sample size, age, chosen techniques, best practices, and performance values. The authors found that the random forest algorithm (RFA) was the most commonly used ML technique, ensuring the best results and accuracy. The authors also suggested that AI techniques can provide valuable information to researchers and clinicians in areas such as personalized treatment, diagnosis, and prognosis, given the heterogeneity of the data obtained from anxiety patients. Furthermore, Lotfi et al. [15] proposed a system for monitoring and managing anxiety among young people using machine learning. The authors describe the development of a mobile application that collects data on users' anxiety levels and uses machine-learning algorithms to predict and manage anxiety. Their system aims to provide an effective and efficient approach to identifying and managing anxiety among young people. Their research concludes that the proposed system improves the quality of life of young people and facilitates the early detection and management of anxiety disorders. Furthermore, Kumar et al. [16], developed a new supervised learning-based prediction model, an Anxious Depression (AD) prediction model, for efficiently predicting anxious depression in real-time tweets. Based on the user's posting patterns and linguistic cues, the feature set is defined using a five-tuple vector <w, t, f, s, c>. The representation of each entry is introduced as follows:

< w: word >: The presence or absence of the anxiety-related word.

< t: timing >: More than two posts during odd night hours.

< f: frequency >: More than thirty posts in an hour.

< s: sentiment >: On average, more than 25% of posts over 30 days with negative polarity.

< c: contrast >: The presence of a polarity contrast of more than 25% in posts within the past twenty-four hours.

Moreover, Susanto et al. [17] proposed a fuzzy-logic-based model to predict mathematics anxiety in students. The model was constructed using a dataset of 470 Turkish university students who completed a mathematics-anxiety scale and a demographic questionnaire. The study examined the relationship between mathematics anxiety and various demographic factors, such as gender, age, and major. The model was constructed using a Mamdani-type fuzzy inference system, and its performance was evaluated using various statistical measures. The results showed that the fuzzy model can accurately predict mathematics anxiety in students and can be useful for educators and researchers to identify students at risk of developing mathematics anxiety.

Further, Khullar et al. [18] proposed a method for detecting anxiety based on physiological signals. Their study used an electrocardiogram (ECG) and galvanic skin response (GSR) signals to develop a machine-learning model for anxiety detection. The proposed model uses an ensemble approach that combines multiple machine-learning algorithms, such as decision trees, random forest, and support vector machine. The performance of the proposed model was evaluated using a dataset of 18 subjects, who were exposed to anxiety-inducing stimuli. The results showed that the model accurately detects anxiety based on physiological signals. The authors suggested that the proposed approach can be further explored to better predict anxiety levels.

Furthermore, Devi et al. [19] proposed a method for predicting the anxiety levels of students using an adaptive neuro-fuzzy-inference system (ANFIS). The study used a dataset of 208 undergraduate students who completed the State-Trait Anxiety Inventory (STAI) questionnaire. The proposed ANFIS model considers various demographic and academic factors, such as age, gender, major, academic performance, and STAI score. The performance of the proposed model was evaluated using various statistical measures, such as mean absolute error and root-mean-squared error. The results showed that the proposed ANFIS model achieved high accuracy in predicting the anxiety levels of the students. The authors further suggested that a developed system should be further explored and improved for the better prediction of anxiety.

From the literature review presented above, we can conclude that anxiety symptoms can be different from person to person and that they can also be different across regions. For example, people in Asian countries, such as Pakistan, may show different anxiety symptoms. Therefore, it is necessary to create a fuzzy logic system that can target the Asian continent in particular, in order to predict the disorder correctly. To this end, we will used symptoms widely shown in patients with anxiety in Pakistan to build our fuzzy expert system. Figure 1 represents the core symptoms of anxiety.

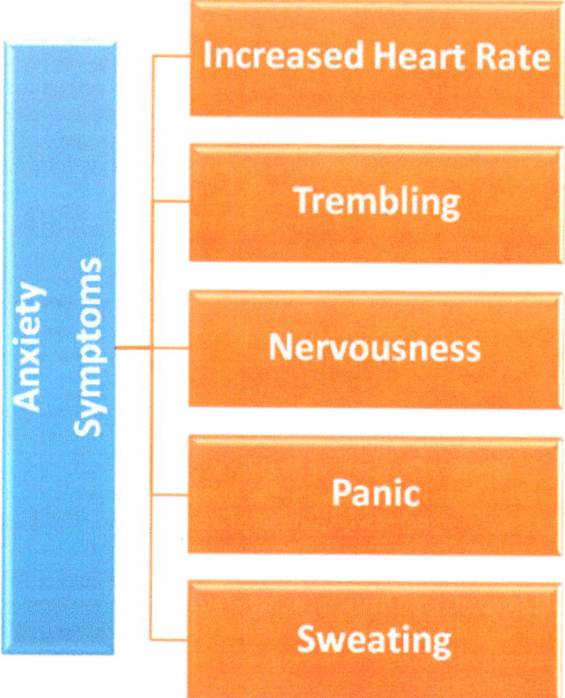

Figure 1. Anxiety symptoms [20].

3. Methodology

3.1. Fuzzy Inference System

A FIS system uses fuzzy logic to map input and output variables based on rules. An expert system can make decisions or predictions based on fuzzy reasoning. The FIS consists of three main components: the fuzzifier, the inference engine, and the defuzzifier. The fuzzifier converts input data, which may be crisp (i.e., exact) or fuzzy (i.e., imprecise), into fuzzy sets representing the degree of membership in a particular category. The inference engine applies a set of rules defined by experts or generated automatically from data to determine the appropriate output based on the input data. Finally, the defuzzifier converts the fuzzy output into a crisp output value. The FIS system can be used in various applications, such as control, decision-making, and pattern-recognition systems. They are particularly useful in studies with high levels of uncertainty or imprecision in their data.

The proposed system uses fuzzy logic as the ideal methodology for decision assistance. Fuzzy logic was chosen because it enables flexible and nuanced decision making through the IF-ELSE-based methodology. It can handle values in points, which conventional rule-based systems cannot. In this study, which focused on predicting anxiety levels, fuzzy logic's capacity to collect and interpret data more granularly was beneficial and in line with the study's goals. Therefore, fuzzy logic was chosen for its usefulness, its compatibility with the study's goals, and its ability to handle values in points.

3.2. Fuzzy Inference System for Prediction of Anxiety Levels

Our proposed system can predict anxiety at different levels (no, low, mild, and high). The fuzzy system's architecture is depicted in Figure 2. The fuzzification process creates fuzzy logic sets from crisp inputs. Based on the input given, the inference engine analyzes the matched set.

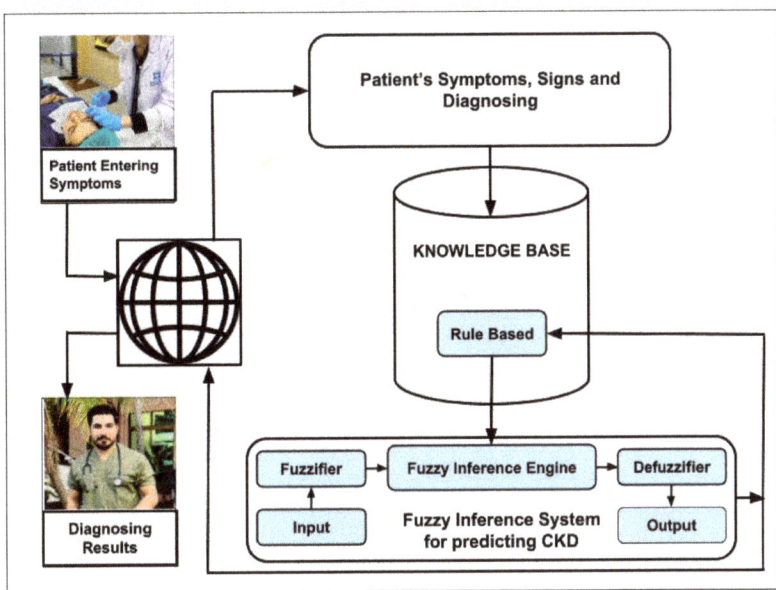

Figure 2. Fuzzy system architecture for predicting anxiety.

The input and output of the fuzzy system are depicted in Figure 3. The fuzzy sets that defuzzification uses as the inference engine transform inputs to produce the results. Figure 4 visually represents the system flowchart, highlighting the sequence of steps and decision points involved in its operation. Figure 4 serves as a useful tool for understanding the system's functionality and identifying potential areas for improvement.

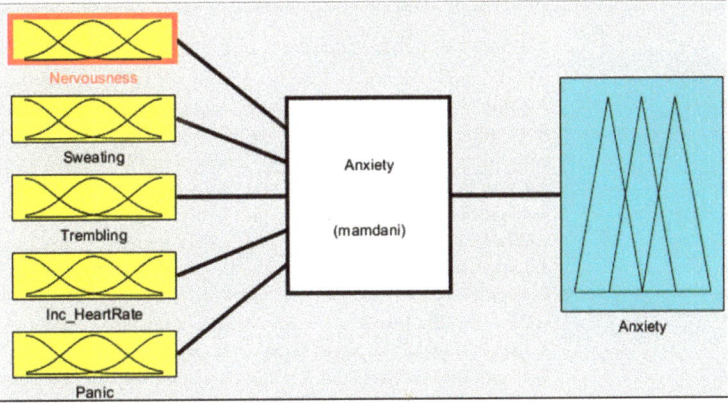

Figure 3. Input and output of the proposed system.

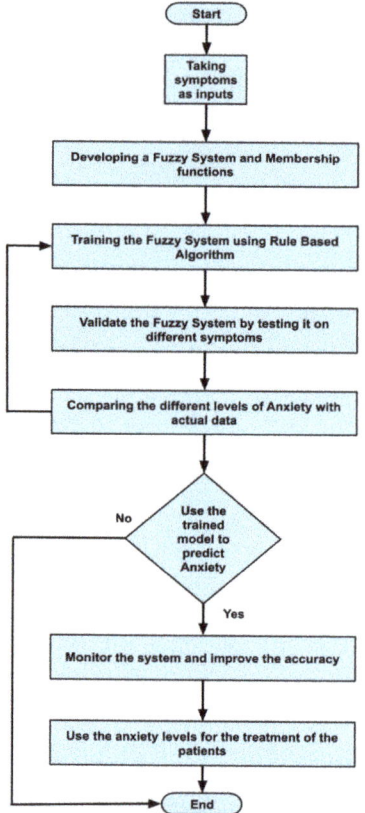

Figure 4. Flowchart of the proposed system.

3.3. Knowledge-Base Input-and-Output Fuzzy System

Five symptoms were taken as input variables and one output variable was used for this system simulation. Fuzzy inputs Nervousness and Panic had three fuzzy terms (low, mild and high), while inputs sweating, trembling, and increased heart rate had two fuzzy terms (no and yes). Creation of FIS rules was executed with the guidance of experts in the field. The rules were defined using a combination of linguistic terms and mathematical expressions. In order to accurately define the rules, a thorough analysis of the available linguistic terms was conducted, and they were mathematically combined to form a comprehensive set of rules. The linguistic terms were categorized based on the number of values they represented to facilitate the creation of these rules. This was achieved by separating the terms with two values from those with three. In this way, we effectively managed the vast array of terms and created precise rules. The rules were defined using a mathematical expression resulting from multiplying the number of linguistic terms with two values by the number of terms with three. This expression yielded a total of 72 rules, each of which was carefully crafted to ensure accuracy and reliability in the FIS system. Overall, creation of these rules was a rigorous and meticulous process involving the input of domain experts and the use of advanced mathematical techniques.

Further details are shown in Table 1. The parameter regarding the fuzzy set describes the range of levels of symptoms. As we already know, the range is from 0–1, which means that when the symptom has three values, the range is divided into three equal sizes. If the symptom has two values, the range is divided into two. Furthermore, the symptom level is shown based on the range value provided.

Table 1. Ranges of fuzzy systems for predicting anxiety.

Input and Output Variables	Linguistic Term	Parameters of Fuzzy Sets
Nervousness	{"Low", "Mild", "High"}	{[0, 0, 0.33], [0, 0.33, 0.66], [0.33, 0.66, 0.99]}
Sweating	{"No", "Yes"}	{[0, 0, 0.5, 0.51], [0.5, 0.51, 1, 1]}
Trembling	{"No", "Yes"}	{[0, 0, 0.5, 0.51], [0.5, 0.51, 1, 1]}
Increased Heart Rate	{"No", "Yes"}	{[0, 0, 0.5, 0.51], [0.5, 0.51, 1, 1]}
Panic	{"Low", "Mild", "High"}	{[0, 0, 0.33], [0, 0.33, 0.66], [0.33, 0.66, 0.99]}
Anxiety	{"No", "Low", "Mild", "High"}	{[0, 0, 0.25], [0, 0.25, 0.50], [0.25, 0.50, 0.75], [0.50, 0.75, 1]}

3.4. Membership Functions

Membership functions are essential in fuzzy logic as they represent an element's degree of membership in a fuzzy set. This is important because fuzzy sets have continuous boundaries, which differ from crisp sets with clear boundaries. Membership functions enable the representation of uncertainty and imprecision in the input data, facilitating more flexible and nuanced decision-making. They are commonly used to perform mathematical operations on a given dataset and are applicable across different platforms. Table 2 describes the membership functions employed in the proposed system.

Table 2. Membership functions.

Input	Membership Functions	Graphical Representation of Membership Functions
Nervousness	$\mu_{low}(x) = \left\{ \frac{0.33-x}{0.33} \; 0 \leq x \leq 0.33 \right\}$ $\mu_{mild}(x) = \left\{ \begin{array}{l} \frac{x}{0.33} \; 0 \leq x \leq 0.33 \\ \frac{0.66-x}{0.33} \; 0.33 \leq x \leq 0.66 \end{array} \right\}$ $\mu_{high}(x) = \left\{ \begin{array}{l} \frac{x}{0.66} \; 0 \leq x \leq 0.66 \\ \frac{0.99-x}{0.33} \; 0.66 \leq x \leq 0.99 \end{array} \right\}$	

Table 2. *Cont.*

Input	Membership Functions	Graphical Representation of Membership Functions
Sweating	$\mu_{no}(y) = \begin{cases} \frac{y-0}{0} & 0 \leq y \leq 0.51 \\ \frac{0.52-y}{0.01} & 0.51 \leq y \leq 0.52 \end{cases}$ $\mu_{yes}(y) = \begin{cases} y - 0.51 & 0 \leq y \leq 0.51 \\ \frac{1-y}{0} & 0.51 \leq y \leq 0.52 \end{cases}$	
Trembling	$\mu_{no}(t) = \begin{cases} \frac{t-0}{0} & 0 \leq t \leq 0.5 \\ \frac{0.51-t}{0.01} & 0.5 \leq t \leq 0.51 \end{cases}$ $\mu_{yes}(t) = \begin{cases} t - 0.5 & 0 \leq t \leq 0.5 \\ \frac{1-t}{0} & 0.5 \leq t \leq 0.51 \end{cases}$	
Increased Heart Rate	$\mu_{no}(u) = \begin{cases} \frac{u-0}{0} & 0 \leq u \leq 0.49 \\ \frac{0.5-u}{0.01} & 0.49 \leq u \leq 0.50 \end{cases}$ $\mu_{yes}(u) = \begin{cases} u - 0.49 & 0 \leq u \leq 0.49 \\ \frac{1-u}{0} & 0.49 \leq u \leq 0.5 \end{cases}$	
Panic	$\mu_{low}(x) = \begin{cases} \frac{0.35-x}{0.35} & 0 \leq x \leq 0.35 \end{cases}$ $\mu_{mild}(x) = \begin{cases} \frac{x}{0.35} & 0 \leq x \leq 0.35 \\ \frac{0.7-x}{0.35} & 0.35 \leq x \leq 0.7 \end{cases}$ $\mu_{high}(x) = \begin{cases} \frac{x}{0.7} & 0 \leq x \leq 0.7 \\ \frac{1.05-x}{0.35} & 0.7 \leq x \leq 1.05 \end{cases}$	

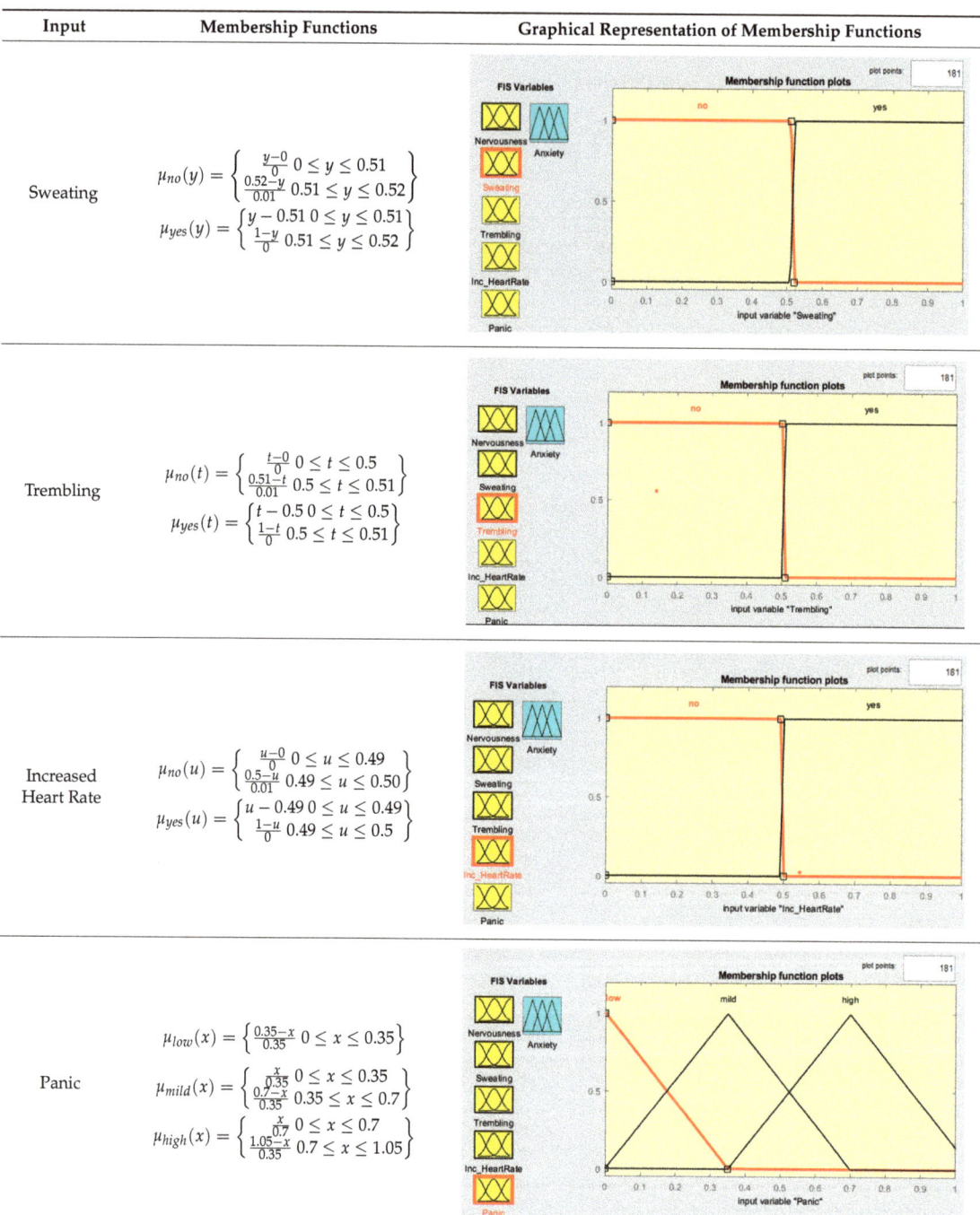

Table 2. Cont.

Input	Membership Functions	Graphical Representation of Membership Functions
Anxiety	$\mu_{no}(z) = \left\{ \frac{0.25-z}{0.25}\ 0 \leq z \leq 0.25 \right\}$ $\mu_{low}(z) = \left\{ \begin{array}{l} \frac{z}{0.25}\ 0 \leq z \leq 0.25 \\ \frac{0.5-z}{0.25}\ 0.25 \leq z \leq 0.5 \end{array} \right\}$ $\mu_{mild}(z) = \left\{ \begin{array}{l} \frac{z}{0.5}\ 0 \leq z \leq 0.5 \\ \frac{0.75-z}{0.25}\ 0.5 \leq z \leq 0.75 \end{array} \right\}$ $\mu_{high}(z) = \left\{ \begin{array}{l} \frac{z}{1}\ 0 \leq z \leq 1 \\ \frac{1-z}{0.25}\ 0.75 \leq z \leq 1 \end{array} \right\}$	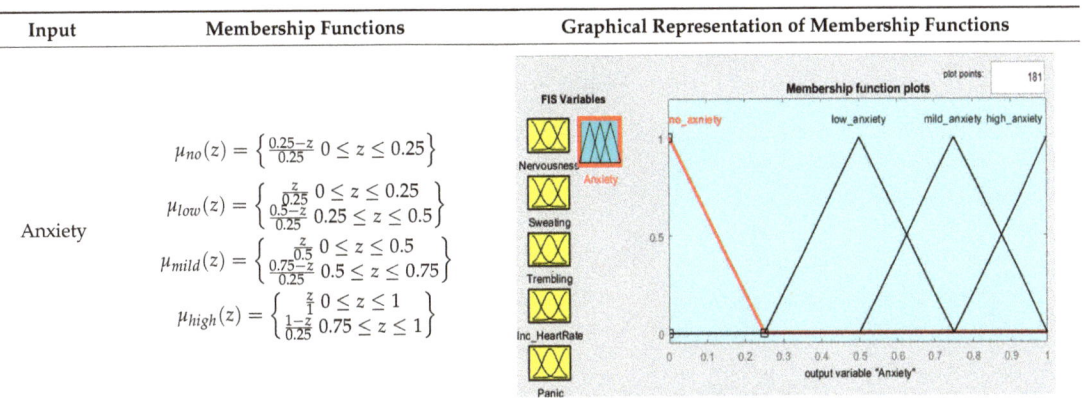

3.5. Defuzzification

Defuzzification is the process of converting fuzzy logic sets and related membership degrees into a calculable or quantifiable output in crisp logic. Fuzzy logic is a form of reasoning that deals with imprecise or uncertain information, while crisp logic deals with precise or certain information. Defuzzification takes the fuzzy input and output values from a fuzzy set and converts them into a crisp value that can be used for decision making. There are several methods for defuzzification, including the centroid method, mean-of-the-maximum method, and weighted-average method. Each method has advantages and disadvantages, and the choice of method depends on the specific application and the desired level of accuracy.

Figures 5–7 show graphical representations of the defuzzification. Figure 5 shows the control surface; the yellowish color represents the anxiety level, bluish represents the sweating, and greenish represents the nervousness in the patient. We can see that anxiety levels gradually increased after giving two predictors, sweating and nervousness. Figure 6 shows the control surface, in which the yellowish color represents the anxiety levels, bluish represents the panic, and light-blue color represents increased heart rate in any patient. We can see that after giving another two predictors, panic and increased heart rate, the anxiety level reached a new value. Figure 7 shows the control surface, in which the yellowish color represents the anxiety levels, bluish color represents the trembling, and the greenish color represents panic in any patient. Here, we can see that the anxiety level drastically increased after giving symptoms of trembling and panic. This illustrates how different symptoms can be mapped to different fuzzy sets and how defuzzification can convert these fuzzy sets into crisp values for decision making.

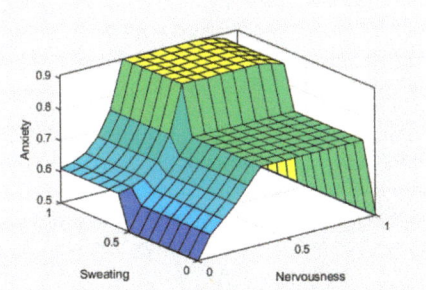

Figure 5. Surface inputs (X—Sweating and Y—Nervousness) and output (Anxiety).

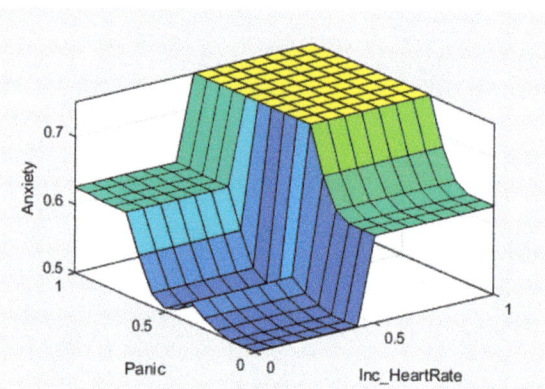

Figure 6. Surface inputs (X–Panic and Y—Increased Heart Rate) and output (Anxiety).

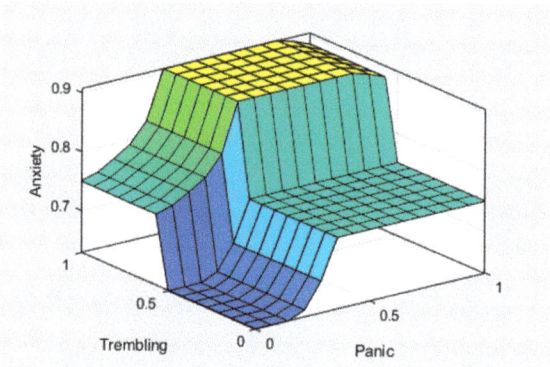

Figure 7. Surface inputs (X—trembling and Y—panic) and output (anxiety).

4. Demonstration and Evaluation

The demonstration-and-evaluation phase consisted of three activities. The experiment results generated using MATLAB were discussed in the first activity, while the second activity featured the prediction of anxiety by the fuzzy expert system. The third activity featured a detailed description of the comparative evaluation of the proposed system and a discussion.

4.1. Experimental Results

MATLAB is a widely used tool in scientific computing, image processing, data visualization, and numerical computation. In this study, it was used to calculate the experimental results. The figures, specifically Figures 8–11, illustrate the use of MATLAB in the determination of anxiety levels based on certain factors, such as nervousness, sweating, trembling, increased heart rate, and panic.

Figure 8 outlines the criteria for assessing anxiety levels without any or with low factors. If nervousness, sweating, trembling, increased heart rate, and panic were absent or low, the anxiety level was considered as "No."

Figure 9 outlines the criteria for assessing anxiety levels when some factors are present. If nervousness is low, sweating is absent, trembling is present, increased heart rate is not present, and panic is mild, then the anxiety level is considered "Low."

Figure 10 outlines the criteria for assessing anxiety levels when several factors are present. If nervousness is mild, sweating, trembling, and increased heart rate are present, and panic is mild, then the anxiety level is considered "Mild."

Figure 11 outlines the criteria for assessing anxiety levels when most factors are high. Specifically, if nervousness, sweating, trembling, increased heart rate, and panic are all high, then the anxiety level is considered "High." In summary, MATLAB was utilized to calculate the experimental results. Figures 8–11 demonstrate the use of MATLAB in the determination of the anxiety levels based on specific factors, such as nervousness, sweating, trembling, increased heart rate, and panic. We set the range from 0 to 1. If we changed the value of 'No' or any other symptom to "Mild" or "High," the anxiety levels changed accordingly. This is because the levels were solely based on the input ranges. The figures below demonstrate how the different range values affected the anxiety levels.

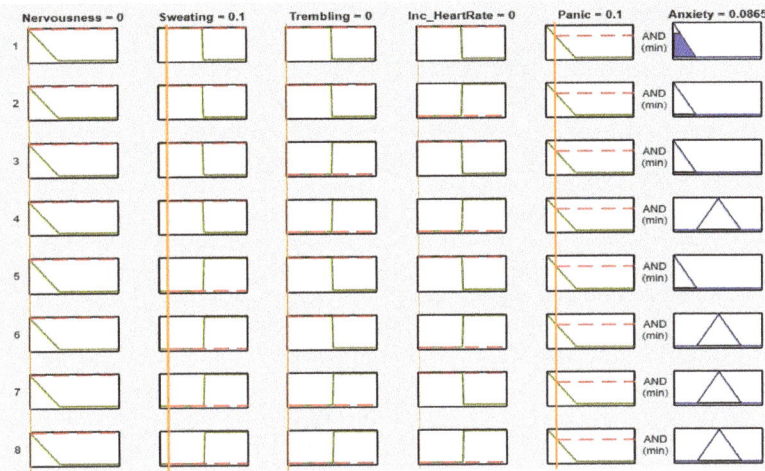

Figure 8. Lookup diagram for no anxiety.

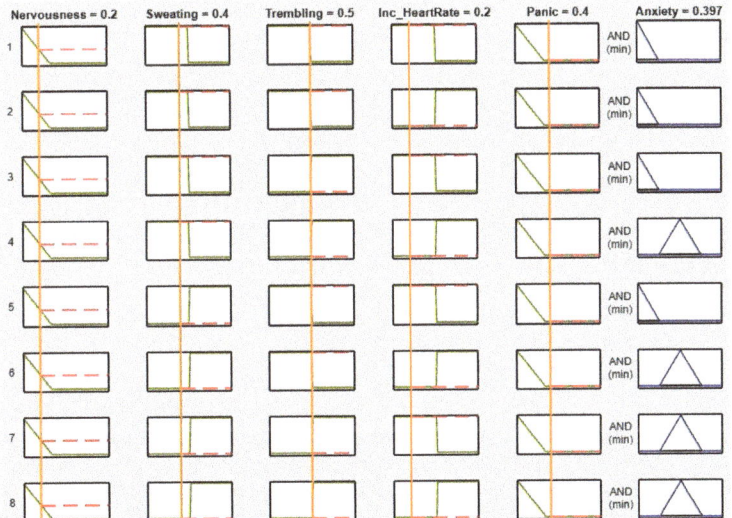

Figure 9. Lookup diagram for low anxiety.

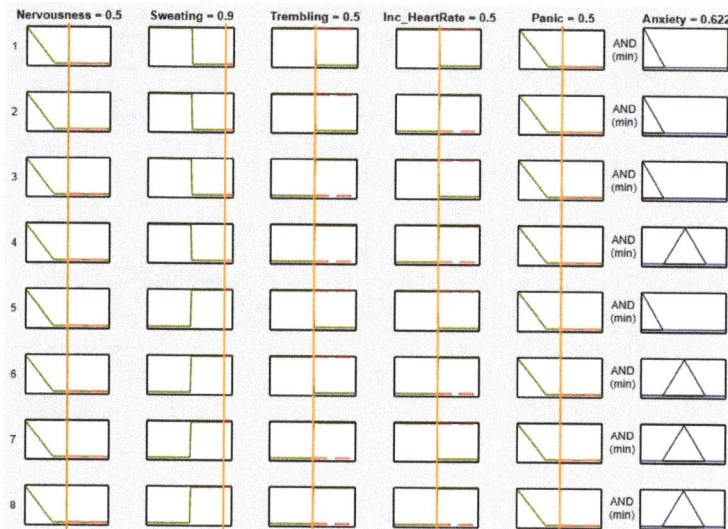

Figure 10. Lookup diagram for mild anxiety.

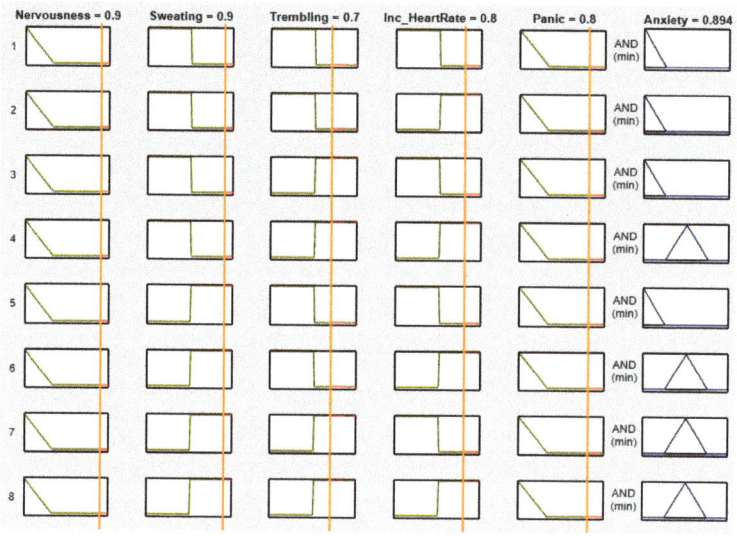

Figure 11. Lookup diagram for high anxiety.

4.2. Anxiety-Prediction Fuzzy Expert System

The anxiety-prediction system uses various anxiety-related symptoms as inputs from the user. These symptoms include nervousness, sweating, trembling, increased heart rate, and panic. By analyzing these inputs, the system can predict the user's anxiety levels, which can be categorized into four levels: no anxiety, low anxiety, mild anxiety, and high anxiety. The proposed system can be visualized as shown in Figure 12, which demonstrates how the inputs from the user are processed and used to determine the user's anxiety levels. The system is designed to provide a quick and accurate assessment of the user's anxiety level, which can be used to inform treatment decisions or track progress over time.

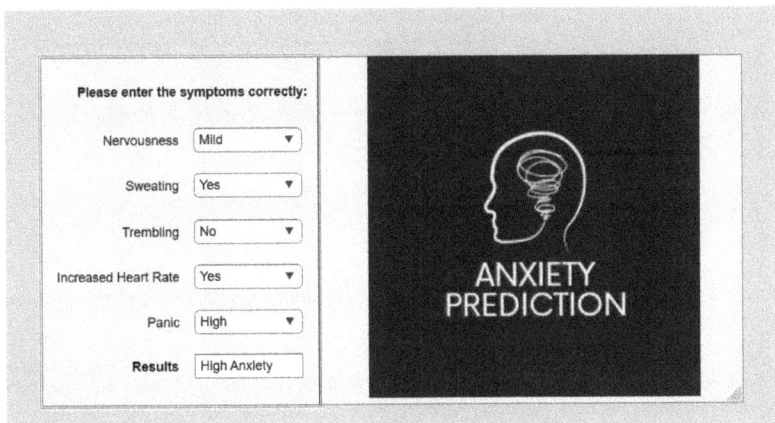

Figure 12. Proposed anxiety-prediction system.

4.3. Comparative Evaluation of the Proposed System

The data were collected from the research center of Shalimar Hospital in Lahore, Pakistan. The data were organized during the examination of patients in the outpatient department (OPD). Doctors diagnosed the disorder by collecting the symptoms, which were subsequently used to evaluate the accuracy and results of our system. Table 3 presents a comprehensive assessment of the performance of our proposed anxiety-prediction system by comparing the results obtained from the experts' opinions, medical reports, and our system. To evaluate the accuracy of our system, we tested it on the patient data from Shalimar Hospital. We compared the results obtained from our system with the doctors' reviews and the patients' medical reports. By comparing the results from the expert opinions, medical reports, and our proposed system, we identified any discrepancies and assessed the overall accuracy of our system. The comparison also helped to identify areas in which the system may need further improvement. Our proposed system showed an accuracy of 87%. We determined the accuracy by comparing our results with the medical records, using the data for 15 patients. Our proposed system correctly predicted the anxiety levels of 13 patients. We estimated the accuracy using the formula $(100/15 \times 13)$. Table 3 provides a comprehensive assessment of the overall performance of our proposed system, which can be used to inform future developments and improve the system's accuracy.

Patient 1 has high anxiety according to the expert opinion and medical reports; our system also accurately established that this patient has high anxiety. Furthermore, patient 2 has high anxiety according to the expert opinion and medical reports, and our system also accurately predicted the high anxiety of this patient. Patient 3 has no anxiety according to the expert opinion and medical reports, which our system also accurately predicted. Patient 4 has low anxiety according to the expert opinion and our proposed system, while the medical report suggests that the patient has mild anxiety. Patient 5 has no anxiety according to the expert opinion, the medical report, and our proposed system. The data were collected in a random manner. We evaluated a total of 15 patients in Table 3. A total of three patients showed high anxiety, five showed low anxiety, five showed no anxiety, and two showed mild anxiety.

Table 3. Comparative evaluation.

Patients	Expert Opinions	Medical Reports	Proposed System
Patient 1	High Anxiety	High Anxiety	High Anxiety
Patient 2	High Anxiety	High Anxiety	High Anxiety
Patient 3	No Anxiety	No Anxiety	No Anxiety
Patient 4	Low Anxiety	Mild Anxiety	Low Anxiety
Patient 5	No Anxiety	No Anxiety	No Anxiety
Patient 6	Low Anxiety	Low Anxiety	Low Anxiety
Patient 7	Low Anxiety	Low Anxiety	Low Anxiety
Patient 8	Low Anxiety	Low Anxiety	Low Anxiety
Patient 9	No Anxiety	No Anxiety	No Anxiety
Patient 10	Mild Anxiety	Mild Anxiety	Mild Anxiety
Patient 11	No Anxiety	No Anxiety	No Anxiety
Patient 12	High Anxiety	High Anxiety	High Anxiety
Patient 13	High Anxiety	High Anxiety	Mild Anxiety
Patient 14	Low Anxiety	Low Anxiety	Low Anxiety
Patient 15	No Anxiety	No Anxiety	No Anxiety

We evaluated our system by comparing its results with expert opinions and the results from medical reports. The expert opinion represented the doctor's diagnosis before any medical results were reported. The medical reports represented the results from the GAD test, which is essentially a test to examine anxiety disorder. After comparing the results of our system with expert opinions and medical reports, we determined the accuracy of our system. Table 3 compares three results from our system, the expert opinions, and the medical reports. It displays the anxiety levels predicted by each source for each patient. The proposed system exhibited an accuracy of 87%, accurately predicting the anxiety level of 13 out of 15 patients. We also compared these results with previous research on the same issue. This comparison with previous research indicated that the proposed system outperforms the models developed by Boukhechba et al. [21], Dhyani et al. [22], and Borse et al. [23] in terms of accuracy. The comparison of the results from different sources helped to identify discrepancies and assess the system's overall accuracy. This study's findings can be used to inform future developments and improve the accuracy of the system.

5. Answers to Research Questions

The answers to the research questions were formulated at the beginning of this research.

RQ1. Which symptoms can lead to anxiety in a patient?

Anxiety can manifest in many ways and the symptoms can vary from person to person. Some common anxiety symptoms include nervousness and panic, excessive sweating, trembling, increased heart rate, and restlessness.

In Asian countries, such as Pakistan, where there are high stress levels due to various factors, such as financial instability, political turmoil, and societal pressure, anxiety can be a prevalent issue. Additionally, cultural factors, such as the stigmatization of mental health issues and a lack of awareness and education, can further exacerbate the problem.

RQ2. How much accuracy can be obtained by predicting anxiety at certain levels?

Th prediction of anxiety at certain levels can be challenging due to the complex nature of the disorder and the variability of symptoms from person to person. Anxiety can occur in short-term and long-term forms and present differently depending on the individual. It is important to carefully select the symptoms to use as indicators of the disorder to predict anxiety at different levels accurately. A proposed system was developed in a research study to predict anxiety at four levels. The system used a combination of patient symptoms and medical reports to predict anxiety levels. The study's results were then compared to patients' medical reports and expert opinions. The system was found to have an accuracy of 87% in predicting anxiety levels. This suggests that the system accurately identified the different anxiety levels in the studied patients.

It should be noted that the system's accuracy may vary depending on the population studied and the specific symptoms used as indicators of the disorder. Further research is needed to validate the system in different populations and to improve its accuracy. However, the results of this study provide a promising start in the development of effective tools for predicting anxiety at different levels.

6. Conclusions and Future Work

This paper proposed an expert system that utilizes fuzzy inference techniques to provide an objective and reliable method for detecting anxiety levels in patients early. The proposed system's success in achieving an accuracy of 87% an Asian country, Pakistan, is noteworthy and provides hope for the early-stage diagnosis and treatment of anxiety disorders. The proposed system is an innovative approach that aims to predict the levels of anxiety in patients by using a combination of fuzzy enhanced rule-based sets and patient symptoms as inputs. Our results indicate that the system has a level of accuracy that is comparable to an expert opinion, which can potentially assist clinicians in diagnosing and treating anxiety disorders. However, a limitation of this work is that it only predicted anxiety and did not predict other psychological disorders, such as depression. In future work, the system's scope can be expanded to predict the outputs based on the symptoms of both anxiety and depression, thereby providing a more comprehensive assessment of a patient's mental health. By addressing the limitations of this work and incorporating more symptoms and relevant information, the system's accuracy can be further improved, making it more useful to medical professionals in providing accurate assessments and treatment recommendations for patients. Overall, this expert system represents a promising tool for the early detection and treatment of anxiety disorders.

Author Contributions: Methodology, M.H.; Software, M.R.; Validation, H.N.A.; Data curation, M.R. and H.N.A.; Writing—original draft, M.H.; Writing—review & editing, A.A.A. and H.A.A.; Visualization, H.A.A. and A.A.A., funding A.A.A. and H.A.A. All authors have read and agreed to the published version of the manuscript.

Funding: This work was supported by Princess Nourah bint Abdulrahman University Researchers Supporting Project number (PNURSP2023R 308), Princess Nourah bint Abdulrahman University, Riyadh, Saudi Arabia.

Institutional Review Board Statement: Not applicable.

Informed Consent Statement: Informed consent was obtained from all subjects involved in the study.

Data Availability Statement: Not applicable.

Acknowledgments: Authors would like to give thanks for the support of Princess Nourah bint Abdulrahman University Researchers Supporting Project number (PNURSP2023R 308), Princess Nourah bint Abdulrahman University, Riyadh, Saudi Arabia.

Conflicts of Interest: The authors declare no conflict of interest.

References

1. Yuen, K.F.; Wang, X.; Ma, F.; Li, K.X. The psychological causes of panic buying following a health crisis. *Int. J. Environ. Res. Public Health* **2020**, *17*, 3513. [CrossRef] [PubMed]
2. Daas, P.J.; Puts, M.J.; Buelens, B.; van den Hurk, P.A. Big data as a source for official statistics. *J. Off. Stat.* **2015**, *31*, 249. [CrossRef]
3. Chalhoub, Z.; Koubeissy, H.; Fares, Y.; Abou-Abbas, L. Fear and death anxiety in the shadow of COVID-19 among the Lebanese population: A cross-sectional study. *PLoS ONE* **2022**, *17*, e0270567. [CrossRef] [PubMed]
4. Jehi, T.; Khan, R.; Dos Santos, H.; Majzoub, N. Effect of COVID-19 outbreak on anxiety among students of higher education; A review of literature. *Curr. Psychol.* **2022**. [CrossRef] [PubMed]

5. Charlson, F.; van Ommeren, M.; Flaxman, A.; Cornett, J.; Whiteford, H.; Saxena, S. New WHO prevalence estimates of mental disorders in conflict settings: A systematic review and meta-analysis. *Lancet* **2019**, *394*, 240–248. [CrossRef] [PubMed]
6. Minervini, G.; Franco, R.; Marrapodi, M.M.; Mehta, V.; Fiorillo, L.; Badnjević, A.; Cervino, G.; Cicciù, M. The Association between COVID-19 Related Anxiety, Stress, Depression, Temporomandibular Disorders, and Headaches from Childhood to Adulthood: A Systematic Review. *Brain Sci.* **2023**, *13*, 481. [CrossRef] [PubMed]
7. Hebert, E.A.; Dugas, M.J. Behavioral experiments for intolerance of uncertainty: Challenging the unknown in the treatment of generalized anxiety disorder. *Cogn. Behav. Pract.* **2019**, *26*, 421–436. [CrossRef]
8. Waqas, A.; Raza, N.; Lodhi, H.W.; Muhammad, Z.; Jamal, M.; Rehman, A. Psychosocial factors of antenatal anxiety and depression in Pakistan: Is social support a mediator? *PLoS ONE* **2015**, *10*, e0116510. [CrossRef] [PubMed]
9. Aqeel, M.; Abbas, J.; Shuja, K.H.; Rehna, T.; Ziapour, A.; Yousaf, I.; Karamat, T. The influence of illness perception, anxiety and depression disorders on students mental health during COVID-19 outbreak in Pakistan: A web-based cross-sectional survey. *Int. J. Hum. Rights Healthc.* **2022**, *15*, 17–30. [CrossRef]
10. Nemesure, M.D.; Heinz, M.V.; Huang, R.; Jacobson, N.C. Predictive modeling of depression and anxiety using electronic health records and a novel machine learning approach with artificial intelligence. *Sci. Rep.* **2021**, *11*, 1–9. [CrossRef] [PubMed]
11. Schwalbe, N.; Wahl, B. Artificial intelligence and the future of global health. *Lancet* **2020**, *395*, 1579–1586. [CrossRef] [PubMed]
12. Kumar, Y.; Koul, A.; Singla, R.; Ijaz, M.F. Artificial intelligence in disease diagnosis: A systematic literature review, synthesizing framework and future research agenda. *J. Ambient Intell. Humaniz. Comput.* **2022**, 1–28. [CrossRef] [PubMed]
13. Lee, E.E.; Torous, J.; De Choudhury, M.; Depp, C.A.; Graham, S.A.; Kim, H.-C.; Paulus, M.P.; Krystal, J.H.; Jeste, D.V. Artificial intelligence for mental health care: Clinical applications, barriers, facilitators, and artificial wisdom. *Biol. Psychiatry Cogn. Neurosci. Neuroimaging* **2021**, *6*, 856–864. [CrossRef] [PubMed]
14. Altintaş, E.; Aksu, Z.U.; Demir, Z.G. Machine Learning Techniques for Anxiety Disorder. *Avrupa Bilim Teknol. Derg.* **2021**, *31*, 365–374. [CrossRef]
15. Lotfi, F.; Rodić, B.; Bogdanović, Z. A System For Monitoring And Managing The Anxiety Among The Young People Using Machine Learning. In Proceedings of the E-business Technologies Conference Proceedings, Vancouver, BC, Canada, 20–21 October 2022; Volume 2, pp. 91–94.
16. Kumar, A.; Sharma, A.; Arora, A. Anxious depression prediction in real-time social data. *arXiv* **2019**, arXiv:1903.10222. [CrossRef]
17. Susanto, H.P.; Purnamasari, M.I. Constructing a Fuzzy Model to Predict Math Anxiety. In Proceedings of the 5th International Conference on Current Issues in Education (ICCIE 2021), Yogyakarta, Indonesia, 12–13 November 2021; Atlantis Press: Amsterdam, The Netherlands, 2022; pp. 25–30.
18. Khullar, V.; Tiwari, R.G.; Agarwal, A.K.; Dutta, S. Physiological signals based anxiety detection using ensemble machine learning. In *Cyber Intelligence and Information Retrieval: Proceedings of CIIR 2021*; Springer: Singapore, 2022; pp. 597–608.
19. Devi, S.; Kumar, S.; Kushwaha, G.S. An adaptive neuro fuzzy inference system for prediction of anxiety of students. In Proceedings of the 2016 Eighth International Conference on Advanced Computational Intelligence (ICACI), Chiang Mai, Thailand, 14–16 February 2016; pp. 7–13.
20. Schäfer, J.Ö.; Naumann, E.; Holmes, E.A.; Tuschen-Caffier, B.; Samson, A.C. Emotion regulation strategies in depressive and anxiety symptoms in youth: A meta-analytic review. *J. Youth Adolesc.* **2017**, *46*, 261–276. [CrossRef] [PubMed]
21. Boukhechba, M.; Chow, P.; Fua, K.; Teachman, B.A.; Barnes, L.E. Predicting social anxiety from global positioning system traces of college students: Feasibility study. *JMIR Ment. Health* **2018**, *5*, e10101. [CrossRef] [PubMed]
22. Dhyani, M.; Kushwaha, G.S.; Kumar, S. A novel intuitionistic fuzzy inference system for sentiment analysis. *Int. J. Inf. Technol.* **2022**, *14*, 3193–3200. [CrossRef]
23. Borse, S.; Pardeshi, K.; Jha, P.; Gangurde, S.; Vidhate, P.; Borse, L.; Gulecha, V. An Artificial Intelligence Enabled Clinical Decision Support System in Mental Health Disorder. *J. Posit. Sch. Psychol.* **2022**, *6*, 2371–2378.

Disclaimer/Publisher's Note: The statements, opinions and data contained in all publications are solely those of the individual author(s) and contributor(s) and not of MDPI and/or the editor(s). MDPI and/or the editor(s) disclaim responsibility for any injury to people or property resulting from any ideas, methods, instructions or products referred to in the content.

 healthcare

Article

Identifying the Barriers to Acceptance of Blockchain-Based Patient-Centric Data Management Systems in Healthcare

Ibrahim Mutambik [1,*], John Lee [2], Abdullah Almuqrin [1] and Zahyah H. Alharbi [3]

1. Department of Information Science, College of Humanities and Social Sciences, King Saud University, P.O. Box 11451, Riyadh 11437, Saudi Arabia; aalmogren@ksu.edu.sa
2. School of Informatics, The University of Edinburgh, 10 Crichton St., Edinburgh EH8 9AB, UK; john.lee@ed.ac.uk
3. Department of Management Information Systems, College of Business Administration, King Saud University, P.O. Box 28095, Riyadh 11437, Saudi Arabia; zalharbi@ksu.edu.sa
* Correspondence: imutambik@ksu.edu.sa

Abstract: A number of recent studies have shown that wastage and inefficiency are a significant problem in all global healthcare systems. One initiative that could radically improve the operational efficiency of health systems is to make a paradigm shift in data ownership—that is, to transition such systems to a patient-centric model of data management by deploying blockchain technology. Such a development would not only make an economic impact, by radically cutting wastage, but would deliver significant social benefits by improving patient outcomes and satisfaction. However, a blockchain-based solution presents considerable challenges. This research seeks to understand the principal factors, which act as barriers to the acceptance of a blockchain-based patient-centric data management infrastructure, in the healthcare systems of the GCC (Gulf Cooperation Council) countries. The study represents an addition to the current literature by examining the perspectives and views of healthcare professionals and users. This approach is rare within this subject area, and is identified in existing systematic reviews as a research gap: a qualitative investigation of motivations and attitudes among these groups is a critical need. The results of the study identified 12 key barriers to the acceptance of blockchain infrastructures, thereby adding to our understanding of the challenges that need to be overcome in order to benefit from this relatively recent technology. The research is expected to be of use to healthcare authorities in planning a way forward for system improvement, particularly in terms of successfully introducing patient-centric systems.

Keywords: blockchain; blockchain barriers; blockchain in healthcare; future of healthcare blockchain; GCC countries

1. Introduction

In most countries today, the efficiency of healthcare systems is critical. Yet few, if any, nations can claim that there is not a high level of wastage inherent within their existing health infrastructure [1]. According to a 2023 report by the Peter G. Peterson Foundation, for example, as much as 25% of healthcare expenditure in the US alone is considered wasteful, which amounted, in 2021, to some USD 1.2 trillion annually [2,3], while various other studies have found that similar problems beset all OECD countries [4,5].

One of the principal reasons for this inefficiency lies in the complex and multi-dimensional interactions between various (healthcare system) stakeholders, ranging from primary care actors (clinical practitioners) and the patients themselves, to hospitals and national governments [4,5]. The result, without exception, has been the deployment of technology in a disparate and non-integrated way, by using different digital platforms for different parts of the system. This leads to a number of serious issues, resulting from an inability to find and exchange patient data across multiple health providers and related organisations [6], including increased costs, high error rates, knowledge mismanagement, poor quality of care [7] and higher mortality rates [6]. Yet, creating an integrated system retrospectively has proved a formidable challenge.

In the past few years, however, a powerful, cost-efficient, and—in principle—technically viable solution has emerged: blockchain. This is essentially a form of distributed (decentralised) database, which means that the database on which transactions are recorded is shared by several participants (called nodes) in the blockchain. Such a decentralised infrastructure allows for a faster, more secure transaction process, which all participants can trust.

The evidence is clear [4,8,9] that blockchain offers a major opportunity to develop healthcare systems across the world to better meet evolving societal and economic development goals. However, while the technology is being increasingly implemented [6,10–12], it is, according to the World Economic Forum, still "massively under-utilised in global healthcare" [12]. There have been a number of studies that have investigated why this is the case. One such study, for example, highlighted the importance of perceived security in encouraging the adoption of blockchain [13–15], while others have found that a government's published position on the technology has a significant effect on adoption levels [13,14,16].

These studies have contributed significantly to our knowledge, indicating that acceptance of the technology by various stakeholders remains an impediment to adoption. However, these studies have all had a relatively narrow focus on specific issues. There are few studies that have examined the barriers to blockchain adoption by considering a wider range of general factors, such as administrative challenges, usability issues, future-proofing, and regulatory frameworks, from the direct perspective of professionals involved, either internally or externally, in the healthcare system [17–19]. Existing systematic reviews have also identified a research gap in this area. For example, Tandon et al. [20] conducted an extensive systematic review of the literature on blockchain in healthcare specifically, and noted among their recommendations:

based on the gaps explicated from the review, implications arise for the need to broader disciplinary and particularly methodological coverage of the research to advance the current understanding of the field. For example, research based on survey methodology or interviews can enhance the understanding of e.g., the challenges and barriers inhibiting user adoption. (p. 20)

Similarly, Yang et al. [21], addressing a business context but identifying issues that arise equally in healthcare, observe that

there is also a lack of empirical studies examining the incentives leading business organizations to invest in and adopt blockchain technology. Indeed, knowledge about the reasons for adopting and using blockchain technology in private and public organizations is rather scarce. We suggest that future studies investigate the motivations associated with blockchain adoption and how these motivations influence how blockchain initiatives are implemented and managed in companies. (p. 4480)

Insights into the nature of barriers and inhibitions to user adoption that arise from the attitudes, motivations, and understandings of participants in the specific area of activity are exactly the motivation behind this study, as such insights will support healthcare providers in providing more holistic and efficient systems, yielding a broad range of societal benefits.

This study aims to address this issue by identifying key challenges/barriers to the acceptance of a blockchain-based patient-centric data management system, as a first step towards the development of effective solutions. To achieve this, this study analysed (using thematic analysis) the data collected from semi-structured interviews with 30 professionals working in the healthcare systems of the countries of the Gulf Cooperation Council (GCC). This study explored the research question (see below) across the GCC region, as opposed to a single country, for a number of reasons. Principal among these reasons was the fact that the GCC countries face several similar challenges in the healthcare field, such as a disease burden that includes communicable and non-communicable diseases, mental health issues, and accidental injuries. Addressing this issue will require an innovative and cost-effective strategy [22]. Given that the GCC countries each have government-funded healthcare models, and that the GCC Charter, formulated in 1981, states that the region's nations should 'coordinate and integrate' on joint economic and social ventures [23], it was decided that the study could meaningfully and usefully encompass all nations in the region. As the GCC countries share many cultural, social, and economic characteristics [22,23], the results of the study are expected to be broadly generalisable, though with some limitations. This will be discussed further at a later stage in this report.

A qualitative approach was adopted, as implied by Tandon et al. [20], in order to gain informed, expert opinion which reflects the true, on-the-ground problems that face healthcare providers in implementing a blockchain approach to healthcare infrastructure. This is a crucial precursor to potential future quantitative research. In particular, the paper seeks to provide insights related to the research question:

RQ: What are the main user acceptance barriers to adopting blockchain-based patient-centric data management systems in the GCC healthcare sector?

This paper is structured as follows. Section 2 provides a review of the background, in terms of blockchain-based patient-centric data management systems in healthcare, against which we construct our study. Following this, the research methodology, results, and discussion are presented. Lastly, the conclusions of the study are provided.

2. Theoretical Background

There are conflicting claims in the existing literature about the potential benefits that could be delivered by the adoption of blockchain-based ICT solutions in healthcare environments. Many researchers make broad and substantial claims for these benefits. These range from improved data and record management [24,25] to more effective containment of pandemic instances, such as COVID-19 [26,27]. Further, there exist several meta-studies of the blockchain for healthcare development literature, which identifies a wide range of potential use cases for the technology. In general, however, these use cases can be grouped into three main areas: clinical trial improvement, pharmaceutical supply chains, and data ownership (specifically, patient-centred data management systems). We will briefly discuss these in turn.

According to research in [28], the cost of clinical trials approximately doubled between 2008 and 2019. There are a number of reasons for this, such as the inflationary cost of materials, the increasing complexity of regulations, and higher costs associated with collecting, managing, and analysing data. Some of these issues could be entirely eliminated through the implementation of blockchain technology. In fact, blockchain-based systems could effectively revolutionise clinical trial management, by enabling the automation of aspects such as site pre-screening, protocol approvals (such as patient consent, etc.), participant enrolment and monitoring, data collection and compliance, and analysis and reporting [29].

Automating these aspects of clinical trials in a secure, trusted, transparent, and auditable way would radically improve the efficiency of the process. The savings, in terms of both cost and time, could be invested in improving patient care and the development of innovations that would move the healthcare industry forward [30].

There are claims that supply chains, too, could be radically improved through the use of blockchain. Consider, for example, prescription drugs. These are, according to the US National Library of Medicine, the largest counterfeit market in the world, estimated to be worth up to USD 432 billion [31,32]. The WHO (World Health Organisation) reports [33] that up to 10% of medical products worldwide are counterfeit or substandard [34,35], resulting in the death, globally, of more than a million patients every year due to counterfeit pharmaceuticals.

The probability of totally eliminating such criminal activity in the short term is low. However, it is perfectly realistic to reduce the level of crime significantly, through better monitoring and control of supply chains. It is argued that this can be achieved through the use of blockchain. ID (identification) markers can embedded on all products, facilitating full and accurate accountability at every stage of the supply chain by using smart contracts, which are programs stored on blockchain that execute automatically when predetermined conditions are met [6], allowing stakeholders to resolve issues concerning product provenance directly between themselves, without the need for unreliable middlemen [36]. This makes the supply chain not merely less vulnerable to corruption, but also more cost efficient.

The third area where it is argued that blockchain could deliver major benefits is by changing the current model of data ownership to patient-centric data management systems. By using blockchain, the need for third-party data management could be eliminated, allowing individuals to have more control over their medical information through specific apps. This has several significant benefits, such as allowing patients to more easily access test results and medication lists, manage appointments, and contact health and care staff when needed [6]. Such an approach would not merely improve patient outcomes, but would significantly and positively impact system efficiency in terms of the use of resources. In fact, if properly implemented, patient-centric data management has the potential to produce the best possible outcomes in a healthcare context [6,37,38]. For these reasons, patient-centric systems are considered by many to be the 'holy grail' of healthcare [39], offering a full restructuring of health systems [6,37,38] and delivering a wide range of economic and social benefits.

While there are several important studies that examine the possibility of blockchain as a basis for patient-centric models [6,11,37], these are all proposals based on theoretical constructs or existing technical knowledge; they are in several respects controversial and do not address the question of whether the technology will be regarded as acceptable from the perspectives of specific and relevant professionals. The present study is different in that it addresses this key aspect.

While many benefits of using blockchain in the healthcare sector are widely proclaimed, a range of factors combine to make the adoption of the technology less straightforward and slower than proponents would like [6,40,41]. One of these factors is that blockchain is still considered to be in its relatively early stages of development, and some significant challenges remain concerning privacy, security and scalability [42]. Another challenge that would need to be overcome, before blockchain can be successfully implemented in healthcare on a wide scale, is the large volume of data concerned. According to a 2020 OECD report [4], storing full electronic medical records or genetic data on blockchain would be costly and inefficient because of the constraints of replicating the blockchain across every network node [43], which undermines the potential of blockchain as a basis for patient-centric data management. The challenges of using very large volumes of data on blockchain have also been identified in other contexts, such as supply chains [6,44,45], which have to be set against the potential advantages of blockchain, such as combatting counterfeiting in this area.

Other problems which arise in the implementation of blockchain include "a lack of standardisation, accessibility, ownership, and change management" [46]. Possibly one of the most significant issues, however, is the need for interoperability between different blockchains. This is because healthcare infrastructures are complex, and would typically involve a number of different blockchain ecosystems. These ecosystems would need to talk seamlessly to each other. However, while interoperability is a significant challenge, viable solutions are beginning to emerge [6,44].

The challenges described above have not completely prevented the adoption of blockchain in healthcare. In fact, the technology is beginning to find acceptance in healthcare environments for a range of administrative and clinical use cases, such as the management of medical records, remote patient monitoring, pharmaceutical supply chains, insurance claims, and data analytics [47]. Yet, as already noted, the proclaimed potential of blockchain technology is far from being realised. This is because one of the most important areas in which blockchain can contribute to social development by helping to revolutionise healthcare is in the third category mentioned above: developing patient-centred data management systems. While there have been some attempts to build patient-centric systems, most have not been as successful as originally hoped [48]. The reasons for this lack of success are unclear, though there is some evidence that inhibiting factors include concerns—by patients, administrators, and clinical staff—over data confidentiality, integrity, and availability [49]. Although there have been a few other studies of the lack of success of similar blockchain projects in the education sector, these have either focused on a specific, pre-identified, adoption barrier [50,51] or have been metastudies that address difficulties only at a very general level [52]. They therefore shed little light on the issues that face the healthcare sector, in terms of blockchain adoption, and specifically the challenges involved in delivering patient-centric systems.

It is this gap in knowledge that this paper attempts to fill. It aims to better understand the challenges that health authorities must overcome if blockchain solutions are to be successful at scale in creating patient-centred data management models. We deploy a qualitative methodology that collects data from a range of stakeholders—administrators, clinical staff, and patients—and analyse these data to identify key issues. While it is true that there is a considerable number of papers on blockchain, most focus narrowly on potential benefits [25,29,53,54]. There are few studies that deploy a qualitative methodology to explore the concerns of the full base of stakeholders. This paper aims to address this.

3. Research Method

The aim of this research is to identify the principal barriers to the acceptance of blockchain in healthcare. In particular, we seek to gain insights into acceptance barriers that inhibit the adoption of blockchain-based patient-centric data management systems in the GCC. While there is likely to be a range of significant barriers to adoption in any healthcare environment, there are several aspects of the GCC context (as they share a similar culture and society) that may produce unique and specific barriers. These include, in particular:

- Cultural Factors: Most GCC healthcare systems will use their own unique and familiar record-keeping and data-sharing practices. There may therefore be a strong resistance to disrupting these processes.
- Regulatory Environment: Blockchain technology may not easily align with the regulatory frameworks of the GCC, particularly in terms of data protection and cross-border data exchange.
- Technological Readiness: Blockchain systems may require technical infrastructures and levels of digital literacy among the (staff) population which may not yet exist, and will take significant time to develop.
- Distrust: As a new technology, blockchain may engender distrust in healthcare professionals and patients in the GCC, which may be difficult to overcome.

This study set out to identify any barriers to acceptance that may arise from these factors, and others. To achieve this aim, we analysed semi-structured interviews from a sample (N = 30) of professionals and potential service users, using thematic analysis, in order to identify patterns in participants' views and attitudes [55–57]. Several key areas relating to blockchain, in the context of healthcare, were covered, including:

- The nature and benefits of blockchain in healthcare;
- Internal and external attitudes towards the ideological and practical aims of delivering patient-centric data systems;
- Aspects of departmental/organisational culture that could hinder blockchain initiatives in the field of data ownership;
- Perceived barriers, either internal or external, to the development of blockchain applications;
- The implications of slow progress towards patient-centric data management systems.

The interviews were conducted in two phases:

- Structure and format: Prior to carrying out the interviews, primary open-ended questions were developed to provide the basis for follow-up. All questions were

constructed in such a way as to avoid 'leading' the interviewee, which can result in bias.
- The interview processes: The time given to each interview was approximately the same (1 h). Consistency and validity were promoted through the use of field notes.

To help ensure that the proposed questions would deliver the depth and scope of the data required for the study [55,58–60], three pilot interviews were conducted before interviewing the full sample. The pilot process indicated that the relevant areas of investigation (described above) were adequately covered.

3.1. Sampling

This study employed a combination of purposeful and snowball [58,61,62] sampling techniques. Such a combination allowed the researchers to benefit from the advantages of both techniques [63–65]. While purposeful sampling was used to identify participants, who were highly informed on the topic in question, which helped to ensure that data was accurate and relevant, the snowball technique allowed the researchers to enlarge the sample to other, similarly expert, participants. This helped this study achieve richer and more comprehensive results.

The majority of the sample were professionals in the healthcare system with a clear perspective of the benefits and implications of introducing blockchain processes in delivering patient-centric systems, though it was also important to gather the views of other key stakeholders. Potential participants were as follows:

- Responsible for the development and/or management of administrative or clinical departments within national healthcare systems;
- Experienced in the design and implementation of blockchain projects;
- Other stakeholders in a patient-centric data management system (individuals not necessarily professionally involved with healthcare delivery).

The initial (purposeful) process of selecting participants was carried out by identifying appropriately qualified professionals on LinkedIn. After a comprehensive review of profiles, 70 invitations to participate were made, outlining the purpose of the study and emphasising its ethical construction. This resulted in 20 suitably qualified professionals volunteering to participate in the study. These professionals then recommended a further 10 potential participants. The final sample size was 30. All were based in GCC countries, and each country was represented by an (approximately) equal number of participants, in order to minimise the possibility of bias towards any country's perspective (Table 1).

Table 1. Summary of details of interviewees.

Participant No.	Job Title	Years of Experience	Country
P1	Professor of Cryptology	16	UAE
P2, P3	Academic Advisor in Blockchain	6	Bahrain
P4	Post-doctoral Researcher in Blockchain Systems	4	Bahrain
P5	Clinical Practice Manager	13	Oman
P5, P6	Lecturer in Information Technology	11	Saudi
P7	Professor in Computer Engineering	7	Saudi
P8	Multidisciplinary Team Leader in Healthcare	6	Oman
P9	Clinical Practice Manager	8	UAE
P10	Hospital CEO	10	Bahrain
P10, P11	Pharmaceutical Product Manager	7	Kuwait
P12	University IT Developer	5	UAE
P13	Healthcare Manager	3	Qatar

Table 1. Cont.

Participant No.	Job Title	Years of Experience	Country
P14	Clinical Operations Manager	3	Kuwait
P15	Project Manager, IT	4	Saudi
P16	Professor in Informatics	14	Oman
P17	Associate Professor in Software/IT	7	Oman
P18	Ph.D. Student in Distributed Ledger Technology	4	Qatar
P19	Clinical Operations Manager	11	UAE
P20, P21	Senior Healthcare Scientist	12	Saudi
P22	Professor in Computer Engineering	7	Kuwait
P23	Professor of Blockchain Technology	5	Qatar
P24	Professor in Informatics	14	Saudi
P25, P26	Director of Healthcare Services	12	Bahrain
P27	University IT Manager	4	Kuwait
P28	Director of Healthcare Services	10	Saudi
P29, P30	Hospital CEO	17	Qatar

3.2. Data Collection

The 30 interviews which form the basis of this study were conducted over a two-month period. All original interviews were conducted in Arabic, and analysed in Arabic, with excerpts translated into English for the purposes of this report. Due to the diversity of geographic locations, interviews were carried out using online meeting platforms (Zoom, etc.) and were recorded using the platform's tools.

All appropriate ethical guidelines were followed in conducting the research. Before each interview, the purpose and scope of the study were explained to the participant, and it was emphasised that there were no correct or incorrect answers to any question. Written assurances of full confidentiality were also provided, together with an explanation of how the data would be used. Each participant was informed that they could withdraw their participation at any time, and no part of any interview was recorded without full and explicit permission from the interviewee. All participation was voluntary, and no incentive, financial or otherwise, was offered to any participant.

Each interview contained 17 questions, and the process was meticulously designed to ensure that the questions were both rigorous and relevant to our study's goals (all the questions are presented in Appendix A, together with a description of how the questions were designed). As the interviews were semi-structured, the questions were open-ended and conformed to a predetermined thematic framework [62]. This framework covered the broad areas described above. Example questions were the following:

- What benefits do you feel could be gained from deploying blockchain applications in healthcare?
- Do you feel that patient-centric data management is a desirable policy objective? Why?
- How realistic is any plan to implement patient-centric data infrastructures within your departmental framework?
- How do you think (health) service users will respond to patient-centric data management?
- Could failure to implement patient-centric data management have long-term consequences, either good or bad?

Following the interviews, each recording was matched against the field notes by two independent researchers. This helped to ensure the accuracy and consistency of data in preparation for the coding and analysis phase.

3.3. Data Analysis

As noted above, the study used thematic analysis [66] to identify themes in the interview data. The steps involved were as follows:

- Initial coding: Segment-by-segment coding to identify similarities in interview content.
- Focused coding: Codes thought to be of particular significance are grouped to form patterns.
- Theme search: Focused codes were recorded to facilitate the identification of groups that shared attributes of meaning. These sub-themes were then examined for higher-level commonalities to form main themes.
- Theme identification: This used (a) internal homogeneity (meaningful coherence within themes) and (b) external homogeneity (a clear and identifiable distinction between themes) to identify themes [58].

4. Results

As a result of the analysis, 4 main themes and 12 sub-themes connected to barriers to the adoption of blockchain (in healthcare) emerged (Table 2).

Table 2. Main themes and sub-themes.

Main (Barrier) Themes	Sub-Themes
Administrative and Management	Knowledge and Skills Recruitment Funding and Financial Support Management Commitment Security
User Perspectives	Ease of Use Privacy and Data Use
Future Proofing	Scalability Interoperability Standardisation Sustainability
Regulatory Issues	Compliance Governance and Liability

4.1. Administrative and Management

This theme contains four sub-themes. We will look at each in turn.

4.1.1. Knowledge and Skills Recruitment

The participants expressed a wide variety of concerns about the specialist knowledge and skills required to successfully implement blockchain solutions. However, there was general agreement among all participants, including those not professionally involved in healthcare, that the advanced and immature nature of blockchain would represent a particular challenge in terms of recruitment. According to one participant, for example:

> Blockchain is still relatively new on the scene, and it's a complex area. I realise that there are quite a few suitably qualified people around, but most of these have been swept up by private and high-paying enterprises. My guess would be that it would prove quite a challenge to find the number and type of staff required over the short or medium term.

Another participant emphasised the difficulties of recruitment, saying:

> It will be tough enough to find the technical development staff needed, but there's also the senior project management people to worry about. They need to be highly experienced in the strengths and limitations of blockchain technologies, and they're few and far between at the moment.

One participant, who is not a healthcare professional, commented:

I'm not an expert in blockchain, but I know it's what makes cryptocurrencies such as bitcoin work, so I would imagine there's a high demand for blockchain specialists in the crypto sector. That might make it hard for the healthcare industry to find the right people.

It is clear from these comments that the participants consider the practical challenge of system development to be a significant issue in itself, quite apart from any ideological or policy concerns.

4.1.2. Funding and Financial Support

One sub-theme that was easily identified from the interview data was connected to cost. Several participants pointed out that blockchain projects are expensive to fund, and, as the technology is relatively new and untried in healthcare, projections of return on investment are not easy to make. Other interviewees mentioned that the nature of blockchain, which requires considerable computing power, could make systems expensive to run, so careful thought would have to be given to how they would be funded on an ongoing basis. A typical comment on this point was as follows:

Using blockchain to build a patient-centric data management system sounds good idea on the face of it, but it could be surprisingly expensive to operate. A huge amount of information and a high blockchain transaction level is involved, which means it is likely to be very expensive to fund the consensus protocols and cryptographic needs of the underlying chain.

Several participants emphasised the cost element of the recruitment difficulties discussed in the skills section above. For example:

Specialist developers in DLTs [Distributed Ledger Technologies] such as blockchain are becoming more common, but they are still relatively rare animals. Most of them are either working in research, or are in relatively high paying posts with private companies. Health authorities will need to be prepared to pay high rates, if they want properly qualified staff.

Other interviewees pointed out the implications of immature technologies such as blockchain. One said:

You have to remember that, as blockchain is still relatively new and developing, the initial cost of the project is just the start. After that, there's likely to be an ongoing requirement for infrastructure upgrades as the technology improves, as well as the addition of new features and functions to meet the needs of the healthcare system as societal requirements change.

Overall, it seems that the nature of the technology, together with the relative scarcity of qualified project managers and developers, makes cost a significant issue.

4.1.3. Management Commitment

Despite its many benefits, the nascent and unproven nature of blockchain technology is the source of significant levels of concern at the level of senior management and policy makers. But without a clear and positive commitment at these levels, to both the concept of patient-centric systems and the underlying blockchain technology, adoption of the technology is likely to be slow. As one interviewee put it:

Projects such as that under discussion require buy-in across the board, and this simply won't happen unless there is unequivocal backing from senior management. Personally, I don't see that backing from my own employer, but my experience may not be typical.

In fact, this view *was* quite typical, as several participants expressed a similar opinion, though some were more explicit about the causes of management concern. For example:

There is certainly a lot of management hesitancy about moving forward with blockchain, and I think a lot of this is connected to personal risk. New technologies like this can easily fail due to unexpected problems, and the cost of project failure, however that's defined, can be high at a political, institutional and personal level. Few senior managers are willing to take that risk.

Other participants supported this view. One commented:

I think that management, in general terms, is playing it safe at the moment. The stakes are high, and they're waiting to see how other initiatives work out and whether they can learn from them.

4.1.4. Security

Although the consensus mechanisms and decentralised nature of blockchain mean that records are immutable and tamperproof, there remain questions over security, at both the institutional and individual levels, which can represent significant concerns to management and authorities. Nearly all of the participants felt that security, in one way or another, was a serious issue for the concept of patient-centric data management systems. According to one participant, for example:

Despite blockchain's famously high security levels, many people would be surprised at how vulnerable it can be. 51% attacks, for example, are well known in the Bitcoin arena and they can happen in any blockchain context. These give cybercriminals control over the system, which could be a disaster, particularly in a public service context.

Another interviewee remarked:

Because of a lack of standardisation, blockchain's—and especially public chain systems—don't necessarily meet all of the regulatory security and privacy requirements such as the GDPR and similar frameworks. I think this needs to be resolved before blockchain could safely be used in healthcare systems.

While many participants commented on the institutional risks of blockchain, others pointed to risks at the personal stakeholder level. As one participant put it:

Most members of the public will be unaware of the importance of protecting their private key. This means they could either mislay them or use weak keys—either way, it could lead to quite serious problems at a systemic and individual level.

4.2. User Perspectives

This theme contains two sub-themes. We will look at each in turn.

4.2.1. Ease of Use

While the majority of people today are comfortable with engaging with digital systems, there remains a very significant minority for whom digital technology is a challenge. Some do not engage at all. However, a healthcare system, by its nature, is fully inclusive. This means that, if the benefits of blockchain-based patient-centred systems are to be fully realised, they must be easy to engage with. This is not only true at the service user level, but at the clinical and administrative level also. However, many of the study's participants expressed doubts as to whether the required levels of ease of use would be achieved. One participant said:

In the end, a system like this is all about the patients. Unless they engage with it, it will fail. This means the interface must be ultra-easy to use, and access issues such as private keys must be simple to understand. My worry is, that a lot of people will find themselves unable or unwilling to use the system.

Another interviewee made a similar point:

Ease of use, in my mind, is critical. Unless the apps are simple and straightforward, engagement levels might be low. This would mean traditional record-keeping systems would have to be fully maintained, and that could negate, or significantly reduce, the benefits of a blockchain approach.

While the importance of engagement of service users is critical, it is also essential that the system is accessible and easy for managers, administrators, and clinicians. The following remark by an interviewee was typical:

> It may be 2023, but a lot of professionals in healthcare are addicted to paper-based systems. The success of a patient-centric data management system will need something of a culture shift. For this to happen, the benefits will need to be clear, and the system extremely easy to use. I'm not convinced that either of those conditions would be true in the short term.

4.2.2. Privacy and Data Use

Administrative errors threaten the safety of thousands of patients worldwide every year. Although blockchain could realistically contribute to reducing these errors significantly, there remain a number of systemic security vulnerabilities. Some of these have been touched upon in the security subtheme (Section 4.1.1). However, there is a related, but separate, issue that was highlighted by many of the study's participants. This issue is the problem of user perceptions of privacy and appropriate data use. As one participant commented:

> Although blockchain infrastructures offer high privacy levels, I would say that the large majority of the public don't realise this. To them, an open system which gives them control over their own data could cause real concern. I think the authorities would have to run comprehensive educational campaigns to correct this impression, if people are to accept the system.

The problem of inaccurate user perceptions, resulting from a lack of blockchain awareness extends to other areas related to privacy, such as data abuse. One participant, a professional outside of the healthcare field, commented:

> While, in reality, traditional record keeping systems can also be abused, there is a danger that the public would see a more apparently open system as presenting higher risk. This could present a problem in terms of adoption levels.

This view was echoed by another participant, who commented:

> It's ironic, really, that the privacy protections of blockchain, such as ZKPs [Zero-Knowledge Proofs], can deliver higher levels of privacy than anything we've known before, yet the average member of the public probably wouldn't see it like that. I think there needs to be a broader understanding of how blockchain works before systems based on the technology will be accepted by the public.

In summary, most participants felt that there was a wide gap between reality and public perception in terms of data privacy, and closing this gap represents a significant challenge.

4.3. Future Proofing

This theme contains four sub-themes. We will look at each in turn.

4.3.1. Scalability

The scalability of blockchain networks is an important issue in many applications. This is particularly true with large public blockchains, which tend to have large amounts of data. This means there is a large number of transactions (changes to data). As each transaction must be validated through peer-to-peer verification, the network must have high scalability (the ability to support an increasing transaction load, as well as more network nodes). This represents a key challenge for the healthcare sector. One participant, for example, pointed out that:

> One problem is that the system under discussion is fundamentally public facing, which means it not only must be easy to use, but it has to be real-time. Unless users see changes pretty much instantly, they are unlikely to engage with the system. However, not all blockchains can deliver the required scalability.

Another participant made a similar point:

> In theory, the system proposed could consist of small private blockchains, which would help to maintain scalability and speed. But this would be expensive. In practise, the chances are that scalability, or the lack of it, would present a major challenge to healthcare providers.

Scalability could also present issues where payments are concerned, according to several participants. For example:

In itself, transaction speed might not be a major problem in many contexts, such as data management. But the system will also need to allow for payment processes in some situations, such as private consultancies, and this could be a much more serious challenge. Scalability of the system will be essential.

This sub-theme of the results highlights that, even though blockchain has been used in real-world applications for over a decade, scalability issues may still hinder growth and innovation.

4.3.2. Interoperability

Historically, blockchain technology has evolved in a siloed environment, which has led to a number of issues that have hindered the adoption of blockchain. One of these issues is a lack of interoperability—different blockchains do not easily 'talk' to other systems. Several participants identified this as a potential problem. As one interviewee put it:

It's crucial that blockchains used in a healthcare system can interoperate, and communicate with legacy systems. As things stand, this ability is, though improving, still limited, and could present real challenges, in terms of enabling healthcare systems based on blockchain to adapt to future needs easily and cost effectively.

Another participant emphasised the same point, but was a little more positive about emerging solutions:

There's no doubt that interoperability has been major factor in Decisions whether or not to implement solutions based on blockchain. It's still an important factor, but at least realistic and viable solutions are now beginning to emerge, such as Over ledger from Quant [Network], and Ark.

It is important to note that interoperability is not only about blockchain-to-blockchain communications. To be effective, blockchains must also communicate with other IT systems and applications within the healthcare infrastructure. This point was made by a number of participants. For example:

If a patient-centric system is going to deliver on its potential, it needs to be able to exchange data with a wide range of other systems, internal and external. I'm not convinced that this ability exists right now, without compromising core attributes such as security and privacy.

In short, while interoperability is improving rapidly, it is still a significant concern and may limit options in the future.

4.3.3. Standardisation

Another significant issue is the lack of standards—each different blockchain has been developed with its own set of protocols. This has played a large part in inhibiting mass adoption. One participant remarked:

Standards are necessary for any technology to succeed at a global, or even national, level. This is no different for blockchain. It's true that the landscape is beginning to change, But standards take time to develop and we're not there yet. This means that there are real risks involved with Investing huge amounts in blockchain-based healthcare systems.

The observation that blockchain standards are beginning to emerge was made by several participants. However, most still had caveats. According to one:

It's good to see that several industry alliances and other bodies are now collaborating to create global blockchain standards, but we're still some way away from a framework that will trigger mass adoption of the technology.

The issues of standardisation and interoperability, though separate, are also closely connected. As one participant said:

It's true that blockchain technology offers a range of major benefits over other data exchange solutions, but the lack of industry standards could cause serious interoperability issues.

4.3.4. Sustainability

Most blockchains are energy intensive to run. The implementation of systems based on the technology may therefore not be consistent with the commitment of most countries to become environmentally sustainable and energy efficient. For example, all 193 members of the UN's Sustainable Development Goals (SDGs), which include climate and clean energy targets for 2030. This issue was a clear concern among the study's participants. One interviewee, for example, commented:

The energy consumption issue is a real dilemma. Most blockchains operate on a proof-of-work basis which needs a lot of energy, which can be hard to justify, given current concerns over climate change and so on. The alternative is proof-of-stake models, but although they're less energy-intensive, they're much more complex and can lead to system vulnerabilities.

The above comment was repeated in various forms. For example, another participant said:

Implementing blockchain could easily end up being a decision that the authorities might regret, due to the energy demands of the system. I seem to remember that it was the power requirements of blockchains that led to Tesla reversing their decision to allow payment with Bitcoin in 2021. That surely tells us something about the difficulties involved.

Another participant recognised the sustainability issue, but was more positive:

Some experts think that the types of blockchain suited to use in healthcare have far lower energy requirements than those of Bitcoin. But, even if this is the case, combatting the negative perceptions of blockchain energy consumption won't be easy.

The technology is evolving rapidly, and it remains to be seen whether healthcare blockchains will encounter energy usage issues. If suitable solutions are identified, it will be imperative to combat the negative perceptions of current blockchain energy consumption.

4.4. Regulatory Issues

This theme contains two sub-themes. We will look at each in turn.

4.4.1. Compliance

The implementation of blockchain in healthcare must comply with relevant national, and in some cases, regional (e.g., GCC, EU, US) regulatory frameworks. These frameworks vary according to country/region and therefore make different demands of the system. Due to blockchain's relatively immature technical nature, compliance is not always easy to ensure. This was highlighted by several participants. For example:

I believe most GCC countries now have an equivalent to the EU's [European Union] GDPR [General Data Protection Regulation], which contains a right to be forgotten. However, this is incompatible with blockchain's immutability. I'm not sure how this will be resolved.

Another participant said:

While most regulatory frameworks don't yet have a standard for decentralised identity, this is something that will almost certainly emerge soon. But until we know its precise form, it will be difficult to know how easy compliance will be. It might require processes within the blockchain system to be reengineered.

The problem of complying with the requirements of most regulatory frameworks to anonymise personal (sensitive) information was raised by several interviewees. For example:

The main difficulty is establishing whether the information held on the [blockchain] system is considered sensitive. And even when this is known, there are question-marks over whether the information can be anonymised in a way that's compliant with regulations such as GDPR.

4.4.2. Governance and Liability

It is important to establish a clear and robust governance model concerning interactions between parties involved with the blockchain network. In a healthcare system, these parties would include the network operator and its participants. While this is not, in general terms, especially difficult in a technical or legal sense, the relatively novel nature of blockchain could present a challenge. As one participant phrased it:

Blockchain poses a range of different risks connected to issues such as security, confidentiality and data protection. However, these risks aren't, as yet, fully understood in practical context, so the attribution of liability needs to be carefully analysed.

Another participant made a similar point:

It is important to establish the legal structure, liability and governance model of the blockchain systems, so that all rules, rights and obligations are legally clear. This is critical to ensuring everyone involved, from service users to administrators can use the system without concern.

While most blockchain healthcare systems will fulfil internal national needs, there are situations where they may be required to function across borders. This could pose a number of complex jurisdictional issues which require careful consideration. One participant said:

If blockchain nodes are located in different countries or regions, it's not clear which legal framework would be relevant. Local laws may apply, but there is some confusion over exactly what laws would be enforceable and how they would be enforced.

4.5. Comparison of IT and Clinical Perspectives

A further analysis of the data allows for a comparison between the perspectives of the IT specialists whose focus is on the management and delivery of IT systems and processes and the clinical practice managers whose focus is on clinical provision. This is based on the themes and sub-themes identified through the analysis, which are then traced through the data. This enables us to list concerns and prospects recognised by these two groups, which are related yet importantly different in certain respects, as follows.

4.5.1. IT Specialist Perspective

- Scalability: IT specialists must consider how well a blockchain solution can scale to support a large number of transactions. For example, public blockchains like Bitcoin and Ethereum can process a limited number of transactions per second, which may not be suitable for a large healthcare organisation. Private or consortium blockchains, on the other hand, might provide better scalability.
- Integration with Existing Systems: IT specialists have to ensure that the blockchain solution can integrate seamlessly with existing health IT systems. This involves considering issues like data migration, user training, and system maintenance.
- Regulatory Compliance: IT specialists must ensure that the blockchain solution complies with regulations related to health data, such as the Health Insurance Portability and Accountability Act.
- Blockchain Type: IT specialists need to decide between public, private, or consortium blockchains. Public blockchains are open to anyone and are secured by decentralization, but they might not be suitable for sensitive health data due to privacy concerns.

- Private and consortium blockchains, which are only accessible to invited participants, might be a better choice for healthcare applications.
- Smart Contracts: IT specialists are likely to be interested in the potential of smart contracts, which are self-executing contracts with the terms of the agreement directly written into lines of code. In healthcare, smart contracts could automate many processes, such as claims adjudication in health insurance.
- Data Standardisation: To ensure interoperability, IT specialists need to consider standardising the data stored in the blockchain.

4.5.2. Clinical Practice Manager Perspective

- Patient Engagement: Clinical Practice Managers are likely to view blockchain as a tool for enhancing patient engagement. For example, blockchain could enable patients to control who has access to their health data, thereby promoting a more patient-centric approach to healthcare.
- Collaboration with other Healthcare Providers: Blockchain could facilitate collaboration by creating a shared, immutable record of patient data that can be accessed by different healthcare providers. This could lead to more coordinated and efficient patient care.
- Cost Implications: Clinical Practice Managers must consider the cost implications of implementing blockchain technology. This includes not only the costs involved in the technology itself, but also training costs, maintenance costs, and potential cost savings from improved efficiency.
- Staff Training and Adaptation: Clinical Practice Managers would need to plan for staff training to ensure that all staff members understand how to use the new system.
- Patient Empowerment: Clinical Practice Managers may see blockchain as a way to empower patients. With blockchain, patients could have more control over their health data, deciding who can access it and what they can do with it.
- Improved Care Coordination: Blockchain could improve care coordination by providing a single, up-to-date, and immutable record of a patient's health history. This could help all providers involved in a patient's care stay on the same page.
- Operational Efficiency: Clinical Practice Managers may be interested in how blockchain could improve operational efficiency. For example, blockchain could speed up the claims process in health insurance by reducing the need for intermediaries.
- Change Management: Implementing a new technology like blockchain would involve significant change. Clinical Practice Managers would need to manage this change carefully, ensuring that staff are trained, also that relationships with external agencies and providers are developed and maintained, that patients are in certain respects trained, informed, and appropriately advised, and that workflows are updated as necessary.

4.5.3. Reflection

To aid the process of consolidating these outcomes, a further exercise was conducted in which an IT specialist and a practice manager were interviewed and encouraged to reflect further on the issues, on the nature and sources of their expectations surrounding blockchain technologies, and on underlying factors critical to success.

They emphasised that the majority of clinical practice managers typically undergo a profound process of culture change management. This process, which often occurs before, during, and after the implementation of new systems or practices, significantly shapes their comprehension of and approach to implementing novel technologies or changes. Their perspectives are formulated in the context of a variety of factors, including their hands-on professional experiences, continuous training and personal development, interactions with colleagues and industry experts, and adherence to regulatory standards and guidelines.

In the context of healthcare settings, IT professionals serve as a pivotal force in evaluating and deploying new technologies such as blockchain. These professionals are equipped with the technical expertise to decipher the complexities of blockchain technology, its potential applications, and the prerequisites for its successful implementation.

The successful integration of blockchain technology within healthcare systems is not a task that rests solely on the shoulders of IT professionals. It demands a concerted and cooperative effort from clinical practice managers, IT professionals, and other significant stakeholders. Clinical practice managers bring to the table indispensable insights into the functional needs of the clinic. Their understanding of day-to-day operations, patient interactions, and workflow dynamics is critical to shaping the implementation strategy.

On the other hand, IT professionals employ their specialist knowledge to architect the most effective technical solutions that meet these needs. Their expertise ensures that the blockchain technology is tailored to fit seamlessly within the existing system, enhancing efficiency without disrupting established workflows.

The professionals see this harmonious collaboration as the backbone of a successful implementation strategy. It ensures that blockchain technology is not just technically integrated, but also woven into the fabric of the clinic's operations in a manner that maximizes benefits, to serve the clinic and its patients optimally.

Overall, we seem to observe here that the complementary concerns of these professional groups emphasise the need, above all, for clear articulation and excellent communication between them. This will be especially critical as the implications of introducing a new technology become clearer. For example, it was remarked that the technology should enhance efficiency "without disrupting established workflows"; but, in general, it cannot be expected that established workflows will, or always should, remain unaffected. This may well at times be inconsistent with the goal of enhancing efficiency. Working out what kinds of changes will be inevitable, which should be embraced and which should be resisted, is a process that strongly depends on the collaboration, but may test the harmonious relationship quite severely.

5. Discussion

This study has explored the barriers to the use of blockchain-based infrastructures in healthcare, with particular emphasis on the delivery of patient-centric data management systems, which leads to a focus on issues such as data ownership and privacy. A sample of 30 professionals (healthcare and non-healthcare) were interviewed, and the resulting data was subjected to thematic analysis. In total, 12 specific themes relating to barriers were identified, categorised under four main themes: Administrative and Management, User Perspectives, Future Proofing, and Regulatory Issues. In light of these themes, a comparison was carried out between the perspectives of professionals from the IT domain, on the one hand, and clinical practice management on the other.

The results of this study identified two main themes which were of equal concern to the participants: administrative and management issues, and future-proofing issues (Table 2). All of the sub-themes that were identified within these main themes are common to any kind of information technology that might be used in this area, but assume particular forms and significance in relation to the (relatively) new and unproven nature of blockchain technology. This is a result of the fact that, while blockchain is widely recognised as having the vast potential to drive social and economic development, by revolutionising many business and public sector processes, it remains a developing technology and raises questions about many of these key issues, and in particular security risk, cost, scalability, stability and flexibility [43,67]. This study also found that issues connected to end-user perceptions and regulation were also a concern. For example, most participants felt that the public perception of data privacy was significantly different from the reality, and this gap would not be easy to eliminate. This finding reflects those of a number of other studies, such as in [68,69].

On some barriers, such as Knowledge and Skills Recruitment, there was particularly strong agreement. Such a consensus suggests that the practical issue of system development will be a significant challenge, separate from any political or doctrinal concerns. This is consistent with the findings of other studies and relevant skills and recruitment sources. Although a similar issue could arise in relation to other technologies, it is apparently perceived as especially serious for the case of blockchain technologies, due to the expected competition for skills from areas such as cryptocurrencies.

Another area in which the participants were in strong agreement was management commitment. A lack of commitment at senior levels will inevitably have a negative impact on the adoption levels of any technology [70,71], and this lack of support for blockchain technologies was highlighted by the majority of participants. This view is supported by some other sources [72], who argue that there are often powerful vested interests that oppose the wider developmental impact of blockchain technology and may do everything in their power to hinder its implementation.

However, the general evidence from the wider field of literature is not in total agreement with these observations. While there is some evidence that senior clinicians and other medical personnel are sceptical of the technology [73,74], there is a considerable amount of de facto evidence that senior tiers of management and policymakers are strongly in favour of adoption—this evidence lies in the fact that blockchain is being increasingly deployed in the healthcare sector [75,76]. However, this evidence from the literature applies only to the general field of healthcare, not to the specific use case of patient-centric data management systems. Further, all participants in the current study were from GCC countries, and cultural and political factors may have played a role in biasing the results. Further research is needed here.

Some sub-themes in the study showed a divergence of views. The barrier of interoperability, for example, was seen by all participants to be a major ICT development issue, but some participants saw it as insurmountable over the short and medium term, while others saw it as less critical, as there is evidence of viable solutions already emerging. The literature is inconclusive on this point. Interoperability is often difficult even in cases where there are mature standards, due to differences in implementation and the problems of updating as technologies and standards evolve; the present state of blockchain technology does not make this any easier. However, while there is a broad consensus that blockchain interoperability remains a key issue [77], there are mixed views on the degree to which it is a problem. There are now several examples of large-scale, real-world blockchain projects that depend critically on interoperability, such as LACChain, which is a pan-regional blockchain programme across 12 countries in Latin America and the Caribbean, which are designed to drive social and economic development, promote financial inclusion and sustainability, and create new efficiencies through digitisation [78]. In this case, interoperability of multiple blockchains has been provided through a solution from Quant Network—a solution that is 'blockchain agnostic' and can be applied to a wide range of industries, including healthcare. However, there is also an argument that true interoperability between blockchains is impossible [79] and that the solutions mentioned must therefore relax the formal definition of a blockchain in some way. It remains to be seen whether such a relaxation may be shown to be of particular concern in a healthcare data context.

Standardisation, which is a related but separate issue to interoperability, was also identified as a barrier by the participants, though several observed that various standards bodies are beginning to develop international frameworks. Nonetheless, the lack of standardisation remains an issue that inhibits the development of blockchain-based systems in healthcare, as well as other sectors. This is because effective healthcare, and particularly patient-centric data management, can only be delivered through the integration and data sharing of multiple different ecosystems and legacy systems, and this requires the development and implementation of middleware and interfaces [45,80,81]. This development process can only be effective if the blockchain platforms are developed to a common set of standards [82]. One can observe that organisations such as the Institute of Electrical and Electronics Engineers (IEEE) [83] and the European Commission [84] are working to develop and promote standards as quickly as possible, but their uptake and overall effectiveness are not yet clear.

Another issue that was identified as a barrier, but with different degrees of resolvability by the participants, was scalability. Although many participants saw scalability as a barrier to be addressed if blockchain is to fulfil its developmental potential, some pointed out that if a system is built around private blockchains, then it could be less of an issue [85]. Other interviewees, however, remarked that the use of private blockchains is more complex and expensive, and so may prove impractical in many contexts. Overall, the study confirmed the findings of other research that has looked at the issue of scalability in more general health (and other) contexts [86,87].

Because of the nature of the kind of development under discussion in this study (patient-centric data management), security is a particular concern. Nearly all of the participants felt that, despite the inherent security strengths of blockchain, this remained a serious issue. This reflects the findings of other studies, which have highlighted that blockchain infrastructures are vulnerable [88,89]. One participant mentioned the possibility of 51% of attacks, which are recognised as a significant risk [90]. However, we note that the nature of the risk is not necessarily clear. In practice, these attacks, along with Sybil attacks, seem to be a risk to the integrity of the organisation of the blockchain; they might allow arbitrary data to be added or removed, and thus completely disrupt the operation of systems. This may not, e.g., allow the attacker to read any personal patient-related information directly (since such information is usually stored off the blockchain), but may allow it to be accessed indirectly if the blockchain controls access to the patient data. We conclude that security, and in particular the perception of security and risk, is among the main concerns/barriers that need to be addressed.

Today, the issue of environmental sustainability forms a key part of the CSR (Corporate Social Responsibility) policies of nearly all organisations. It is therefore a critical consideration for healthcare authorities. However, most blockchains are energy intensive to run, which has implications for sustainability. In fact, it has been reported that in some sectors, such as education, lack of sustainability is the prime reason for low adoption rates of blockchain [42]. The results of this study indicate that it remains a key development issue in healthcare, although some participants were of the view that solutions were emerging. It should be noted that this study was confined to the implications of blockchain for environmental (as opposed to social and economic) sustainability—future studies on its implications for social and economic sustainability development would be useful.

In any context, the issues of regulatory compliance and governance are important. But they are particularly important in public service contexts such as healthcare. In general terms, this may be considered a complex, but uncontentious, issue. However, the relatively novel setting of blockchain applications presents significant challenges, as there is currently no 'settled law' of blockchain [12]. In other words, regulatory frameworks and national/international governance models are yet to be fully defined. This issue was stressed by several participants in the current study.

Overall, the result of this study suggests that the barriers to implementing a blockchain-based, patient-centric data management system fall into two categories: those that were supported by all participants as being a major challenge to the adoption of the technology, and those which, while still a significant challenge, are seen as possibly having viable, short-term solutions. For example, issues such as skills recruitment and management commitment were shown to be ongoing problems without an obvious or immediate resolution, while it is believed that interoperability and scalability can currently be addressed, at least to some extent. This has been demonstrated by use cases in a number of sectors, including healthcare [86,87].

Our comparison of perceptions among different groups suggests that their concerns tend to be complementary: one way of looking at this is that they worry about the things that fall into their own area of responsibility and can do something about them, while they leave other concerns to others. This is a rational approach, but without excellent lines of communication may carry the risk that issues are addressed without full consultation, leading to conflict, dissatisfaction, loss of trust, and potentially a fall in the level of acceptance of the whole technology strategy.

6. Conclusions

Today, improving healthcare for citizens is a key development policy objective of countries around the world. One route to achieving this, which is attracting attention from an increasing number of governments, is to change the data ownership model by transitioning to a patient-centric system. Such an approach can not only improve patient outcomes and satisfaction, but increase the efficiency of healthcare resources. However, the design and delivery of such systems is a complex issue. While a possible solution is to deploy blockchain technology, this is still in its relatively early stages of development. Its application can present significant challenges to ICT development and introduction—challenges that differ from use case to use case.

This study seeks to identify the key barriers to the acceptance of blockchain-based, patient-centric data management systems, as an important first step in achieving widespread adoption. Unlike the few existing studies on the topic, this research examines the perspectives of professionals (healthcare and non-healthcare) in order to identify and prioritise relevant challenges. Our approach addresses a research gap identified by existing reviews [21]. The results are expected to be of interest to healthcare bodies looking to improve the effectiveness and efficiency of their service delivery infrastructures.

This study identified 12 barriers, which were classified into 4 main themes (Administrative and Management, User Perspectives, Future Proofing, and Regulatory Issues). Of these main themes, two (Administrative and Management and Future Proofing) were of most concern to participants, though the sub-themes (barriers) within these main themes varied in their significance and potential impact on blockchain adoption. Some of the barriers were perceived by participants to have no short-term solution, while others, though still significant, had short-term workarounds or resolutions. Overall, the two most challenging barriers for the immediate term were found to be (a) skills recruitment and (b) management commitment. Any of these barriers will need to be addressed in an environment of full and open consultation across different stakeholder groups, to promote and enhance a "harmonious collaboration" that will be crucial to maintaining acceptance across the organisation.

These findings have some important theoretical and practical implications for GCC healthcare systems. GCC countries are, for example, actively working to improve their healthcare infrastructures, and the deployment of distributed ledger (blockchain) technology could play a key role in achieving objectives such as improved outcomes and system efficiency. However, the adoption of such technology by users is critical to success, so understanding barriers to acceptance is key. The identification of these barriers will not only help policymakers and healthcare providers transition to blockchain-based systems effectively, but will also help to ensure that new systems and processes are fully aligned with the needs and expectations of GCC populations.

We believe that our study will contribute significantly to the understanding of the issue of acceptance barriers and help pave the way for more effective implementation strategies for blockchain technology in the region's healthcare sector.

It is important to bear in mind that the study reported here was carried out in the GCC countries. These all have advanced healthcare systems, comparable with those found in other developed countries, but the cultural context varies in some respects, and this could affect the perceptions of participants. Other research has noted cultural differences in this region in, for example, attitudes to information privacy, data, and security, and in approaches to these issues among management and policy makers [91–93]. We believe that many of the findings we report would be similar in other advanced healthcare contexts. A fuller comparative study would be extremely valuable but, as yet, the literature does not offer comparable findings for other contexts. Our study is necessarily limited, but we hope to inspire and inform further research that will allow investigation of potential regional or cultural specificities in more detail.

Although the study sought to obtain generalisable conclusions through its methodology and sampling, it should be noted that the sample (interviewees) consisted of professionals from GCC countries, and therefore may have been affected by cultural issues. Further research with a broader sample would provide further insights. Our current study focused on qualitative analysis to explore the barriers to acceptance of blockchain-based patient-centric data management systems in healthcare, and we believe this is an essential precursor to potential further study using a more quantitative methodology. Statistical data needs to be acquired and interpreted in a context where, for example, questionnaires can be designed with a clear view of how participants will understand them. Well-informed quantitative research can then provide insights with potential value to explore comparative analysis across different societies and cultures.

Author Contributions: Conceptualization, I.M., J.L., A.A. and Z.H.A.; methodology, I.M., A.A. and Z.H.A.; validation, I.M., J.L., A.A. and Z.H.A.; formal analysis, I.M., J.L., A.A. and Z.H.A.; writing—original draft preparation, I.M., J.L., A.A. and Z.H.A.; writing—review and editing I.M., J.L., A.A. and Z.H.A. All authors have read and agreed to the published version of the manuscript.

Funding: This research was funded by the Researchers Supporting Project number (RSP2024R233), King Saud University, Riyadh, Saudi Arabia.

Institutional Review Board Statement: This study was conducted according to the guidelines of the Declaration of Helsinki, and approved by the Institutional Review Board (Human and Social Researches) of King Saud University (approval code KSU-18-242, date 27 November2022).

Informed Consent Statement: Informed consent was obtained from all subjects involved in the study.

Data Availability Statement: Data are available on request due to restrictions of privacy.

Acknowledgments: The authors are grateful for the facilities and other support given by the Researchers Supporting Project number (RSP2024R233), King Saud University, Riyadh, Saudi Arabia.

Conflicts of Interest: The authors declare no conflict of interest.

Appendix A

Interview Questions: As is conventional for semi-structured interviews, this study's interviews were designed with open questions to provide the researcher with the opportunity to explore particular themes or responses further. Much consideration was given to the question design, to ensure that they provided information that was relevant to the study's goals. The key guiding principles of interview design were as follows:

Consistency with Research Objectives: Every question was designed to help identify and understand the barriers to the acceptance of blockchain-based patient-centric data management systems.

Theoretical Relevance: All questions were designed to leverage the existing knowledge base (from the current literature) to ensure they were consistent with a solid theoretical framework built around general factors such as technology adoption and resistance to change, as well as specific factors such as data management using blockchain systems.

Expert Assessment: Before being used in interviews, all prototype questions were assessed by several experts with experience in blockchain technology, healthcare systems, and the cultural/regulatory landscape of the GCC. This was important in ensuring that all questions were clear, relevant, and likely to provide the richness of response required by the study's analysis.

Pilot Test: Before carrying out actual interviews for the study, several pilot interviews were conducted to test clarity and quality of response. Following these pilot interviews, a number of refinements were made, in order to remove potential ambiguity.

The researchers believe that the rigorous process described above resulted in a set of questions that yielded valuable contributions to answering the study's RQ.

The questions used in the interviews were as follows:

1. What are the key benefits of blockchain technology for healthcare?
2. How can blockchain be used to improve the security of patient data?
3. How can blockchain be used to improve the interoperability of patient data?
4. How can blockchain be used to improve the efficiency of patient care?
5. How can blockchain be used to improve the transparency of patient care?
6. How can blockchain be used to improve the accountability of healthcare providers?
7. What are some specific examples of how blockchain is being used in healthcare today?
8. What are the challenges of deploying blockchain applications in healthcare?
9. How do you think blockchain will impact the future of healthcare?
10. What is patient-centric data management?
11. What are the benefits of patient-centric data management?
12. How can policy be used to promote patient-centric data management?
13. What are some specific examples of how policy has been used to promote patient-centric data management?
14. Do you think patient-centric data management is a desirable policy objective? Please give reasons for your answer.
15. What are the challenges of implementing patient-centric data infrastructures within a healthcare department?
16. What are the benefits of implementing patient-centric data infrastructures within a healthcare department?
17. How would you go about implementing patient-centric data infrastructures within a healthcare department?

References

1. Chandra, A.; Staiger, D.O. Identifying Sources of Inefficiency in Healthcare. *Q. J. Econ.* **2020**, *135*, 785–843. [CrossRef]
2. Shrank, W.H.; Rogstad, T.L.; Parekh, N. Waste in the US Health Care System. *JAMA* **2019**, *322*, 1501. [CrossRef] [PubMed]
3. Peter, G. Peterson Foundation. Almost 25% of Healthcare Spending is Considered Wasteful. Here's Why. Available online: https://www.pgpf.org/blog/2023/04/almost-25-percent-of-healthcare-spending-is-considered-wasteful-heres-why (accessed on 12 September 2023).
4. Gavurova, B.; Kocisova, K.; Sopko, J. Health System Efficiency in OECD Countries: Dynamic Network DEA Approach. *Health Econ. Rev.* **2021**, *11*, 40. [CrossRef] [PubMed]
5. Ozcan, Y.A. *Health Care Benchmarking and Performance Evaluation*; Springer: Boston, MA, USA, 2008; Volume 120. [CrossRef]
6. Khan, D.; Jung, L.T.; Hashmani, M.A. Systematic Literature Review of Challenges in Blockchain Scalability. *Appl. Sci.* **2021**, *11*, 9372. [CrossRef]
7. Reis, Z.S.N.; Maia, T.A.; Marcolino, M.S.; Becerra-Posada, F.; Novillo-Ortiz, D.; Ribeiro, A.L.P. Is There Evidence of Cost Benefits of Electronic Medical Records, Standards, or Interoperability in Hospital Information Systems? Overview of Systematic Reviews. *JMIR Med. Inform.* **2017**, *5*, e26. [CrossRef] [PubMed]
8. Ma, Y.; Fang, Y. Current Status, Issues, and Challenges of Blockchain Applications in Education. *Int. J. Emerg. Technol. Learn. (iJET)* **2020**, *15*, 20. [CrossRef]
9. Zhang, S.; Lee, J.-H. Analysis of the Main Consensus Protocols of Blockchain. *ICT Express* **2020**, *6*, 93–97. [CrossRef]
10. Adere, E.M. Blockchain in Healthcare and IoT: A Systematic Literature Review. *Array* **2022**, *14*, 100139. [CrossRef]
11. Singh, D.; Monga, S.; Tanwar, S.; Hong, W.-C.; Sharma, R.; He, Y.-L. Adoption of Blockchain Technology in Healthcare: Challenges, Solutions, and Comparisons. *Appl. Sci.* **2023**, *13*, 2380. [CrossRef]
12. World Economic Forum. Is Blockchain the Solution for Failing Global Healthcare? Available online: https://www.weforum.org/agenda/2022/09/blockchain-solution-for-failing-global-healthcare/ (accessed on 2 September 2023).
13. Guru, A.; Mohanta, B.K.; Mohapatra, H.; Al-Turjman, F.; Altrjman, C.; Yadav, A. A Survey on Consensus Protocols and Attacks on Blockchain Technology. *Appl. Sci.* **2023**, *13*, 2604. [CrossRef]
14. Alanazi, R.; Bahari, G.; Alzahrani, Z.A.; Alhaidary, A.; Alharbi, K.; Albagawi, B.S.; Alanazi, N.H. Exploring the Factors behind Nurses' Decision to Leave Clinical Practice: Revealing Causes for Leaving and Approaches for Enhanced Retention. *Healthcare* **2023**, *11*, 3104. [CrossRef]
15. Alabdulatif, A.; Khalil, I.; Saidur Rahman, M. Security of Blockchain and AI-Empowered Smart Healthcare: Application-Based Analysis. *Appl. Sci.* **2022**, *12*, 11039. [CrossRef]
16. Schuetz, S.; Venkatesh, V. Blockchain, Adoption, and Financial Inclusion in India: Research Opportunities. *Int. J. Inf. Manag.* **2020**, *52*, 101936. [CrossRef]

17. Umrao, D.; Rakshe, D.S.; Prakash, A.J.; Korde, S.K.; Singh, D.P. A Comparative Analysis of the Growing Role of Blockchain Technology in the Healthcare Sector. In Proceedings of the 2022 2nd International Conference on Advance Computing and Innovative Technologies in Engineering (ICACITE), Greater Noida, India, 28–29 April 2022; IEEE: Piscataway, NJ, USA, 2022; pp. 1599–1603. [CrossRef]
18. Fernandez-Quilez, A. Deep Learning in Radiology: Ethics of Data and on the Value of Algorithm Transparency, Interpretability and Explainability. *AI Ethics* **2023**, *3*, 257–265. [CrossRef]
19. Gupta, S.; Rhyner, J. Mindful Application of Digitalization for Sustainable Development: The Digitainability Assessment Framework. *Sustainability* **2022**, *14*, 3114. [CrossRef]
20. Tandon, A.; Dhir, A.; Islam, A.K.M.N.; Mäntymäki, M. Blockchain in Healthcare: A Systematic Literature Review, Synthesizing Framework and Future Research Agenda. *Comput. Ind.* **2020**, *122*, 103290. [CrossRef]
21. Li, Y.; Marier-Bienvenue, T.; Perron-Brault, A.; Wang, X.; Paré, G. Blockchain Technology in Business Organizations: A Scoping Review. In Proceedings of the 51st Hawaii International Conference on System Sciences, Waikoloa Village, HI, USA, 2–6 January 2018. [CrossRef]
22. Khoja, T.; Rawaf, S.; Qidwai, W.; Rawaf, D.; Nanji, K.; Hamad, A. Health Care in Gulf Cooperation Council Countries: A Review of Challenges and Opportunities. *Cureus* **2017**, *9*, e1586. [CrossRef]
23. Arabic Casa. Similarities and Differences between the 6 GCC Countries. Arabic Casa. Available online: https://arabiccasa.com/similarities-and-differences-between-6-gcc-countries/ (accessed on 12 March 2023).
24. Yaqoob, I.; Salah, K.; Jayaraman, R.; Al-Hammadi, Y. Blockchain for Healthcare Data Management: Opportunities, Challenges, and Future Recommendations. *Neural Comput. Appl.* **2022**, *34*, 11475–11490. [CrossRef]
25. Attaran, M. Blockchain Technology in Healthcare: Challenges and Opportunities. *Int. J. Healthc. Manag.* **2022**, *15*, 70–83. [CrossRef]
26. Alsaed, Z.; Khweiled, R.; Hamad, M.; Daraghmi, E.; Cheikhrouhou, O.; Alhakami, W.; Hamam, H. Role of Blockchain Technology in Combating COVID-19 Crisis. *Appl. Sci.* **2021**, *11*, 12063. [CrossRef]
27. Bahbouh, N.; Basahel, A.; Sendra, S.; Abi Sen, A.A. Tokens Shuffling Approach for Privacy, Security, and Reliability in IoHT under a Pandemic. *Appl. Sci.* **2022**, *13*, 114. [CrossRef]
28. Rathore, A. Getting a Handle on Clinical Trial Costs. Available online: https://www.clinicalleader.com/doc/getting-a-handle-on-clinical-trial-costs-0001 (accessed on 23 March 2023).
29. Wu, T.-C.; Ho, C.-T.B. Blockchain Revolutionizing in Emergency Medicine: A Scoping Review of Patient Journey through the ED. *Healthcare* **2023**, *11*, 2497. [CrossRef] [PubMed]
30. Benchoufi, M.; Ravaud, P. Blockchain Technology for Improving Clinical Research Quality. *Trials* **2017**, *18*, 335. [CrossRef]
31. O'Hagan, A.; Garlington, A. Counterfeit Drugs and the Online Pharmaceutical Trade, a Threat to Public Safety. *Foresic. Res. Criminol. Int. J.* **2018**, *6*, 906–912. [CrossRef]
32. Ofori-Parku, S.S. Fighting the Global Counterfeit Medicines Challenge: A Consumer-Facing Communication Strategy in the US Is an Imperative. *J. Glob. Health* **2022**, *12*, 03018. [CrossRef]
33. World Health Organization. *Substandard and Falsified Medical Products*; World Health Organization: Geneva, Switzerland, 2018.
34. El-Dahiyat, F.; Fahelelbom, K.M.S.; Jairoun, A.A.; Al-Hemyari, S.S. Combatting Substandard and Falsified Medicines: Public Awareness and Identification of Counterfeit Medications. *Front. Public Health* **2021**, *9*, 754279. [CrossRef] [PubMed]
35. Glass, B. Counterfeit Drugs and Medical Devices in Developing Countries. *Res. Rep. Trop. Med.* **2014**, *11*, 11–22. [CrossRef] [PubMed]
36. Prause, G. Smart Contracts for Smart Supply Chains. *IFAC-PapersOnLine* **2019**, *52*, 2501–2506. [CrossRef]
37. Zhuang, Y.; Sheets, L.R.; Chen, Y.-W.; Shae, Z.-Y.; Tsai, J.J.P.; Shyu, C.-R. A Patient-Centric Health Information Exchange Framework Using Blockchain Technology. *IEEE J. Biomed Health Inform.* **2020**, *24*, 2169–2176. [CrossRef]
38. Tareen, F.N.; Alvi, A.N.; Malik, A.A.; Javed, M.A.; Khan, M.B.; Saudagar, A.K.J.; Alkhathami, M.; Abul Hasanat, M.H. Efficient Load Balancing for Blockchain-Based Healthcare System in Smart Cities. *Appl. Sci.* **2023**, *13*, 2411. [CrossRef]
39. Dalton, J.; Chambers, D.; Harden, M.; Street, A.; Parker, G.; Eastwood, A. Service User Engagement in Health Service Reconfiguration: A Rapid Evidence Synthesis. *J. Health Serv. Res. Policy* **2016**, *21*, 195–205. [CrossRef] [PubMed]
40. Kamišalić, A.; Turkanović, M.; Mrdović, S.; Heričko, M. A Preliminary Review of Blockchain-Based Solutions in Higher Education. In *Learning Technology for Education Challenges*; Springer: Cham, Switzerland, 2019; pp. 114–124. [CrossRef]
41. Khezr, S.; Moniruzzaman, M.; Yassine, A.; Benlamri, R. Blockchain Technology in Healthcare: A Comprehensive Review and Directions for Future Research. *Appl. Sci.* **2019**, *9*, 1736. [CrossRef]
42. Park, J. Promises and Challenges of Blockchain in Education. *Smart Learn. Environ.* **2021**, *8*, 33. [CrossRef]
43. Jabarulla, M.Y.; Lee, H.-N. Blockchain-Based Distributed Patient-Centric Image Management System. *Appl. Sci.* **2020**, *11*, 196. [CrossRef]
44. Caldarelli, G.; Ellul, J. Trusted Academic Transcripts on the Blockchain: A Systematic Literature Review. *Appl. Sci.* **2021**, *11*, 1842. [CrossRef]
45. Xi, P.; Zhang, X.; Wang, L.; Liu, W.; Peng, S. A Review of Blockchain-Based Secure Sharing of Healthcare Data. *Appl. Sci.* **2022**, *12*, 7912. [CrossRef]
46. Poquiz, W.A. Blockchain Technology in Healthcare: An Analysis of Strengths, Weaknesses, Opportunities, and Threats. *J. Healthc. Manag.* **2022**, *67*, 244–253. [CrossRef]

47. Elangovan, D.; Long, C.S.; Bakrin, F.S.; Tan, C.S.; Goh, K.W.; Yeoh, S.F.; Loy, M.J.; Hussain, Z.; Lee, K.S.; Idris, A.C.; et al. The Use of Blockchain Technology in the Health Care Sector: Systematic Review. *JMIR Med. Inform.* **2022**, *10*, e17278. [CrossRef]
48. Royle, J.; Jones, R. Patient Centric Healthcare—What's Stopping Us? In *Patient Centric Blood Sampling and Quantitative Bioanalysis*; Wiley: Hoboken, NJ, USA, 2023; pp. 1–15. [CrossRef]
49. El-Gazzar, R.; Stendal, K. Blockchain in Health Care: Hope or Hype? *J. Med. Internet Res.* **2020**, *22*, e17199. [CrossRef] [PubMed]
50. Arndt, T.; Guercio, A. Blockchain-Based Transcripts for Mobile Higher-Education. *Int. J. Inf. Educ. Technol.* **2020**, *10*, 84–89. [CrossRef]
51. Hidrogo, I.; Zambrano, D.; Hernandez-de-Menendez, M.; Morales-Menendez, R. Mostla for Engineering Education: Part 1 Initial Results. *Int. J. Interact. Des. Manuf. (IJIDeM)* **2020**, *14*, 1429–1441. [CrossRef]
52. Risius, M.; Spohrer, K. A Blockchain Research Framework. *Bus. Inf. Syst. Eng.* **2017**, *59*, 385–409. [CrossRef]
53. Lutfiani, N.; Aini, Q.; Rahardja, U.; Wijayanti, L.; Nabila, E.A.; Ali, M.I. Transformation of Blockchain and Opportunities for Education 4.0. *Int. J. Educ. Learn.* **2021**, *3*, 222–231. [CrossRef]
54. Lavorgna, L.; Russo, A.; De Stefano, M.; Lanzillo, R.; Esposito, S.; Moshtari, F.; Rullani, F.; Piscopo, K.; Buonanno, D.; Brescia Morra, V.; et al. Health-Related Coping and Social Interaction in People with Multiple Sclerosis Supported by a Social Network: Pilot Study with a New Methodological Approach. *Interact. J. Med. Res.* **2017**, *6*, e10. [CrossRef]
55. Mann, C.; Stewart, F. *Internet Communication and Qualitative Research: A Handbook for Researching Online*; Sage Publications Ltd.: London, UK, 2000.
56. Ritchie, J.; Lewis, J. *Qualitative Research Practice: A Guide for Social Science Students and Researchers*; Sage: London, UK, 2003. [CrossRef]
57. Strauss, A.L.; Corbin, J.M. *Basics of Qualitative Research, Techniques and Procedures for Grounded Theory*; Sage Publications Ltd.: Thousand Oaks, CA, USA, 1998.
58. Patton, M.Q. *Qualitative Research and Evaluation Methods.*; Sage: London, UK, 2002.
59. Hyde, K.F. Recognising Deductive Processes in Qualitative Research. *Qual. Mark. Res. Int. J.* **2000**, *3*, 82–90. [CrossRef]
60. King, N.; Horrocks, C. *Interviews in Qualitative Research*; Sage: Los Angeles, CA, USA, 2010.
61. Silverman, D. *Doing Qualitative Research*; Sage Publications Ltd.: London, UK, 2006.
62. Jamshed, S. Qualitative Research Method-Interviewing and Observation. *J. Basic Clin. Pharm.* **2014**, *5*, 87. [CrossRef]
63. Creswell, J.W.; Clark, V.L.P. *Designing and Conducting Mixed Methods Research*; Sage: London, UK, 2011.
64. Johnson, R.B.; Onwuegbuzie, A.J. Mixed Methods Research: A Research Paradigm Whose Time Has Come. *Educ. Res.* **2004**, *33*, 14–26. [CrossRef]
65. Gorard, S. *Quantitative Methods in Educational Research: The Role of Numbers Made*; Continuum: London, UK, 2001.
66. Braun, V.; Clarke, V. Using Thematic Analysis in Psychology. *Qual. Res. Psychol.* **2006**, *3*, 77–101. [CrossRef]
67. Levis, D.; Fontana, F.; Ughetto, E. A Look into the Future of Blockchain Technology. *PLoS ONE* **2021**, *16*, e0258995. [CrossRef]
68. Fatima, N.; Agarwal, P.; Sohail, S.S. Security and Privacy Issues of Blockchain Technology in Health Care—A Review. In *ICT Analysis and Applications*; Springer Nature: Singapore, 2022; pp. 193–201. [CrossRef]
69. Esmaeilzadeh, P.; Mirzaei, T. The Potential of Blockchain Technology for Health Information Exchange: Experimental Study from Patients' Perspectives. *J. Med. Internet Res.* **2019**, *21*, e14184. [CrossRef] [PubMed]
70. Farooque, M.; Jain, V.; Zhang, A.; Li, Z. Fuzzy DEMATEL Analysis of Barriers to Blockchain-Based Life Cycle Assessment in China. *Comput. Ind. Eng.* **2020**, *147*, 106684. [CrossRef]
71. Raimundo, R.; Rosário, A. Blockchain System in the Higher Education. *Eur. J. Investig. Health Psychol. Educ.* **2021**, *11*, 276–293. [CrossRef]
72. da Cunha, P.R.; Soja, P.; Themistocleous, M. Blockchain for Development: A Guiding Framework. *Inf. Technol. Dev.* **2021**, *27*, 417–438. [CrossRef]
73. Esmaeilzadeh, P. Benefits and Concerns Associated with Blockchain-Based Health Information Exchange (HIE): A Qualitative Study from Physicians' Perspectives. *BMC Med. Inform. Decis. Mak.* **2022**, *22*, 80. [CrossRef]
74. Hau, Y.S.; Lee, J.M.; Park, J.; Chang, M.C. Attitudes Toward Blockchain Technology in Managing Medical Information: Survey Study. *J. Med. Internet Res.* **2019**, *21*, e15870. [CrossRef] [PubMed]
75. Haleem, A.; Javaid, M.; Singh, R.P.; Suman, R.; Rab, S. Blockchain Technology Applications in Healthcare: An Overview. *Int. J. Intell. Netw.* **2021**, *2*, 130–139. [CrossRef]
76. Wang, Q.; Qin, S. A Hyperledger Fabric-Based System Framework for Healthcare Data Management. *Appl. Sci.* **2021**, *11*, 11693. [CrossRef]
77. Reegu, F.A.; Abas, H.; Hakami, Z.; Tiwari, S.; Akmam, R.; Muda, I.; Almashqbeh, H.A.; Jain, R. Systematic Assessment of the Interoperability Requirements and Challenges of Secure Blockchain-Based Electronic Health Records. *Secur. Commun. Netw.* **2022**, *2022*, 1–12. [CrossRef]
78. Allende, M. *LACChain Framework for Permissioned Public Blockchain Networks: From Blockchain Technology to Blockchain Networks*; Pardo, A., Da Silva, M., Eds.; Inter-American Development Bank: Washington, DC, USA, 2021. [CrossRef]
79. Lafourcade, P.; Lombard-Platet, M. About Blockchain Interoperability. *Inf. Process. Lett.* **2020**, *161*, 105976. [CrossRef]
80. Astill, J.; Dara, R.A.; Campbell, M.; Farber, J.M.; Fraser, E.D.G.; Sharif, S.; Yada, R.Y. Transparency in Food Supply Chains: A Review of Enabling Technology Solutions. *Trends Food Sci. Technol.* **2019**, *91*, 240–247. [CrossRef]

81. Abutaleb, R.A.; Alqahtany, S.S.; Syed, T.A. Integrity and Privacy-Aware, Patient-Centric Health Record Access Control Framework Using a Blockchain. *Appl. Sci.* **2023**, *13*, 1028. [CrossRef]
82. Chen, G.; Xu, B.; Lu, M.; Chen, N.-S. Exploring Blockchain Technology and Its Potential Applications for Education. *Smart Learn. Environ.* **2018**, *5*, 1. [CrossRef]
83. IEEE BLOCKCHAIN. Standards. Available online: https://innovate.ieee.org/subscriptions-for-ieee-standards-and-related-content/ (accessed on 6 December 2023).
84. European Commission. Urbanisation Worldwide. Available online: https://knowledge4policy.ec.europa.eu/foresight/topic/continuing-urbanisation/urbanisation-worldwide_en (accessed on 23 September 2023).
85. Hillman, V.; Ganesh, V. Kratos: A Secure, Authenticated and Publicly Verifiable System for Educational Data Using the Blockchain. In Proceedings of the 2019 IEEE International Conference on Big Data (Big Data), Los Angeles, CA, USA, 9–12 December 2019; IEEE: Piscataway, NJ, USA, 2019; pp. 5754–5762. [CrossRef]
86. Siyal, A.A.; Junejo, A.Z.; Zawish, M.; Ahmed, K.; Khalil, A.; Soursou, G. Applications of Blockchain Technology in Medicine and Healthcare: Challenges and Future Perspectives. *Cryptography* **2019**, *3*, 3. [CrossRef]
87. McGhin, T.; Choo, K.-K.R.; Liu, C.Z.; He, D. Blockchain in Healthcare Applications: Research Challenges and Opportunities. *J. Netw. Comput. Appl.* **2019**, *135*, 62–75. [CrossRef]
88. Guo, H.; Yu, X. A Survey on Blockchain Technology and Its Security. *Blockchain: Res. Appl.* **2022**, *3*, 100067. [CrossRef]
89. Cernian, A.; Tiganoaia, B.; Sacala, I.; Pavel, A.; Iftemi, A. PatientDataChain: A Blockchain-Based Approach to Integrate Personal Health Records. *Sensors* **2020**, *20*, 6538. [CrossRef] [PubMed]
90. Juricic, V.; Radosevic, M.; Fuzul, E. Creating Student's Profile Using Blockchain Technology. In Proceedings of the 2019 42nd International Convention on Information and Communication Technology, Electronics and Microelectronics (MIPRO), Opatija, Croatia, 20–24 May 2019; IEEE: Piscataway, NJ, USA, 2019; pp. 521–525. [CrossRef]
91. Benamati, J.H.; Ozdemir, Z.D.; Smith, H.J. Information Privacy, Cultural Values, and Regulatory Preferences. *J. Glob. Inf. Manag.* **2021**, *29*, 131–164. [CrossRef]
92. Nemati, H.; Wall, J.D.; Chow, A. Privacy Coping and Information-Sharing Behaviors in Social Media: A Comparison of Chinese and U.S. Users. *J. Glob. Inf. Technol. Manag.* **2014**, *17*, 228–249. [CrossRef]
93. Milberg, S.J.; Smith, H.J.; Burke, S.J. Information Privacy: Corporate Management and National Regulation. *Organ. Sci.* **2000**, *11*, 35–57. [CrossRef]

Disclaimer/Publisher's Note: The statements, opinions and data contained in all publications are solely those of the individual author(s) and contributor(s) and not of MDPI and/or the editor(s). MDPI and/or the editor(s) disclaim responsibility for any injury to people or property resulting from any ideas, methods, instructions or products referred to in the content.

Review

A Systematic Literature Review of Health Information Systems for Healthcare

Ayogeboh Epizitone [1,*], Smangele Pretty Moyane [2] and Israel Edem Agbehadji [3]

1. ICT and Society Research Group, Durban University of Technology, Durban 4001, South Africa
2. Department of Information and Corporate Management, Durban University of Technology, Durban 4001, South Africa
3. Centre for Transformative Agricultural and Food Systems, School of Agricultural, Earth and Environmental Sciences, University of KwaZulu-Natal, Pietermaritzburg 3209, South Africa
* Correspondence: ayogebohe@dut.ac.za; Tel.: +27-(0)73-310-9150

Abstract: Health information system deployment has been driven by the transformation and digitalization currently confronting healthcare. The need and potential of these systems within healthcare have been tremendously driven by the global instability that has affected several interrelated sectors. Accordingly, many research studies have reported on the inadequacies of these systems within the healthcare arena, which have distorted their potential and offerings to revolutionize healthcare. Thus, through a comprehensive review of the extant literature, this study presents a critique of the health information system for healthcare to supplement the gap created as a result of the lack of an in-depth outlook of the current health information system from a holistic slant. From the studies, the health information system was ascertained to be crucial and fundament in the drive of information and knowledge management for healthcare. Additionally, it was asserted to have transformed and shaped healthcare from its conception despite its flaws. Moreover, research has envisioned that the appraisal of the current health information system would influence its adoption and solidify its enactment within the global healthcare space, which is highly demanded.

Keywords: health information system; information system; knowledge management; healthcare

Citation: Epizitone, A.; Moyane, S.P.; Agbehadji, I.E. A Systematic Literature Review of Health Information Systems for Healthcare. *Healthcare* 2023, *11*, 959. https://doi.org/10.3390/healthcare11070959

Academic Editors: Giner Alor-Hernández, Jezreel Mejía-Miranda, José Luis Sánchez-Cervantes and Alejandro Rodríguez-González

Received: 27 February 2023
Revised: 20 March 2023
Accepted: 25 March 2023
Published: 27 March 2023

Copyright: © 2023 by the authors. Licensee MDPI, Basel, Switzerland. This article is an open access article distributed under the terms and conditions of the Creative Commons Attribution (CC BY) license (https://creativecommons.org/licenses/by/4.0/).

1. Introduction

Health information systems (HIS) are critical systems deployed to help organizations and all stakeholders within the healthcare arena eradicate disjointed information and modernize health processes by integrating different health functions and departments across the healthcare arena for better healthcare delivery [1–6]. Over time, the HIS has transformed significantly amidst several players such as political, economic, socio-technical, and technological actors that influence the ability to afford quality healthcare services [7]. The unification of health-related processes and information systems in the healthcare arena has been realized by HIS. HIS has often been contextualized as a system that improves healthcare services' quality by supporting management and operation processes to afford vital information and a unified process, technology, and people [7,8]. Several authors assert this disposition of HIS, alluding to its remarkable capabilities in affording seamless healthcare [9]. Haux [10] modestly chronicled HIS as a system that handles data to convey knowledge and insights in the healthcare environment. Almunawar and Anshari [7] incorporated this construed method to describe HIS to be any system within the healthcare arena that processes data and affords information and knowledge. Malaquias and Filho [11] accentuated the importance of HIS in the same light, highlighting its emergence to tackle the need to store, process, and extract information from the system data for the optimization of processes, enhancing services provided and supporting decision making.

HIS's definition was popularized by Lippeveld [12], and reported to be an "integrated effort to collect, process, report and use health information and knowledge to influence policy-making, programme action and research". Over the course of time, this definition has been adopted and contextualized countlessly by many authors and the World Health Organization (WHO) [3,8,13–15]. Although Haule, Muhanga [8] claimed the definition of HIS varies globally, in actuality, the definition has never changed from its inception, but on the contrary, it has been conceptualized over various contexts. Malaquias and Filho [11] reiterated this definition in the extant literature. These scholars affirmed HIS as "a set of interrelated components that collect, process, store and distribute information to support the decision-making process and assist in the control of health organizations" [11]. The same definition is adopted in this paper, and HIS is construed as "a system of interrelated constituents that collect, process, store and distribute data and information to support the decision-making process, assist in the control of health organizations and enhance healthcare applications". However, it is paramount to note that HIS is broad. In many instances, the definition is of minimal relevance due to its associated incorporation with external applications related to health developments and policy making [16]. Hence, emphasis should not be placed on the definition but on its contribution to all facets of health development.

The current state of HIS is considered to be inadequate despite its numerus deployment of HIS that has been driven by its potential benefit to uplift healthcare and revolutionize its processes [17,18]. The persistence of many constraints and resistance to technology has resulted to the incapacitation of HIS in the attainment of its objectives. The extant literature reveals several challenges in different categories, such as the inadequacy of human resources and technological convergence within the healthcare [18], highlighting the evidence of limitations of HIS that restrict their utilization and deployment within the healthcare. Although several authors identified the unique disposition of HIS in integrating care and unifying the health process, these perspectives seems to be marred by the presence of barriers [17,19]. Garcia, De la Vega [17] alleged that the current HIS deployment is characterized by fragmentation, update instability, and lack of standardization that limit its potential to aid healthcare. Congruently, several authors associated the lack of awareness of HIS potential, the underuse HIS, inadequate communication network, and security and confidentiality concerns among the barriers limiting HIS [20]. Thus, the need for this paper is set forth: to uncover current and pertinent insights on HIS deployment as a concerted effort to strengthen it and augment its healthcare delivery capabilities. This paper comprehensively explores the extant literature systematically with respect to the overarching objective: to ascertain value insights pertaining to HIS holistically from literature synthesis. To achieve this goal, the following research questions are investigated: What has been the development of the HIS since its conception? How has HIS been deployed? Finally, how does HIS enable information and knowledge management in healthcare?

In this paper, an overview HIS from the extant literature in relation to the health sector is presented with associated related work. It is essential to point out that in spite of the surplus of research work conducted on health information systems, there are still many challenges confronting it within the healthcare area that necessitate the need for this study [5]. Therefore, the extant literature is explored in this paper systematically to uncover current and pertinent insights surrounding the deployment of the HIS, an integrated information system (IS) for healthcare. This paper is structured into five sections. The paper commences with an introductory background that presents the contextualization of HIS for healthcare, followed by a methodology that details the method and material used in this study. The next section, which is the discussion, presents the discourse of HIS evolution that highlights its progress to date, its structural deployment, and the information system and knowledge management within the healthcare arena as mediated by HIS. The last part of this study focuses on the conclusion that summarizes the discussion presented in this paper.

2. Material and Method

In this paper, a systematic review is conducted to synthesize the extant literature and analyze the content to ascertain the value disposition of HIS in relation to healthcare delivery. Preceding this review, the used of search engines was employed to retrieve related research publications that fit the study scope and contexts. The main database used was the *Web of Science*. Other databases such as *SCOPUS* and *Google Scholar* were also used to obtain additional relevant work associated with the context. For inclusion criteria, only articles containing references to the keywords HIS, information, healthcare, and related healthcare systems were analyzed scrupulously. Research work that did not have these references, did not constitute a journal or conference-proceeding work, and were not written in the English language were excluded. Figure 1, the PRISMA flow statement, illustrates the methodological phases of this research along with the exclusion and inclusion criteria that were implemented for the study synthesis.

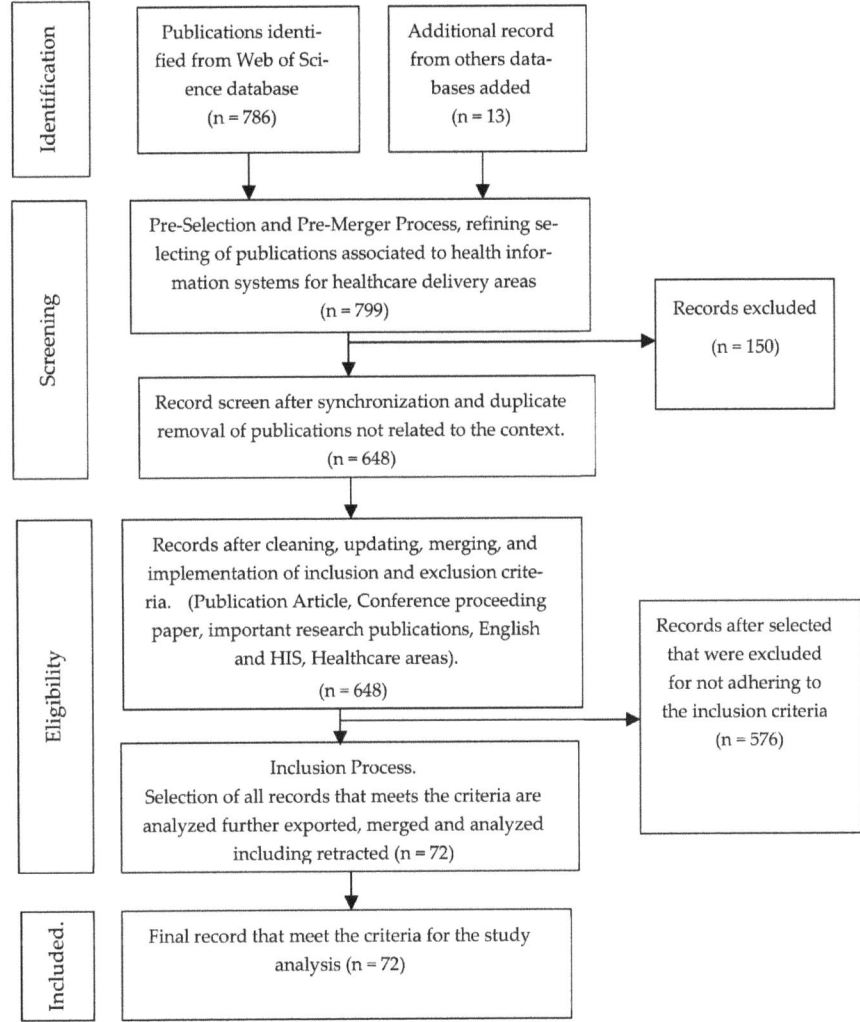

Figure 1. Prisma flow Statement.

3. Discussion

3.1. The Evolution of Health Information Systems

The concept of enhancing healthcare applications has always been the foundation of HIS, which posits that the intercession of information systems with business processes affords better healthcare services [7,21]. According to Almunawar and Anshari [7], many determinants, such as technological, political, social and economic, have enormously influenced the nature of the healthcare industry. The technological determinant, particularly the computerized component, is thought to be deeply ingrained in the enactment and functioning of HIS. According to Panerai [16], this single attribute can be held solely responsible for HIS letdowns rather than its accomplishment.

The ownership of HIS has been contested in the literature, with some authors claiming that HIS belongs to the IT industries [22]. While IT has enabled many developments in various industries, it has also resulted in many dissatisfactions. Recently, there has been an insurgence from many industries, particularly the healthcare industries, who acknowledge the role of IT in optimizing and enhancing health initiatives but want appropriation of their integrated IS. However, according to the definition of HIS, it is presented as "a set of interconnected components that collect, process, store, and distribute information to support decision-making and aid in the control of health organizations"; thus, the disposition of HIS was established. Without bias, the development of HIS was conceived due to unavoidable changes and transformations within the global space.

A good representation and consolidation of this dispute are within the realization that there is a co-existence of different related and non-related components in a system. In this case, the HIS is an entrenched system with several features, including technologies. Panerai [16] supported this notion and theorized HIS to be broad, stating that the relevance of its definition is contextual. In the study, HIS was reiterated as any kind of "structured repository of data, information, or knowledge" that can be used to support health care delivery or promote health development [16]. Thus, maintaining a rigid definition is of minimal practical use because many HIS instances are not directly associated with health development, such as the financial and human resource modules. Moreover, several different HIS examples are categorized according to the functions they are dedicated to serving within the healthcare arena. They highlight the instances of the existence of outliers that are not regarded as the normal HIS even though they contain health determinants data, such as socioeconomic and environmental, which can be used to formulate health policies.

The development of HIS over the years has led many to believe they are solely computer technology. This notion has contributed dramatically to the misconception of the origin of HIS and the lack of peculiarity between the HIS conceptual structure and implemented HIS technology. The literature dates back the origin of HIS, which can be associated with the first record of mortality in the 18th century, revealing their existence to be 200 years or older than the invention of computers [16]. This demonstrates the emergence of digitalized HIS from the availability of commercialized episodes of "electronic medical records" EMR records in the 1970s [23]. Namageyo-Funa, Aketch [24] commended the advancement of technologies in the healthcare arena, recounting the implementation of digitalized HIS that significantly revolutionized the recording and accessing of health information. A study by Lindberg, Venkateswaran [25] highlighted an instance of HIS transition from paper based to digitally based, revealing a streamlined workflow that revolutionized health care applications in the healthcare arena. This HIS transition over the course of time has led to increased adoption of it within the health care arena. Tummers, Tekinerdogan [26] highlighted the landmark of HIS from its transition to digitalization and reported a current trend in healthcare that has now been extended with the inclusion of block chain technology within the healthcare arena. Malik, Kazi [27] assessed HIS adoption in terms of technological, organizational, human, and environmental determinants and reported a variation of different degrees of utilization. Despite these facts, the extant literature maintains the need for a resilient and sustainable HIS for health care applications within the healthcare arena at all levels [18,27,28].

Figure 2 illustrates the successful adoption of HIS amidst the significant determinants of its effectiveness. From the Figure 2, the technological, organizational, human, and environmental determinants are the defining concepts along with individual sub-determinants in each domain that influence HIS adoption. At the technological level, the need for digitalization drives HIS adoption, especially for stakeholders such as clinicians and decision makers. The administrative, management, and planning functions are the driving actors within the organization level that endorse the implementation of HIS. The environmental and human determinants are more concerned with the socio-technical components that have been regarded as complex drivers for HIS adoptions. Perceptions, literacy, and usability are known forces within these categories that necessitate the adoption of HIS in many healthcare arenas.

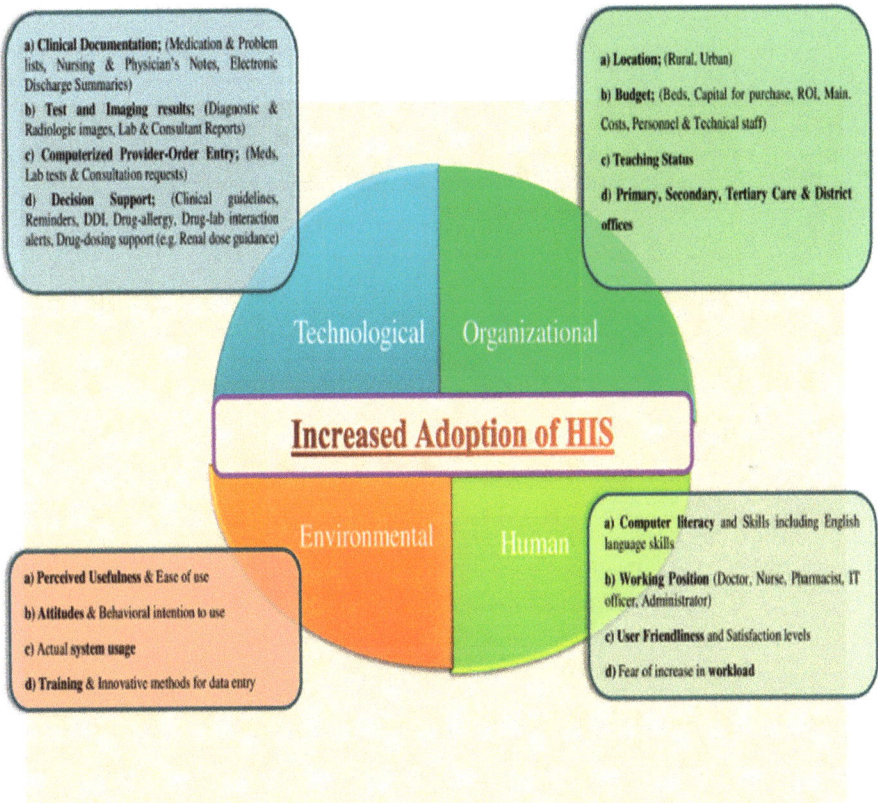

Figure 2. Effective health information system associations with the driving adoption determinants. Source: [27].

3.2. HIS Structural Deployment

HIS's unified front is geared toward assimilating and disseminating health gen to enhance healthcare delivery. HIS consists of different sub-systems that serve several actors within the healthcare arena [29]. These sub-systems are dedicated to specific tasks that perform various functions such as civil registrations, disease surveillance, outbreak notices, interventions, and health information sharing within the healthcare arena. It also supports and links many functions and activities within the healthcare environment, such as recording various data and information for stakeholders, scheduling, billing, and managing. Stakeholders are furnished with health information from diverse HIS scenarios.

These include but are not limited to information systems for hospitals and patients, health institution systems, and Internet information systems. Sligo, Gauld [30] regarded HIS as a panacea within the healthcare ground that improves health care applications. Despite all the limitless capabilities of HIS, it has been reported to be asymmetrical, lacking interactions within subsystems [1,18]. Many decision making methods and policies rely on good health information [31]. According to Suresh and Singh [32], the HIS enables stakeholders such as the government and all other players in the healthcare arena to have access to health information, which influences the delivery of healthcare. The sundry literature further reveals accurate health information to be the foundation of decision making and highlights the decisive role of the human constituent [29,31,33,34].

Furthermore, HIS can be classified into two cogs in today's era: the computer-related constituent that employs ICT-related tools and the non-computer component, which both operate at different levels. These levels include strategic, tactical, and operational. The deployment of HIS at the strategic level offers intelligence functions such as intelligent decision support, financial estimation, performance assessment, and simulation systems [3,35]. At the tactical level, managerial functions are performed within the system, while at the operational level, functions including recording, invoicing, scheduling, administrative, procurement, automation, and even payroll are carried out. Figure 3 shows the three levels within the healthcare system where HIS deployment is utilized.

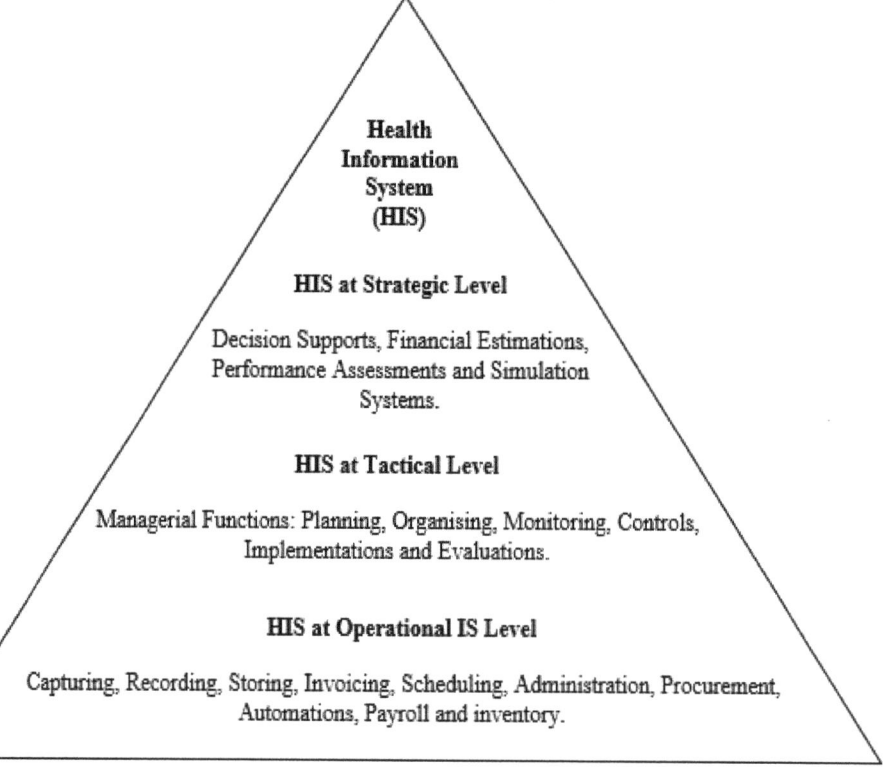

Figure 3. Levels of HIS deployment: source authors.

3.3. Health Information Systems Benefits

HIS, as an interrelated system, houses several core processes and branches in the healthcare arena, affording many benefits. Among these are the ease of access to patients and medical records, reduction of costs and time, and evidence-based health policies and interventions [8,21,36–38]. Several authors revealed the benefits of HIS to be widely known and influential within the healthcare domain [38]. Furthermore, many health organizations are drawn to HIS because of these numerous advantages [22,39]. Moreover, investment in HIS has enabled effective decision making, real-time comprehensive health information for quality health care applications, effective policies in the healthcare arena, scaled-up monitoring and evaluation, health innovations, resource allocations, surveillance services, and enhanced governance and accountability [36,40–42]. Ideally, HIS is pertinent for data, information, and broad knowledge sharing in the healthcare environment. HIS critical features are now cherished due to their incorporation with diverse technology [16,43]. The extant literature reveals the role of HIS to extend beyond its reimbursement. Table 1 presents a summarized extract of various HIS benefits as captured in the literature and some of its core enabling components or instances.

Table 1. HIS core enabling components and its benefits.

Source: Authors	Core Enabling HIS Components	Benefits
Malaquias and Filho [11]	Health ER eHealth mHealth	Ease of access to patient and medical information from records; Cost reduction; Enhance efficiency in patients' data recovery and management; Enable stakeholders' health information centralization and remote access.
Ammenwerth, Duftschmid [44]	eHealth	Upsurge in care efficacy and quality and condensed costs for clinical services; Lessen the health care system's administrative costs; Facilitates novel models of health care delivery.
Tummers, Tobi [45]	HIS	Patient information management; Enable communication within the healthcare arena; Afford high-quality and efficient care.
Steil, Finas [46]	HIS	Enable inter- and multidisciplinary collaboration between humans and machines; Afford autonomous and intelligent decision capabilities for health care applications.
Nyangena, Rajgopal [43]	HIS	Enable seamless information exchange within the healthcare arena.
Sik, Aydinoglu [47]	HIS	Support precision medicine approaches and decision support.

3.4. Information System and Knowledge Management in the Healthcare Arena

The presence of modernized information systems (IS) in the healthcare arena is alleged by scholars to be a congested domain that seldom fosters stakeholders' multifaceted and disputed relationships [48]. On the other hand, it is believed that a significant amount of newly acquired knowledge in the field of healthcare is required for the improvement of health care [49]. Ascertaining and establishing the role of IS and knowledge management is an important step in the development of HIS for healthcare. Flora, Margaret [5] posited that efficient IS and data usage are crucial for an effective healthcare system. Bernardi [50] alleged that the underpinning inkling of a "robust and efficient" HIS enables healthcare stakeholders such as managers and providers to leverage health information to commendably plan and regulate healthcare, which could result in enhanced survival rates. As a result, it is imperative to ground these ideas within the context of the healthcare industry to provide a foundation for developing a robust and sustainable HIS for use in the context of health care applications.

3.4.1. Information System

The assimilation and dissimilation of health information and data within the healthcare system is an important task that influences healthcare outcome. Within the healthcare setting, IS plays a significant role in the assimilation and dissimilation of health information needed by healthcare stakeholders. Many continents endorse the deployment of IS mainly to consolidate mutable information from different sources within the systems. The primary objective for these systems' deployment has been centered on bringing together unique and different components such as institutions, people, processes, and technology in the system under one umbrella [5,51]. An overview of the extant literature reveals that this has rarely been easy, as integration within this system has always been difficult in many contexts. In the context of HIS, many reported the integration phenomena to be problematic, attributing this to the global transformation within the healthcare arena [52,53]. This revolution, coupled with the advancement of the healthcare arena, has resulted in the need for robust allied health IS systems that incorporates different IS and information technology [5,22]. These allied health information systems are necessary to consolidate independent information systems within their healthcare arena use to enhance healthcare applications [54,55]. Organizations in the healthcare arena expect these systems to be sustainable and resilient; however, in order to satisfy these requirements, an integrated information system is needed to unify all independent, agile, and flexible health IS to mitigate challenges for HIS [56].

An aligned HIS that is allied is essential, as it supports health information networks (HIN) that subsequently enhance and improve healthcare applications [44,57]. Thus, many organizations within the healthcare settings are fine-tuning their HIS to be resilient and sustainable. However, the realization of a robust information system within the healthcare arena is challenging and depends on the flow of information as a crucial constituent for suave and efficient functioning [58,59].

3.4.2. Knowledge Management

The process of constructing value and generating a maintainable edge for an industry with capitalization on building, communicating, and knowledge applications procedures to realize set aspirations is denoted as knowledge management [60]. The literature reveals knowledge management as an important contributor to organizational performance through its knowledge-sharing capabilities [61]. In the healthcare industry, there is a high demand for knowledge to enhance healthcare applications [49,62]. Several studies reported that the deployment of knowledge management in the healthcare arena is set to enhance healthcare treatment effectiveness [49,58,61]. Many stakeholders such as governments, World Health Organization (WHO), and healthcare workers rely on the management of healthcare knowledge to complement healthcare applications. According to Kim, Newby-Bennett [61], the focus of knowledge management is to efficaciously expedite knowledge sharing. However, integrating knowledge from different sources is challenging and requires an enabler [61].

The HIS is an indispensable enabler of health knowledge generated from amalgamated health information within the healthcare arena [63–65]. Dixon, McGowan [66] asserted that efficacious modifications in the healthcare arena are made possible by knowledge codification and collaboration from information technologies. Similarly, some authors have pinpointed information and communication technologies within the healthcare arena to be a major determinant in the attainment of a sustainable health system development [58]. The knowledge management relationship with HIS is considered complementary and balanced, as it enables the availability of knowledge that can be shared. The importance of knowledge management is relevant for the realization of an enhanced healthcare application via HIS. Soltysik-Piorunkiewicz and Morawiec [58] claimed that the information society effectively uses HIS as an information system for management, patient knowledge, health knowledge, healthcare unit knowledge, and drug knowledge. The authors herein demonstrated how

HIS facilitates knowledge management in the healthcare sector to improve healthcare applications.

The role of HIS as an integrated IS and key enabler of healthcare knowledge management highlights its potential within the healthcare arena. From the conception of HIS and the records of its evolution, significant achievements have been attained that are demonstrated at different levels of its structural deployment. HIS deployment in several settings of healthcare have positively influenced clinical processes and patients' outcomes [17]. Globally, the need for HIS within the healthcare system is critical in the enhancement of healthcare. Many healthcare actions are dependent on the use of HIS [67–69]. This demand is substantiated by the offerings of HIS in tackling the transformation and digitalization confronting the healthcare system. However, despite the need for HIS and its potential within healthcare, several barriers limit its optimization. Some authors posited the role and involvement of healthcare professionals such as physicians to be important measure that is paramount to decreasing the technical and personal barriers sabotaging HIS deployment [20]. Nonetheless, the design of HIS is accentuated on augmenting health and is considered to be lagging behind in attaining quality healthcare [70].

Although there are equal blessings as well as challenges with HIS deployment, this study appraisal of HIS highlights its capabilities and attributes that enhance healthcare in many ways. From its conception, HIS has evolved significantly to enable the digitalization of many healthcare processes. Its deployment structurally has facilitated many healthcare applications at all levels within the health system where it has been implemented. Many benefits such as ease of access to medical records, cost reduction, data and information management, precision medicine, and autonomous and intelligent decisions have been enabled by HIS deployment. Primarily, HIS is the core enabler of the healthcare information system and knowledge management within the healthcare arena. Ascertaining the attributes and development of HIS is a paramount to driving its implementation and realizing its potential. Many deployments of HIS can be anchored on this study as a reference for planning and executing HIS implementation. The extant literature points out the need for the role of technology such HIS to be ascertained, as little is known in this regard, which as a result has adversely influenced healthcare coordination [19]. Additionally, among the barriers of HIS, the presence of inadequate planning that fails to cater to the needs of those adopting it hinders the optimization of these systems within the healthcare arena [71]. Cawthon, Mion [72] associated the lack of health literacy incorporation in deployed HIS to increased cost and poorer health outcomes. Hence, the insight from this study can be incorporated and associated with HIS initiatives to mitigate these issues. Thus, the findings of this study can be employed to strategize HIS deployment and plans as well as augment its potential to enhance healthcare. Furthermore, the competency of healthcare stakeholders such as patients can be enhanced with the findings of this study that accentuate the holistic representation of HIS in the dissimilation and assimilation of health data and information.

4. Conclusions

In the healthcare information and knowledge arena, assimilation and dissemination is a facet that influences healthcare delivery. The conception and evolution of HIS has positioned this system within the healthcare arena to arbitrate information interchange for its stakeholders. HIS deployment within healthcare has not only enabled information and knowledge management, but it has also enabled and driven many healthcare agendas and continues to maintain a solidified presence within the healthcare space. However, its deployment and enactment globally has been marred and plagued with several challenges that hinder its optimization and defeat its purpose. Phenomena such as the occurrences of pandemics such as COVID-19, which are uncertain, and the advancement of technology that cannot be controlled have caused disputed gradients regarding the positioning of HIS. These phenomena have not only influenced the adoption of HIS but have also limited its ability to be fully utilized. Although much research on HIS has been conducted, the presence of these phenomena and many other inherent challenges such as fragmentation and cost still maintain a constant, prominent presence, which has led to the need for this study.

Consequently, the starting point for this study was to provide insight and expertise regarding the discourse of HIS for healthcare applications. This paper presents current and pertinent insights regarding the deployment of the HIS that, when adopted, can positively aid its employment. This paper investigated the existing HIS literature to accomplish the objective set forth in the introduction. This study's synthesis derived key insights relevant to the holistic view of HIS through a thorough systematic review of the various extant literature on HIS and healthcare. According to the study's findings, HIS are critical and foundational in the drive of information and knowledge management for healthcare. The contribution of HIS to healthcare has been and continues to be groundbreaking since its conception and through its consequent evolution. Nevertheless, despite the presence of some limitations that are external and inherent, it is claimed to have transformed and changed healthcare from the start. Similarly, the evaluation of the current HIS is expected to impact its adoption and strengthen its implementation within the global healthcare space, which is greatly desired. These findings are of great importance to the healthcare stakeholders that directly and indirect interact with HIS. Additionally, scholars and healthcare researchers can benefit from this study by incorporating the findings in future works that plan HIS for healthcare.

Author Contributions: Conceptualization, A.E.; methodology, A.E.; software, A.E.; validation, A.E.; formal analysis, A.E.; investigation, A.E.; resources, A.E.; data curation, A.E.; writing—original draft preparation, A.E.; writing—review and editing, A.E.; visualization, A.E.; supervision, S.P.M. and I.E.A.; project administration, A.E., S.P.M. and I.E.A.; funding acquisition, A.E., S.P.M. and I.E.A. All authors have read and agreed to the published version of the manuscript.

Funding: This research received no external funding.

Institutional Review Board Statement: Not applicable.

Informed Consent Statement: Not applicable.

Data Availability Statement: Not applicable.

Conflicts of Interest: The authors declare there are no conflict of interest.

References

1. Sahay, S.; Nielsen, P.; Latifov, M. Grand challenges of public health: How can health information systems support facing them? *Health Policy Technol.* **2018**, *7*, 81–87. [CrossRef]
2. English, R.; Masilela, T.; Barron, P.; Schonfeldt, A. Health information systems in South Africa. *S. Afr. Health Rev.* **2011**, *2011*, 81–89.
3. Bagayoko, C.O.; Tchuente, J.; Traoré, D.; Moukoumbi Lipenguet, G.; Ondzigue Mbenga, R.; Koumamba, A.P.; Ondjani, M.C.; Ndjeli, O.L.; Gagnon, M.P. Implementation of a national electronic health information system in Gabon: A survey of healthcare providers' perceptions. *BMC Med. Inform. Decis. Mak.* **2020**, *20*, 202. [CrossRef] [PubMed]
4. Berrueta, M.; Bardach, A.; Ciapponi, A.; Xiong, X.; Stergachis, A.; Zaraa, S.; Buekens, P. Maternal and neonatal data collection systems in low- and middle-income countries: Scoping review protocol. *Gates Open Res.* **2020**, *4*, 18. [CrossRef] [PubMed]
5. Flora, O.C.; Margaret, K.; Dan, K. Perspectives on utilization of community based health information systems in Western Kenya. *Pan Afr. Med. J.* **2017**, *27*, 180. [CrossRef]
6. Rachmani, E.; Lin, M.C.; Hsu, C.Y.; Jumanto, J.; Iqbal, U.; Shidik, G.F.; Noersasongko, E. The implementation of an integrated e-leprosy framework in a leprosy control program at primary health care centers in Indonesia. *Int. J. Med. Inform.* **2020**, *140*, 104155. [CrossRef]
7. Almunawar, M.N.; Anshari, M. Health information systems (HIS): Concept and technology. *arXiv* **2012**, arXiv:1203.3923.
8. Haule, C.D.; Muhanga, M.; Ngowi, E. The what, why, and how of health information systems: A systematic review. *Sub Sahar. J. Soc. Sci. Humanit.* **2022**, *1*, 37–43. Available online: http://41.73.194.142/bitstream/handle/123456789/4398/Paper%205.pdf?sequence=1&isAllowed=y (accessed on 1 February 2023).
9. Epizitone, A.; Moyane, S.P.; Agbehadji, I.E. Health Information System and Health Care Applications Performance in the Healthcare Arena: A Bibliometric Analysis. *Healthcare* **2022**, *10*, 2273. [CrossRef]
10. Haux, R. Health information systems–past, present, future. *Int. J. Med. Inform.* **2006**, *75*, 268–281. [CrossRef]

11. Malaquias, R.S.; Filho, I.M.B. Middleware for Healthcare Systems: A Systematic Mapping. In Proceedings of the 21st International Conference on Computational Science and Its Applications, ICCSA 2021, Cagliari, Italy, 13–16 September 2021; Gervasi, O., Murgante, B., Misra, S., Garau, C., Blecic, I., Taniar, D., Apduhan, B.O., Rocha, A.M., Tarantino, E., Torre, C.M., Eds.; Springer Science and Business Media Deutschland GmbH: Cham, Switzerland, 2021; Volume 12957, pp. 394–409. [CrossRef]
12. Lippeveld, T. Routine health information systems: The glue of a unified health system. In Proceedings of the Keynote address at the Workshop on Issues and Innovation in Routine Health Information in Developing Countries, Potomac, MD, USA, 14–16 March 2001.
13. AbouZahr, C.; Boerma, T. Health information systems: The foundations of public health. *Bull. World Health Organ.* **2005**, *83*, 578–583.
14. Bogaert, P.; Van Oyen, H. An integrated and sustainable EU health information system: National public health institutes' needs and possible benefits. *Arch. Public Health* **2017**, *75*, 3. [CrossRef]
15. Bogaert, P.; van Oers, H.; Van Oyen, H. Towards a sustainable EU health information system infrastructure: A consensus driven approach. *Health Policy* **2018**, *122*, 1340–1347. [CrossRef]
16. Panerai, R. *Health Information Systems*; Global Perspective of Heath; Department of Medical Physics, University of Leicester: Leicester, UK, 2014; pp. 1–6.
17. Garcia, A.P.; De la Vega, S.F.; Mercado, S.P. Health Information Systems for Older Persons in Select Government Tertiary Hospitals and Health Centers in the Philippines: Cross-sectional Study. *J. Med. Internet Res.* **2022**, *24*, e29541. [CrossRef]
18. Epizitone, A. Framework to Develop a Resilient and Sustainable Integrated Information System for Health Care Applications: A Review. *Int. J. Adv. Comput. Sci. Appl. (IJACSA)* **2022**, *13*, 477–481. [CrossRef]
19. Walcott-Bryant, A.; Ogallo, W.; Remy, S.L.; Tryon, K.; Shena, W.; Bosker-Kibacha, M. Addressing Care Continuity and Quality Challenges in the Management of Hypertension: Case Study of the Private Health Care Sector in Kenya. *J. Med. Internet Res.* **2021**, *23*, e18899. [CrossRef]
20. Malekzadeh, S.; Hashemi, N.; Sheikhtaheri, A.; Hashemi, N.S. Barriers for Implementation and Use of Health Information Systems from the Physicians' Perspectives. *Stud. Health Technol. Inform.* **2018**, *251*, 269–272.
21. Tossy, T. Major challenges and constraint of integrating health information systems in african countries: A Namibian experience. *Int. J. Inf. Commun. Technol.* **2014**, *4*, 273–279. Available online: https://www.researchgate.net/profile/Titus-Tossy-2/publication/272163842_Major_Challenges_and_Constraint_of_Integrating_Health_Information_Systems_in_African_Countries_A_Namibian_Experience/links/54dca52b0cf28a3d93f8233d/Major-Challenges-and-Constraint-of-Integrating-Health-Information-Systems-in-African-Countries-A-Namibian-Experience.pdf (accessed on 1 February 2023).
22. Vaganova, E.; Ishchuk, T.; Zemtsov, A.; Zhdanov, D. Health Information Systems: Background and Trends of Development Worldwide and in Russia. In Proceedings of the 10th International Joint Conference on Biomedical Engineering Systems and Technologies-Volume 5: HEALTHINF, (BIOSTEC 2017), Porto, Portugal, 21–23 February 2017; pp. 424–428. [CrossRef]
23. Thomas, J.; Carlson, R.; Cawley, M.; Yuan, Q.; Fleming, V.; Yu, F. The Gap Between Technology and Ethics, Especially in Low-and Middle-Income Country Health Information Systems: A Bibliometric Study. *Stud. Health Technol. Inform.* **2022**, *290*, 902–906. [PubMed]
24. Namageyo-Funa, A.; Aketch, M.; Tabu, C.; MacNeil, A.; Bloland, P. Assessment of select electronic health information systems that support immunization data capture—Kenya, 2017. *BMC Health Serv. Res.* **2018**, *18*, 621. [CrossRef] [PubMed]
25. Lindberg, M.H.; Venkateswaran, M.; Abu Khader, K.; Awwad, T.; Ghanem, B.; Hijaz, T.; Morkrid, K.; Froen, J.F. eRegTime, Efficiency of Health Information Management Using an Electronic Registry for Maternal and Child Health: Protocol for a Time-Motion Study in a Cluster Randomized Trial. *JMIR Res. Protoc.* **2019**, *8*, e13653. [CrossRef]
26. Tummers, J.; Tekinerdogan, B.; Tobi, H.; Catal, C.; Schalk, B. Obstacles and features of health information systems: A systematic literature review. *Comput. Biol. Med.* **2021**, *137*, 104785. [CrossRef]
27. Malik, M.; Kazi, A.F.; Hussain, A. Adoption of health technologies for effective health information system: Need of the hour for Pakistan. *PLoS ONE* **2021**, *16*, e0258081. [CrossRef]
28. De Carvalho Junior, M.A.; Bandiera-Paiva, P. Health Information System Role-Based Access Control Current Security Trends and Challenges. *J. Healthc Eng.* **2018**, *2018*, 6510249. [CrossRef] [PubMed]
29. Taye, G. Improving health care services through enhanced Health Information System: Human capacity development Model. *Ethiop. J. Health Dev.* **2021**, *35*, 42–49. Available online: https://www.ajol.info/index.php/ejhd/article/view/210752 (accessed on 1 February 2023).
30. Sligo, J.; Gauld, R.; Roberts, V.; Villa, L. A literature review for large-scale health information system project planning, implementation and evaluation. *Int. J. Med. Inform.* **2017**, *97*, 86–97. [CrossRef]
31. Bosch-Capblanch, X.; Oyo-Ita, A.; Muloliwa, A.M.; Yapi, R.B.; Auer, C.; Samba, M.; Gajewski, S.; Ross, A.; Krause, L.K.; Ekpenyong, N.; et al. Does an innovative paper-based health information system (PHISICC) improve data quality and use in primary healthcare? Protocol of a multicountry, cluster randomised controlled trial in sub-Saharan African rural settings. *BMJ Open* **2021**, *11*, e051823. [CrossRef]
32. Suresh, L.; Singh, S.N. Studies in ICT and Health Information System. *Int. J. Inf. Libr. Soc.* **2014**, *3*, 16–24.
33. Isleyen, F.; Ulgu, M.M. Data Transfer Model for HIS and Developers Opinions in Turkey. *Stud. Health Technol. Inform.* **2020**, *270*, 557–561. [CrossRef] [PubMed]

34. Jeffery, C.; Pagano, M.; Hemingway, J.; Valadez, J.J. Hybrid prevalence estimation: Method to improve intervention coverage estimations. *Proc. Natl. Acad. Sci. USA* **2018**, *115*, 13063–13068. [CrossRef] [PubMed]
35. Sawadogo-Lewis, T.; Keita, Y.; Wilson, E.; Sawadogo, S.; Téréra, I.; Sangho, H.; Munos, M. Can We Use Routine Data for Strategic Decision Making? A Time Trend Comparison Between Survey and Routine Data in Mali. *Glob. Health Sci. Pract.* **2021**, *9*, 869–880. [CrossRef]
36. Kpobi, L.; Swartz, L.; Ofori-Atta, A.L. Challenges in the use of the mental health information system in a resource-limited setting: Lessons from Ghana. *BMC Health Serv. Res.* **2018**, *18*, 98. [CrossRef] [PubMed]
37. Feteira-Santos, R.; Camarinha, C.; Nobre, M.D.; Elias, C.; Bacelar-Nicolau, L.; Costa, A.S.; Furtado, C.; Nogueira, P.J. Improving morbidity information in Portugal: Evidence from data linkage of COVID-19 cases surveillance and mortality systems. *Int. J. Med. Inform.* **2022**, *163*, 104763. [CrossRef]
38. Ker, J.I.; Wang, Y.C.; Hajli, N. Examining the impact of health information systems on healthcare service improvement: The case of reducing in patient-flow delays in a US hospital. *Technol. Forecast. Soc. Chang.* **2018**, *127*, 188–198. [CrossRef]
39. Alahmar, A.; AlMousa, M.; Benlamri, R. Automated clinical pathway standardization using SNOMED CT- based semantic relatedness. *Digital Health* **2022**, *8*, 1–17. [CrossRef] [PubMed]
40. Krasuska, M.; Williams, R.; Sheikh, A.; Franklin, B.; Hinder, S.; TheNguyen, H.; Lane, W.; Mozaffar, H.; Mason, K.; Eason, S.; et al. Driving digital health transformation in hospitals: A formative qualitative evaluation of the English Global Digital Exemplar programme. *BMJ Health Care Inform.* **2021**, *28*, e100429. [CrossRef] [PubMed]
41. Dunn, T.J.; Browne, A.; Haworth, S.; Wurie, F.; Campos-Matos, I. Service Evaluation of the English Refugee Health Information System: Considerations and Recommendations for Effective Resettlement. *Int. J. Environ. Res. Public Health* **2021**, *18*, 10331. [CrossRef] [PubMed]
42. See, E.J.; Bello, A.K.; Levin, A.; Lunney, M.; Osman, M.A.; Ye, F.; Ashuntantang, G.E.; Bellorin-Font, E.; Benghanem Gharbi, M.; Davison, S.; et al. Availability, coverage, and scope of health information systems for kidney care across world countries and regions. *Nephrol. Dial. Transplant.* **2022**, *37*, 159–167. [CrossRef]
43. Nyangena, J.; Rajgopal, R.; Ombech, E.A.; Oloo, E.; Luchetu, H.; Wambugu, S.; Kamau, O.; Nzioka, C.; Gwer, S.; Ndirangu, M.N. Maturity assessment of Kenya's health information system interoperability readiness. *BMJ Health Care Inform.* **2021**, *28*, e100241. [CrossRef]
44. Ammenwerth, E.; Duftschmid, G.; Al-Hamdan, Z.; Bawadi, H.; Cheung, N.T.; Cho, K.H.; Goldfarb, G.; Gulkesen, K.H.; Harel, N.; Kimura, M.; et al. International Comparison of Six Basic eHealth Indicators Across 14 Countries: An eHealth Benchmarking Study. *Methods Inf. Med.* **2020**, *59*, e46–e63. [CrossRef] [PubMed]
45. Tummers, J.; Tobi, H.; Schalk, B.; Tekinerdogan, B.; Leusink, G. State of the practice of health information systems: A survey study amongst health care professionals in intellectual disability care. *BMC Health Serv. Res.* **2021**, *21*, 1247. [CrossRef]
46. Steil, J.; Finas, D.; Beck, S.; Manzeschke, A.; Haux, R. Robotic Systems in Operating Theaters: New Forms of Team-Machine Interaction in Health Care On Challenges for Health Information Systems on Adequately Considering Hybrid Action of Humans and Machines. *Methods Inf. Med.* **2019**, *58*, E14–E25. [CrossRef]
47. Sik, A.S.; Aydinoglu, A.U.; Son, Y.A. Assessing the readiness of Turkish health information systems for integrating genetic/genomic patient data: System architecture and available terminologies, legislative, and protection of personal data. *Health Policy* **2021**, *125*, 203–212. [CrossRef]
48. Bernardi, R.; Constantinides, P.; Nandhakumar, J. Challenging Dominant Frames in Policies for IS Innovation in Healthcare through Rhetorical Strategies. *J. Assoc. Inf. Syst.* **2017**, *18*, 81–112. [CrossRef]
49. Liu, G.; Tsui, E.; Kianto, A. An emerging knowledge management framework adopted by healthcare workers in China to combat COVID-19. *Knowl. Process Manag.* **2022**, *29*, 284–295. [CrossRef]
50. Bernardi, R. Health Information Systems and Accountability in Kenya: A Structuration Theory Perspective. *J. Assoc. Inf. Syst.* **2017**, *18*, 931–958. [CrossRef]
51. Epizitone, A. Critical Success Factors within an Enterprise Resource Planning System Implementation Designed to Support Financial Functions of a Public Higher Education Institution. Master's Thesis, Durban University of Technology, Durban, South Africa, 2021.
52. Ostern, N.; Perscheid, G.; Reelitz, C.; Moormann, J. Keeping pace with the healthcare transformation: A literature review and research agenda for a new decade of health information systems research. *Electron. Mark.* **2021**, *31*, 901–921. [CrossRef]
53. Farnham, A.; Utzinger, J.; Kulinkina, A.V.; Winkler, M.S. Using district health information to monitor sustainable development. *Bull. World Health Organ.* **2020**, *98*, 69–71. [CrossRef] [PubMed]
54. Faujdar, D.S.; Sahay, S.; Singh, T.; Kaur, M.; Kumar, R. Field testing of a digital health information system for primary health care: A quasi-experimental study from India. *Int. J. Med. Inform.* **2020**, *141*, 104235. [CrossRef]
55. Jabareen, H.; Khader, Y.; Taweel, A. Health information systems in Jordan and Palestine: The need for health informatics training. *East. Mediterr. Health J.* **2020**, *26*, 1323–1330. [CrossRef]
56. Ayabakan, S.; Bardhan, I.; Zheng, Z.; Kirksey, K. The Impact of Health Information Sharing on Duplicate Testing. *MIS Q.* **2017**, *41*, 1083–1104. [CrossRef]

57. Mayer, F.; Faglioni, L.; Agabiti, N.; Fenu, S.; Buccisano, F.; Latagliata, R.; Ricci, R.; Spiriti, M.A.A.; Tatarelli, C.; Breccia, M.; et al. A Population-Based Study on Myelodysplastic Syndromes in the Lazio Region (Italy), Medical Miscoding and 11-Year Mortality Follow-Up: The Gruppo Romano-Laziale Mielodisplasie Experience of Retrospective Multicentric Registry. *Mediterr. J. Hematol. Infect. Dis.* **2017**, *9*, e2017046. [CrossRef] [PubMed]
58. Soltysik-Piorunkiewicz, A.; Morawiec, P. The Sustainable e-Health System Development in COVID 19 Pandemic–The Theoretical Studies of Knowledge Management Systems and Practical Polish Healthcare Experience. *J. e-Health Manag.* **2022**, *2022*, 1–12. [CrossRef]
59. Seo, K.; Kim, H.N.; Kim, H. Current Status of the Adoption, Utilization and Helpfulness of Health Information Systems in Korea. *Int. J. Environ. Res. Public Health* **2019**, *16*, 2122. [CrossRef] [PubMed]
60. Mahendrawathi, E. Knowledge management support for enterprise resource planning implementation. *Procedia Comput. Sci.* **2015**, *72*, 613–621.
61. Kim, Y.M.; Newby-Bennett, D.; Song, H.J. Knowledge sharing and institutionalism in the healthcare industry. *J. Knowl. Manag.* **2012**, *16*, 480–494. [CrossRef]
62. Nwankwo, B.; Sambo, M.N. Effect of Training on Knowledge and Attitude of Health Care Workers towards Health Management Information System in Primary Health Centres in Northwest Nigeria. *West Afr. J. Med.* **2020**, *37*, 138–144. [PubMed]
63. Khader, Y.; Jabareen, H.; Alzyoud, S.; Awad, S.; Rumeileh, N.A.; Manasrah, N.; Mudallal, R.; Taweel, A. Perception and acceptance of health informatics learning among health-related students in Jordan and Palestine. In Proceedings of the 2018 IEEE/ACS 15th International Conference on Computer Systems and Applications (AICCSA), Aqaba, Jordan, 28 October–1 November 2018.
64. Benis, A.; Harel, N.; Barak Barkan, R.; Srulovici, E.; Key, C. Patterns of Patients' Interactions With a Health Care Organization and Their Impacts on Health Quality Measurements: Protocol for a Retrospective Cohort Study. *JMIR Res. Protoc.* **2018**, *7*, e10734. [CrossRef]
65. Delnord, M.; Abboud, L.A.; Costa, C.; Van Oyen, H. Developing a tool to monitor knowledge translation in the health system: Results from an international Delphi study. *Eur. J. Public Health* **2021**, *31*, 695–702. [CrossRef] [PubMed]
66. Dixon, B.E.; McGowan, J.J.; Cravens, G.D. Knowledge sharing using codification and collaboration technologies to improve health care: Lessons from the public sector. *Knowl. Manag. Res. Pract.* **2009**, *7*, 249–259. [CrossRef]
67. See, E.J.; Alrukhaimi, M.; Ashuntantang, G.E.; Bello, A.K.; Bellorin-Font, E.; Gharbi, M.B.; Braam, B.; Feehally, J.; Harris, D.C.; Jha, V.; et al. Global coverage of health information systems for kidney disease: Availability, challenges, and opportunitiesfor development. *Kidney Int. Suppl.* **2018**, *8*, 74–81. [CrossRef]
68. Vicente, E.; Ruiz de Sabando, A.; García, F.; Gastón, I.; Ardanaz, E.; Ramos-Arroyo, M.A. Validation of diagnostic codes and epidemiologic trends of Huntington disease: A population-based study in Navarre, Spain. *Orphanet J. Rare Dis.* **2021**, *16*, 77. [CrossRef]
69. Colais, P.; Agabiti, N.; Davoli, M.; Buttari, F.; Centonze, D.; De Fino, C.; Di Folco, M.; Filippini, G.; Francia, A.; Galgani, S.; et al. Identifying Relapses in Multiple Sclerosis Patients through Administrative Data: A Validation Study in the Lazio Region, Italy. *Neuroepidemiology* **2017**, *48*, 171–178. [CrossRef] [PubMed]
70. De Sanjose, S.; Tsu, V.D. Prevention of cervical and breast cancer mortality in low- and middle-income countries: A window of opportunity. *Int. J. Womens Health* **2019**, *11*, 381–386. [CrossRef] [PubMed]
71. Aung, E.; Whittaker, M. Preparing routine health information systems for immediate health responses to disasters. *Health Policy Plan.* **2013**, *28*, 495–507. [CrossRef] [PubMed]
72. Cawthon, C.; Mion, L.C.; Willens, D.E.; Roumie, C.L.; Kripalani, S. Implementing routine health literacy assessment in hospital and primary care patients. *Jt. Comm. J. Qual. Patient Saf.* **2014**, *40*, 68–76. [CrossRef]

Disclaimer/Publisher's Note: The statements, opinions and data contained in all publications are solely those of the individual author(s) and contributor(s) and not of MDPI and/or the editor(s). MDPI and/or the editor(s) disclaim responsibility for any injury to people or property resulting from any ideas, methods, instructions or products referred to in the content.

Review

Internet-Based Healthcare Knowledge Service for Improvement of Chinese Medicine Healthcare Service Quality

Xiaoyu Wang [1], Yi Xie [2,*], Xuejie Yang [2] and Dongxiao Gu [2]

[1] The Department of Pharmacy, Anhui University of Traditional Chinese Medicine, Hefei 230031, China; xywang0551@ahtcm.edu.cn
[2] The School of Management, Hefei University of Technology, Hefei 230009, China; xuejie_y@mail.hfut.edu.cn (X.Y.); dongxiaogu@yeah.net (D.G.)
* Correspondence: yixie928@mail.hfut.edu.cn

Abstract: With the development of new-generation information technology and increasing health needs, the requirements for Chinese medicine (CM) services have shifted toward the 5P medical mode, which emphasizes preventive, predictive, personalized, participatory, and precision medicine. This implies that CM knowledge services need to be smarter and more sophisticated. This study adopted a bibliometric approach to investigate the current state of development of CM knowledge services, and points out that accurate knowledge service is an inevitable requirement for the modernization of CM. We summarized the concept of smart CM knowledge services and highlighted its main features, including medical homogeneity, knowledge service intelligence, integration of education and research, and precision medicine. Additionally, we explored the intelligent service method of traditional Chinese medicine under the 5P medical mode to support CM automatic knowledge organization and safe sharing, human–machine collaborative knowledge discovery and personalized dynamic knowledge recommendation. Finally, we summarized the innovative modes of CM knowledge services. Our research will guide the quality assurance and innovative development of the traditional Chinese medicine knowledge service model in the era of digital intelligence.

Keywords: quality of healthcare service; internet-based health service; Chinese medicine; healthcare knowledge service

Citation: Wang, X.; Xie, Y.; Yang, X.; Gu, D. Internet-Based Healthcare Knowledge Service for Improvement of Chinese Medicine Healthcare Service Quality. *Healthcare* 2023, 11, 2170. https://doi.org/10.3390/healthcare11152170

Academic Editors: Giner Alor-Hernández, Alejandro Rodríguez-González, Jezreel Mejía-Miranda and José Luis Sánchez-Cervantes

Received: 30 May 2023
Revised: 21 July 2023
Accepted: 27 July 2023
Published: 31 July 2023

Copyright: © 2023 by the authors. Licensee MDPI, Basel, Switzerland. This article is an open access article distributed under the terms and conditions of the Creative Commons Attribution (CC BY) license (https://creativecommons.org/licenses/by/4.0/).

1. Introduction

The acceleration of industrialization, urbanization, and population aging has made health the most pressing issue in the international community. The World Health Organization's 13th General Programme of Work has identified the improvement in population health as its third strategic priority, with the aim of having one billion more people enjoying enhanced health and well-being by the end of 2025 [1]. In the context of today's rapid social development, there has been a notable and ongoing increase in public health awareness. Specifically, people have shifted away from seeking treatment for isolated ailments, and have instead prioritized a more comprehensive approach to health management that spans their entire lifecycle. This approach is characterized by a focus on personalized care that addresses a broad range of factors influencing an individual's health and well-being [2]. The Healthy China 2030 Planning Outline proposes integrating the advantages of Chinese medicine (CM) with health management to provide people with comprehensive, lifelong health services [3]. CM does not have an accurate definition at present, and it is generally considered to be a medical theory system gradually formed and developed through long-term medical practice under the guidance of Chinese cultural thought. Traditional Chinese medicine (TCM) posits that illnesses stem from an imbalance in a person's life-force energy, known as "Qi", and its objective is to reestablish harmony within the individual [4]. Modern CM is a combination of TCM and modern medical technology. CM services have a wide range of applications, making them suitable for the treatment, rehabilitation, and health management of various types of diseases, particularly chronic and geriatric conditions. Numerous studies have confirmed the therapeutic and ameliorative effects of CM on chronic diseases, including cancer [5–7], diabetes [8], cardiovascular diseases [9,10], and neurological disorders [11,12] such as epilepsy, Alzheimer's disease, Parkinson's disease, depression, and cerebral ischemia. Additionally, CM has been shown to be effective in treating skin diseases [13,14], infertility [15], and other ailments. It is a reliable and valuable healthcare option for individuals seeking complementary or alternative therapies. Furthermore, CM has demonstrated its critical role in responding to major epidemics, including SARS and the novel coronavirus (COVID-19) [16–18]. The Statistical Bulletin on the Development of China's Health Care in 2021, released by the National Health Commission of China in July 2022, showed that the total number of medical consultations at CM medical and health institutions nationwide reached 1.2 billion in 2021, representing a 13.7% increase from the previous year [19]. The survey results released by the National Administration of Traditional Chinese Medicine in China in 2021 showed that the health literacy of CM culture has been increasing yearly within five dimensions: basic concepts, healthy lifestyle, suitable methods for the public, cultural knowledge, and the ability to understand information. The percent of the population with health literacy in CM reached 20.69% in 2020 [20]. These figures indicate that increasingly more CM health services are being accepted and popularized by the public. However, despite the vast market and globalization of CM services, the safety and efficacy of CM have been subject to questioning by both domestic and international academic communities and the public because of significant differences between CM and modern medicine in terms of evidence-based practice and quality control. Furthermore, China's CM service industry still faces challenges such as low-quality primary CM services, difficulty in allocating high-quality medical resources at grassroots levels, and obstacles in implementing the tiered medical diagnosis and treatment system.

In recent years, the rapid development of new-generation information technologies such as mobile internet, big data, 5G communication technology, cloud computing, artificial intelligence, and the internet of things has facilitated the intelligentization of society, and the healthcare industry is undergoing continuous transformation and innovation. The concept of precision medicine has been proposed to fulfill the growing demand for better health and quality of life. The medical mode has shifted toward the 5P mode, which consists of preventive, predictive, personalized, participatory, and precision medicine. Academia has extensively and deeply researched the combination of new-generation information technologies with healthcare. Liu et al. conducted research on chronic disease management through deep learning by studying data from video-sharing social media platforms [21]. Bobroske et al. studied the effects of early postoperative intervention on patients' long-term use of opioid drugs by constructing a model of patients using opioid drugs [22]. Hajjar and Alagoz developed a randomized modeling framework that provides an accurate solution algorithm and personalized disease screening decision for chronic disease patients or potential chronic disease patients [23]. In addition, multiple scientific studies have shown the significant role of information technology in promoting health and well-being during the COVID-19 pandemic [24–27].

The application of new-generation information technology in the healthcare industry has also created opportunities for the transformation and upgrading of CM services. The Opinions on Promoting the Inheritance, Innovation and Development of Chinese Medicine, issued by the Central Committee of the Communist Party of China and the State Council in 2019, specifically emphasized the full use of new-generation information technologies such as big data and artificial intelligence in CM services, and promoted the deep integration and development of new-generation information technologies with CM health services. Compared with other disciplines, CM is an empirical science that relies more heavily on experience, and its knowledge is more complex and ambiguous. Currently, scholars and experts have achieved certain results in research on the internationalization, standardization, security, and application of artificial intelligence technologies such as deep learning and case-based reasoning (CBR) in CM [28–30]. However, there are still many gaps in the governance of cross-organizational, multimode, and heterogeneous data, organization of CM case knowledge, dynamic updating of knowledge, human–machine collaboration in CM knowledge discovery, and full-cycle personalized proactive knowledge services, which limit the accuracy and capabilities of CM intelligent knowledge services.

In response to the 5P mode transformation and upgrading of the demand for CM services, scholars have started to focus on research into smart CM knowledge services relying on the new generation of information technology. Therefore, this paper summarizes the concept and main features of smart CM knowledge services, conducts an academic review of the current research in this area, analyzes the smart CM knowledge service mode, and explores innovative management methods for smart CM knowledge services. This has significant implications for promoting the safe, effective, and reasonable clinical application of CM, leveraging its unique advantages and benefiting people's health.

2. Concept and Characteristics of Smart CM Knowledge Services

Medical services are crucial to people's lives, health, and safety, and rely heavily on experience and knowledge, with high demands for service accuracy. Relying solely on data or intuition-based solutions can often lead to significant risks. Therefore, the medical and health fields require additional support from domain-specific knowledge.

2.1. The Connotation of Smart CM Knowledge Services

In the field of CM, three types of knowledge are commonly encountered: The first type is general medical knowledge of CM, which includes its basic concepts, principles, and laws, as well as knowledge of clinical diagnosis and treatment. CM clinical diagnosis and treatment knowledge mainly covers the philosophical foundation of CM, as well as the basic theories of its understanding of human physiology, diseases, and their prevention and treatment. The second type consists of medical and health case knowledge that contains rich expert knowledge, and the third type is medical and health reasoning knowledge obtained through various intelligent algorithms. To effectively address complex medical and health management decision-making problems with high risk, it is necessary to integrate these three types of knowledge: general medical knowledge, medical and health case knowledge, and medical and health reasoning knowledge [31].

CM intelligent service refers to the utilization of advanced information technologies such as big data, artificial intelligence, and cloud computing to organize, aggregate, analyze, and provide guidance for CM big data, including historical cases, data on well-known doctors and prescriptions, CM literature, health examination data, and internet health data. These services are tailored to specific clinical scenarios and provide accurate personalized and dynamic pharmaceutical services across multiple scenarios, organizations, and devices for the entire lifecycle.

2.2. The Connotation of Smart CM Knowledge Services

CM intelligent knowledge is a critical technology for CM services, which are characterized by the homogenization of medical service, the intelligence of knowledge services, the integration of medical education and research, and the precision of service.

2.2.1. The Homogenization of Medical Service

CM services require doctors to accumulate a vast amount of knowledge and experience. However, there are significant differences in the expertise levels between younger and senior doctors, and between doctors from remote and medically advanced areas, resulting in inconsistent CM services. CM intelligent services can effectively leverage the high-quality resources of renowned Chinese medicine hospitals and enhance the capacity for community and grassroots hospitals to accommodate more patients. Integrating medical resources within a region or professional field achieves sharing of high-quality resources and hierarchical diagnosis and treatment. Furthermore, the formation of uniform clinical pathways, quality standards, and evaluation systems regulates the diagnosis and treatment behavior and service quality of hospitals at all levels. To improve clinical efficacy, Xuzhou Affiliated Hospital of Nanjing University of Traditional Chinese Medicine and Xuzhou Hospital of Traditional Chinese Medicine carry out homogeneous dialectical treatment with CM characteristics.

2.2.2. The Intelligence of Knowledge Services

CM knowledge comes from various sources and is highly fragmented, with a lack of logical connections among data. CM intelligent knowledge service requires the rapid correlation and standardization of massive amounts of cross-organizational CM data to organically organize the "knowledge fragments" in the CM field and form a richly expressive and highly extensible CM knowledge system that interconnects concepts and knowledge points. Establishing digital knowledge and case libraries and using technologies such as the internet, cloud computing, and big data enables CM knowledge browsing, retrieval, editing, navigation, and visualization, providing a comprehensive CM knowledge view for CM workers, decision-makers, managers, health professionals, and the public. Based on new-generation information technologies such as machine learning, it provides precise and intelligent knowledge services, such as similar case matching, evidence-based medicine assistance, data analysis, and knowledge recommendation services, to assist in policy-making, medical research, and clinical decision-making. The Institute of Traditional Chinese Medicine Information of the Chinese Academy of Chinese Medical Sciences has built large-scale knowledge systems, such as the CM knowledge maps for health preservation, clinical knowledge, and characteristic therapy, to provide support for knowledge management, knowledge services, education, and training in the field of CM.

2.2.3. The Integration of Medical Education and Research

Through the CM intelligent knowledge service platform, a scientific, educational, and research collaboration network can be established to promote communication and collaboration among research personnel from different institutions, fields, and levels. By combining CM knowledge with modern technological methods, advanced techniques such as data mining, artificial intelligence, and cloud computing can be used to digitize and standardize the study and service of CM theory and practice, continuously improving the scientific and accurate knowledge of CM. The combination of CM scientific research and education can promote the development of modern CM and the inheritance and innovation of CM theory and practice. The China Academy of Chinese Medical Sciences and Shanghai University of Traditional Chinese Medicine jointly carry out personnel training to promote the simultaneous development of clinical practice and scientific research innovation.

2.2.4. The Precision of Service

Different organizations and roles have significant differences in their knowledge needs, which change with task demands, age, health status, and other factors. CM intelligent knowledge services can perceive and model dynamically changing information in real time, achieving accurate knowledge matching and meeting personalized service needs throughout the entire process. The precision of CM intelligent knowledge services is reflected in providing personalized treatment recommendations and medication plans for patients based on their medical conditions, personal traits, and medication history, which improves the diagnostic and treatment abilities and efficiency of doctors, reduces the cost of trial and error for patients, and minimizes treatment risks [32]. Academician Xiaolin Tong proposed state-target dialectics, which combines the traditional dialectical thinking of CM with modern pharmacological research, aiming to improve the precision of CM.

3. The Evolution Process of CM Knowledge Services

In this section, we employ bibliometric methods to visualize and summarize the research status of intelligent knowledge services in the field of CM. The bibliometric analysis method extracts tacit knowledge from a large amount of literature data by using data analysis tools, and uses statistical methods to analyze and summarize the fundamental nature and development direction of a certain subject [33]. We use CiteSpace and Excel tools to statistically process the literature data and present the results in the form of tables. The CiteSpace tool is used to analyze relationships such as cooperation and co-occurrence, and draw a visual knowledge map.

In CiteSpace, the overall size of a node indicates the frequency at which the node appears. The node is composed of annual rings of different colors, and each annual ring corresponds to a different time zone and is represented by a different color. The lines between the nodes represent the associations between the nodes. Centrality refers to the intermediary role played by a node on information transfer between other nodes. The higher the centrality of a node, the more important the node is in the process of information transmission [34].

3.1. Data Collection

This study used the SCI-E, SSCI, CPCI-S, ESCI, CCR-E, and IC indexes in Web of Science as the data source, and employed advanced search methods to search for #1 and #2, where #1 represents "Chinese medicine" and #2 represents keywords related to intelligent knowledge services. Specifically, #1 is TS = ("traditional Chinese medicine"), and #2 is TS = ("knowledge service" or "intelligent service" or "smart service" or "knowledge discovery" or "knowledge reasoning" or "knowledge recommendation" or "knowledge aggregation" or "knowledge integration" or "knowledge mining" or "knowledge graph" or "knowledge map" or "artificial intelligence" or "knowledge system*" or "knowledge base*"). The time span used for this study was from 2004 to 2023, and the search cutoff date was 10 February 2023. After removing irrelevant parts and duplicates, a total of 1686 relevant articles were retrieved for this time period.

3.2. Time Distribution Map of CM Intelligent Knowledge Service

Figure 1 illustrates the change over time in the number of articles published in the field of CM related to intelligent knowledge services. Between 2004 and 2008, research on CM in the field of intelligent knowledge services was at its initial stage of development. There are relatively few theoretical and methodological research results for CM knowledge in digitization and intelligence, and no more than 20 academic papers are published each year. From 2009 to 2017, the number of research results on CM intelligent knowledge services fluctuated slightly, but the overall trend was a gradual increase, with a more significant growth rate. Between 2018 and 2021, research on CM intelligent knowledge services developed rapidly, with a substantial increase, and the research theory and methods were relatively mature, reaching a peak of 324 articles in 2021. The number of articles published in 2022 decreased slightly. The research results for 2023 are from only January and February, with a relatively small number of retrieved papers, so their analysis is not presented in this article.

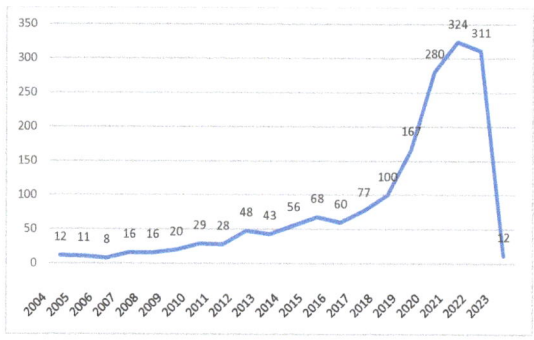

Figure 1. Time-series plot of publication count.

3.3. Space Distribution Analysis

In our study, we analyzed the publication trends of CM knowledge services across 381 institutions. As depicted in Table 1, the average number of articles per institution was 4.42. Notably, the top 20 institutions stood out, collectively publishing 961 articles with an average of 48 articles per institution, surpassing significantly the overall average. Beijing University of Chinese Medicine, Chengdu University of Traditional Chinese Medicine, and China Academy of Chinese Medical Sciences emerged as the three most prolific institutions in terms of publication volume, indicating their prominent position within the field. These institutions also exhibited high centrality. Furthermore, assessing the level of collaboration between research institutions serves as an important indicator for evaluating the research landscape in a specific domain [35]. As shown in Figure 2, it demonstrates that there is close cooperation among the institutions. Figure 3 shows the timeline for institutions to start research on CM knowledge services.

Table 1. List of the top 20 institutions with the number of published articles.

	Year	Number of Published Articles	Institution	Centrality
1	2013	134	Beijing Univ Chinese Med	0.21
2	2013	114	Chengdu Univ Tradit Chinese Med	0.11
3	2013	100	China Acad Chinese Med Sci	0.14
4	2014	79	Guangzhou Univ Chinese Med	0.11
5	2020	59	Shandong Univ Tradit Chinese Med	0.05
6	2013	51	Shanghai Univ Tradit Chinese Med	0.05
7	2019	48	Hosp Chengdu Univ Tradit Chinese Med	0.03
8	2013	46	Tianjin Univ Tradit Chinese Med	0.07
9	2013	46	Chinese Acad Sci	0.12
10	2018	39	Zhejiang Chinese Med Univ	0.04
11	2013	38	Sichuan Univ	0.04
12	2014	34	Capital Med Univ	0.08
13	2016	27	Nanjing Univ Chinese Med	0.02
14	2013	26	Zhejiang Univ	0.04
15	2016	24	Changchun Univ Chinese Med	0.01
16	2020	21	Jiangxi Univ Tradit Chinese Med	0
17	2020	20	Hunan Univ Chinese Med	0.02
18	2019	19	Sun Yat Sen Univ	0.02
19	2013	18	Fudan Univ	0.02
20	2013	18	Kyung Hee Univ	0

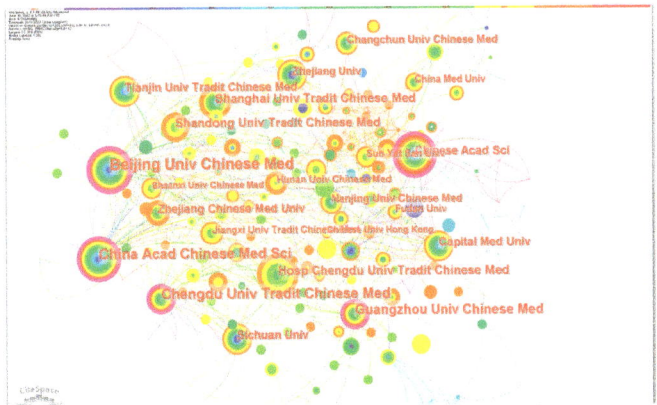

Figure 2. Institutional cooperation network related to CM knowledge service.

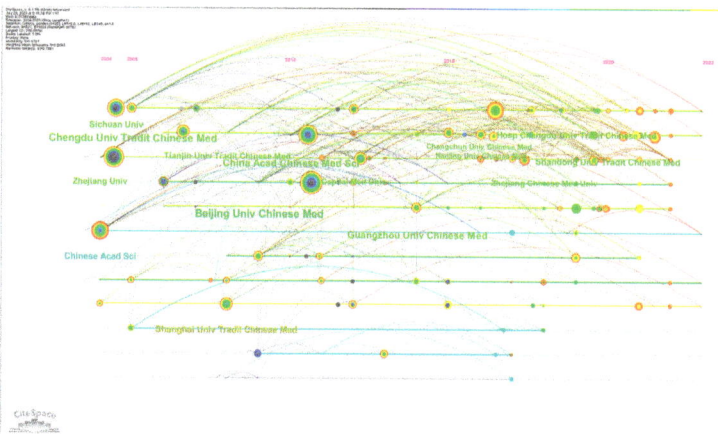

Figure 3. Institutional cooperation timeline chart.

Next, we analyzed the countries/regions that published relevant articles in this field and generated a network of country collaborations, as shown in Figure 4. Therefore, we can find that countries/regions cooperate closely, especially countries with a large number of publications. Figure 5 shows the timeline of countries/regions starting research on CM knowledge services. As shown in Table 2, among the top 20 countries or regions (as shown in the table) of knowledge service research on CM, 8 are in Asia, including China, Taiwan (China), South Korea, India, Singapore, Pakistan, Japan, and Malaysia; 7 in Europe, including the United Kingdom, Germany, Italy, France, Sweden, Scotland, and Romania; in addition, there are the United States and Canada in North America, and Brazil in South America. There is no doubt that China is the country with the largest number of studies on CM, far exceeding other countries and regions, accounting for 69.1% of all publications. Centrality indicates the importance of a node, and among the 20 countries with the largest number of publications, China has the highest centrality, followed by the United States.

Figure 4. Country and region cooperation network related to CM knowledge service.

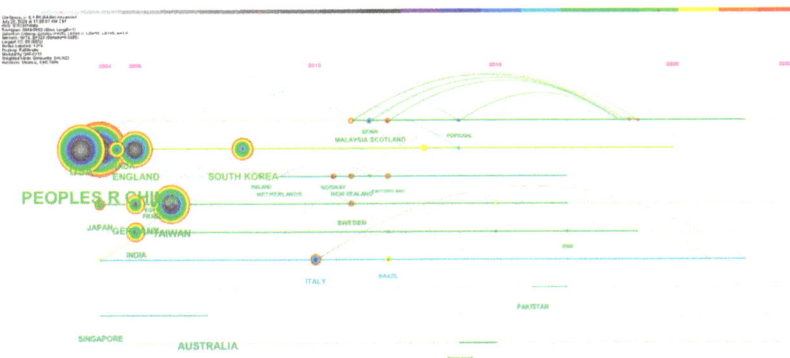

Figure 5. Country and region cooperation timeline chart.

Table 2. List of the top 20 countries/regions with number of published articles.

	Year	Number of Published Articles	Institution	Centrality
1	2013	1270	PEOPLES R CHINA	0.69
2	2013	96	USA	0.26
3	2013	48	AUSTRALIA	0.03
4	2013	42	TAIWAN	0.03
5	2013	37	SOUTH KOREA	0.07
6	2013	28	ENGLAND	0.02
7	2013	25	CANADA	0.01
8	2014	24	GERMANY	0.12
9	2014	20	INDIA	0.19
10	2014	15	SINGAPORE	0
11	2016	14	PAKISTAN	0.09
12	2013	13	JAPAN	0.01
13	2013	12	ITALY	0.19
14	2015	10	MALAYSIA	0.04
15	2018	8	NEW ZEALAND	0
16	2017	8	FRANCE	0.05
17	2016	8	SWEDEN	0.01
18	2013	8	SCOTLAND	0.02
19	2018	7	BRAZIL	0.05
20	2014	7	ROMANIA	0.01

3.4. Evolutionary Analysis of Hot Topics

Keywords are highly concise and general about an article. By analyzing high-frequency keywords, we can understand popular research topics in this field. Important keywords, as shown in Table 3, include "traditional Chinese medicine", "systematic review", "acupuncture", "prevalence", "alternative medicine", "complementary", and "artificial intelligence". It can be seen from Figure 6 that there is a strong connection between keywords, which indicates that most of the research in the field of CM knowledge services is multisubject. Figure 7 shows the timing of keyword co-occurrence. Keywords related to knowledge services include systematic review, knowledge, data mining, etc., indicating that most of the current research on knowledge services in CM is a systematic summary of previous knowledge and knowledge mining. In view of the combination of new technologies and products such as cloud computing and artificial intelligence derived from internet big data and CM, the modernization of CM is accelerating to achieve leapfrog development, which puts forward higher requirements for CM knowledge services. The research results show that the new generation of information technology can be combined with the academic thinking of CM [36] to provide knowledge services in various aspects such as pharmacological analysis [37], auxiliary diagnosis and treatment [38], and optimization of the diagnosis process [39].

Table 3. List of the top 15 keywords with the corresponding frequency.

	Count	Centrality	Year	Keywords
1	478	0.4	2003	traditional Chinese medicine
2	161	0.04	2012	systematic review
3	57	0.04	2010	Chinese medicine
4	56	0.24	2005	acupuncture
5	51	0.07	2006	herbal medicine
6	49	0.01	2011	prevalence
7	48	0.26	2004	alternative medicine
8	45	0.26	2005	complementary
9	43	0.13	2009	therapy
10	42	0.15	2012	oxidative stress
11	41	0.05	2010	artificial intelligence
12	40	0.04	2005	knowledge
13	40	0.06	2007	traditional medicine
14	37	0.01	2003	disease
15	35	0.09	2004	data mining

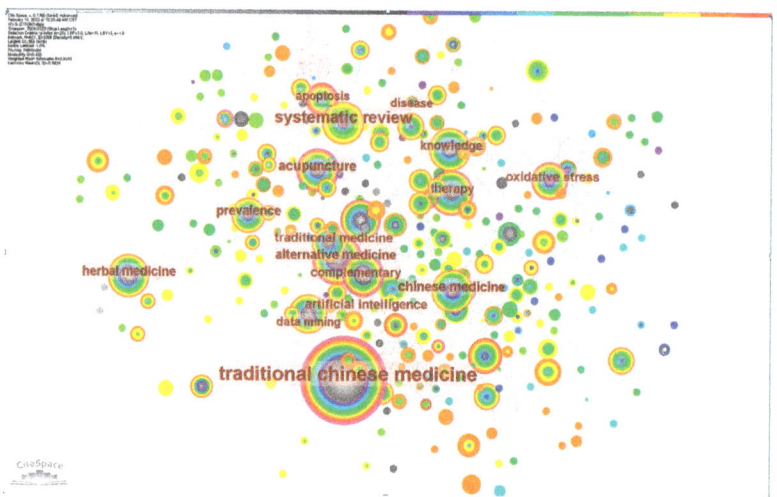

Figure 6. Co-occurrence network of keywords.

By studying the co-occurrence cluster analysis of these keywords, as shown in Figure 8, the research hotspots are mainly focused on the mechanisms of CM, knowledge mining and discovery in CM, application of artificial intelligence in CM, and alternative therapies. As shown in Figure 9, the thematic changes in research on intelligent CM knowledge services over the past 20 years can be observed through keyword clustering and emergent keywords. Strength refers to the burst strength of keywords. It can be seen from the words "complementary" and "alternative medicine" that CM often serves patients as a supplement to modern medicine. Yang et al. pointed out that CM is a good alternative to modern medicine because of its many targets and few side effects [40]. The prominence of words such as "randomized trials", "systems biology", "identification", "antagonistic activity", and "network pharmacology" over the past 20 years indicates the interest of scholars from various countries concerning CM mechanisms and the standardization and internationalization of CM data [28–30]. However, the diagnosis and treatment process of CM is based on the theoretical system of Chinese medicine by examining the condition, determining the type of disease, distinguishing symptoms, and using the method and viewpoint of dialectical treatment to treat the disease. Research on pharmacology or pathology alone is insufficient to cover the knowledge content of CM. Terms such as "knowledge graph", "data mining", "systematic review", and "meta-analysis" indicate that organizing CM knowledge is usually done from a holistic perspective, such as association with CM philosophy, CM physiology, etiology, and pathogenesis. With the vigorous development of the internet, the application of new-generation information technologies such as big data and artificial intelligence in CM diagnosis has made CM diagnosis more quantitative, objective, and standardized [41], and the development of precise CM knowledge services is an inevitable requirement for the modernization of Chinese medicine. At the same time, it is necessary to maintain the dialectical characteristics of CM.

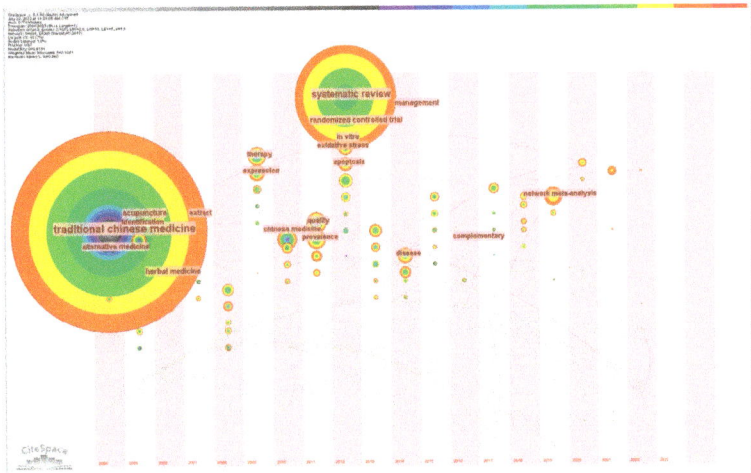

Figure 7. Co-occurrence time chart.

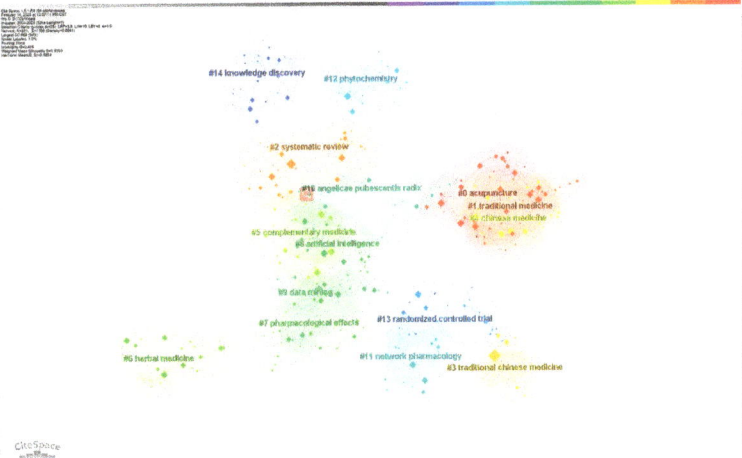

Figure 8. Cluster analysis of keywords.

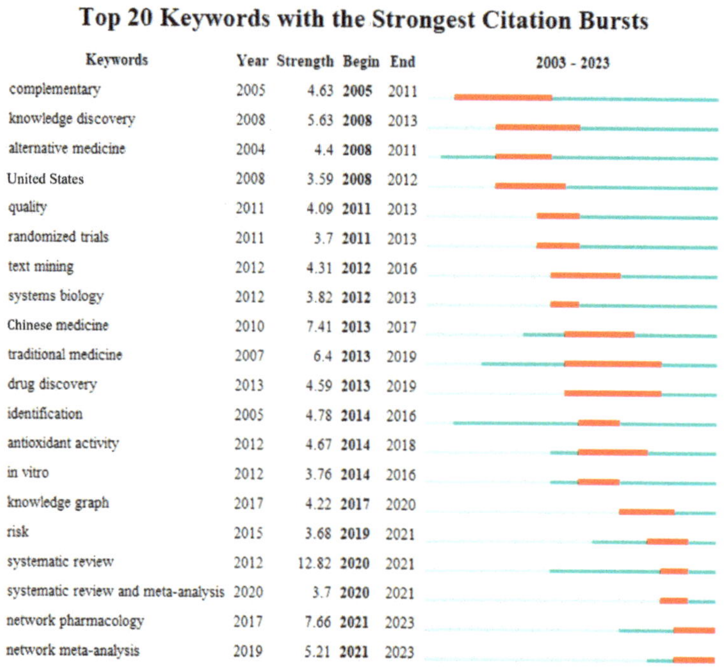

Top 20 Keywords with the Strongest Citation Bursts

Keywords	Year	Strength	Begin	End	2003 - 2023
complementary	2005	4.63	2005	2011	
knowledge discovery	2008	5.63	2008	2013	
alternative medicine	2004	4.4	2008	2011	
United States	2008	3.59	2008	2012	
quality	2011	4.09	2011	2013	
randomized trials	2011	3.7	2011	2013	
text mining	2012	4.31	2012	2016	
systems biology	2012	3.82	2012	2013	
Chinese medicine	2010	7.41	2013	2017	
traditional medicine	2007	6.4	2013	2019	
drug discovery	2013	4.59	2013	2019	
identification	2005	4.78	2014	2016	
antioxidant activity	2012	4.67	2014	2018	
in vitro	2012	3.76	2014	2016	
knowledge graph	2017	4.22	2017	2020	
risk	2015	3.68	2019	2021	
systematic review	2012	12.82	2020	2021	
systematic review and meta-analysis	2020	3.7	2020	2021	
network pharmacology	2017	7.66	2021	2023	
network meta-analysis	2019	5.21	2021	2023	

Figure 9. Keywords with the strongest citation bursts.

4. The Smart CM Services under the 5P Healthcare Mode

The new model of CM service, driven by big data, can address the challenges of CM development under new circumstances. The 5P medical mode mentioned above refers to preventive, predictive, personalized, participatory, and precision medicine. Specifically, preventive refers to the early prevention of disease risks that have not occurred, and predictive refers to predicting the occurrence and development of diseases and uncovering changes in health status. Personalized refers to individualized medicine, including individualized diagnosis and individualized treatment. Participatory means that each individual should be responsible for their personal health and actively participate in disease prevention and health promotion. Precision medicine refers to the practice of personalized multidisciplinary comprehensive treatment. The practice of precision medicine should be a patient-centered, open, medical cognition and practice process that keeps pace with the times and is constantly improving. Compared with general medical knowledge services, there are certain differences in knowledge sources, knowledge systems, and theoretical thinking modes among traditional Chinese medicine knowledge services [42], as shown in Table 4. Under the 5P medical mode, the CM smart service model places the patient at the center of a new medical model that combines CM theory with the 5P medical mode. The CM knowledge smart service model can track changes in the patient's body in real time and adjust the CM implementation plan promptly. By providing a full-cycle smart pharmacy service, including the organization and dynamic updating of medical case knowledge, knowledge generation and discovery based on case reasoning, as well as knowledge service recommendations considering comprehensive utility and diversity, the CM knowledge smart service model provides decision support for CM doctors and helps patients obtain the best treatment plan. This study proposes a basic framework for the CM knowledge smart service model driven by data and knowledge under the 5P medical mode, as illustrated in Figure 10.

Table 4. Comparison between CM knowledge service and general medical knowledge service.

	CM Knowledge Service	General Medical Knowledge Service
Knowledge source	Static knowledge: CM-related academic journals, traditional TCM classics, and guidelines issued by professional TCM organizations. Source of case characteristic data: vision, smell, auscultation, and palpation.	Static knowledge: authoritative sources such as international medical journals, clinical guidelines, and drug registration information. Source of case characteristics data: medical examination report.
Knowledge system	CM knowledge services are mainly based on the theory and practice of CM, including CM, acupuncture, and CM diagnostics.	Based on the modern medical system, including various branches of Western medicine, such as internal medicine, surgery, pediatrics, obstetrics and gynecology, etc.
Theoretical thinking mode	Traditional Chinese medicine emphasizes syndrome differentiation and treatment, and distinguishes the etiology and pathogenesis of diseases through the four diagnostic methods of CM, such as vision, smell, auscultation, and palpation, and then chooses Chinese medicine or acupuncture and other traditional Chinese medicine treatment methods.	Focus on the physiological and pathological mechanisms of diseases, and draw up treatment plans based on large-scale clinical trials.

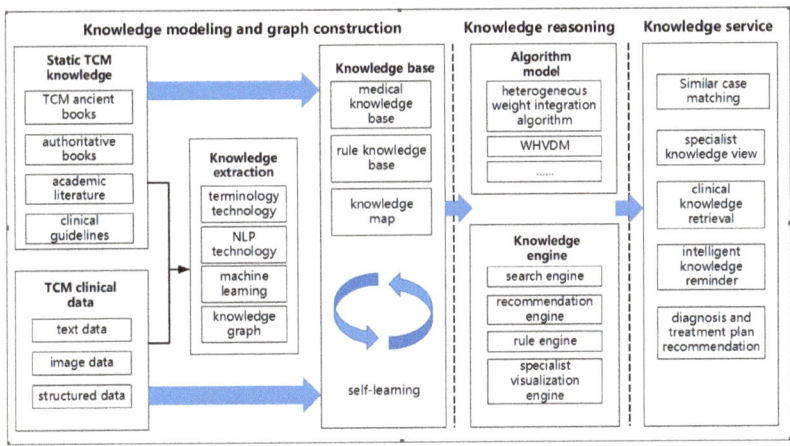

Figure 10. Framework of smart CM knowledge services.

The framework of the CM knowledge intelligent service model includes three parts: knowledge organization, knowledge generation, and knowledge service. Knowledge organization mainly includes case base construction, knowledge modeling, and picture construction. Knowledge generation mainly includes reasoning knowledge, and knowledge service mainly includes case knowledge recommendation, professional knowledge display, and other related services. This model is aimed at improving medical and health decision-making, hospital management, clinical teaching, and scientific research. To achieve this, a medical knowledge base and graph are built using authoritative books, academic literature, clinical pathways, and diagnosis and treatment guidelines. The knowledge base is constantly updated through a self-learning mechanism. Additionally, an example library is created using medical and health big data, and different algorithm models are applied for knowledge generation and discovery in different management decision-making scenarios. Through matching, retrieval, recommendation, reminder, view navigation, and other methods, the framework provides knowledge services to users.

4.1. The Organization of CM Knowledge

Case knowledge is the foundation of CM knowledge, because it contains a vast amount of expert knowledge that is difficult to quantify scientifically but is crucial in providing decision-making information support for the CM medical process. CM doctors use the "observation, listening, questioning, and pulse diagnosis" method to collect clinical data from patients. This approach relies on four aspects of data collection: vision, smell, auscultation, and palpation. By comprehensively analyzing these four aspects of data, CM doctors differentiate syndromes and determine the etiology, location, nature, and pathogenesis of diseases. AI-assisted CM diagnosis relies heavily on data from these four diagnostic methods [43]. For instance, observation can collect data using modern medical imaging techniques (such as CT, MRI, and so on) or in the form of pictures, recording the patient's skin color, facial features, tongue coating, and tongue quality, among others. Listening records patient data information in text, such as the patient's bad breath, body odor, sweat odor, and so on. Questioning records the inquiry information in text or audio. Pulse diagnosis records data through wearable devices or other sensors, or by manually taking the patient's pulse and recording text data. Given the vast amount of heterogeneous data from multiple sources in CM case knowledge, it is essential to effectively organize and manage medical and health case knowledge to achieve fine-grained management and accurate services. Moreover, CM data involves patients' private personal information and medical institutions' business secrets, prompting the need to ensure the security and confidentiality of data. Cross-domain CM data security sharing is an urgent problem that needs to be addressed.

4.1.1. CM Case Knowledge Organization Based on Key Clinical Features Extraction

In the context of CM medical and health management decision-making, CM experts rely on clinical pathways, diagnosis guidelines, and disease consensus to determine the main characteristic attributes, conclusions, and solution categories of cases. They then use a CM case automatic generation algorithm, which integrates natural language processing and key information extraction, to organize CM case knowledge. To evaluate the quality of CM cases, the system establishes a two-stage dual evaluation mechanism of "storage-use" + "quality-usability." Only cases with good evaluations from doctors can enter the case bases, and high-quality cases can be further classified to build rare-disease case bases and well-known-doctor case bases to meet the knowledge service needs for different CM scenarios. The flowchart for building the CM case base is shown in Figure 11.

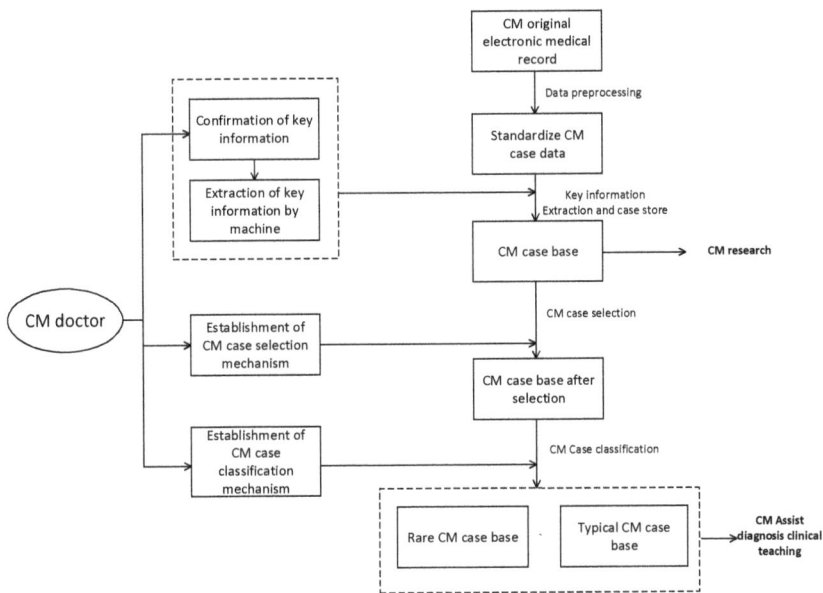

Figure 11. Flowchart for building the case base.

To extract key feature information, natural language processing methods are used based on authoritative disease knowledge provided by doctors. A pretrained medical field word vector dictionary is then used to obtain a word vector matrix of unstructured text data. This matrix is then input into multiple prebuilt segmenters to obtain segmented sentence sequences, which are in turn input into multiple prebuilt part-of-speech taggers to obtain part-of-speech tagging results. Using these results, the key information is obtained, which is then fused and matched based on expert experience to obtain the key case information and form case knowledge.

The first loss function during the pretraining process of the built segmenter is

$$Loss1 = \frac{1}{P}\sum_{i=1}^{P}\left(1 - h_{true}^{(c^{(p)})}\right) \quad (1)$$

where $h_{true}^{(c^{(p)})}$ represents the probability value corresponding to the correct character label, $h_{true}^{(c^{(p)})} \in [0,1]$; P represents the total number of characters; and p represents the pth character.

The second loss function during the training process of the prebuilt part-of-speech tagger is

$$Loss2 = \frac{1}{Q}\sum_{i=1}^{Q}\left(1 - e_{true}^{(w^{(q)})}\right) \quad (2)$$

where $e_{true}^{(w^{(q)})}$ is the probability value of the correct part-of-speech tag, $e_{true}^{(w^{(q)})} \in [0,1]$; Q represents the number of words after sentence segmentation; and q represents the qth word after segmentation.

The calculation of the overall loss function is as follows:

$$Loss = Loss1 + Loss2 \quad (3)$$

The CM knowledge organization system is capable of summarizing and generalizing high-quality case evaluation criteria from historical evaluation data. Through reinforcement learning methods, it automatically learns and forms high-quality case evaluation rules while controlling the case size and continuously injecting high-quality cases and gradually eliminating low-quality cases. This continuous improvement in case quality avoids unlimited expansion and ensures dynamic updates of case knowledge, thereby providing the possibility for large-scale case quality evaluation.

4.1.2. Medical Data Security Sharing Based on Horizontal and Vertical Federated Learning

Cases contain a significant amount of sensitive information, and in recent years, data privacy protection has become an increasingly important concern for society. The exchange of data between enterprises and organizations without user authorization is strictly prohibited. Data sharing between different CM hospitals and clinics is difficult, resulting in various "data islands" of different sizes, which poses significant challenges to AI and machine learning. In the medical and health field, accurate results can be obtained only after analyzing a large amount of data and cases. However, because of the unique nature of medical big data and the differences in information collection systems used by various hospitals, different types of medical data cannot be easily exchanged. Data sharing between hospitals is challenging, and sharing data between other health and elder-care institutions is even more difficult. By sharing user data between different institutions without compromising privacy, more comprehensive analysis data can be obtained, which can assist decision-makers in making informed judgments.

Federated learning is a distributed machine learning algorithm that facilitates secure management of cross-organizational and heterogeneous medical and health data resources. The algorithm involves several key steps. First, based on the first and second data sharing requests initiated by the first institution on the health data resource sharing platform, the second and third institutions that respond to these requests are determined. Next, the relevant parameters of the first model built by the first institution on the data sharing platform are updated using horizontal federated learning algorithms, based on the relevant health data resources of the second institution. The third step involves updating the parameters of the second model built by the first institution on the data resource sharing platform using vertical federated learning algorithms, based on the health data resources of the third institution. In the fourth step, the Shapley value is used to allocate model construction, which helps determine the training results from the first and second models obtained by the first institution. Finally, based on the training results of the first and second models, and in conjunction with the relevant health data resources of the first institution, secure aggregation of medical and health big data across organizations is achieved. The incentive parameters for each participating institution in the allocation modeling process can be expressed by Formula (4).

$$\varphi_i(v) = \sum_{S \subseteq N \setminus \{i\}} \frac{S!(N-S-1)!}{N!} \left(v_{(S \cup \{i\})} - v_S \right) \qquad (4)$$

where $\varphi_i(v)$ represents the incentive parameter of the i-th participating institution; N represents the total number of participating institutions; S represents a subset of the N participating institutions; v_S represents the individual contribution value of the subset S; $v_{(S \cup \{i\})}$ represents the contribution value of the set S; and $N \setminus \{i\}$ represents the subset that does not include the i-th participating institution.

4.2. CBR Method for Health Knowledge Generation and Discovery

CBR is an important subfield of artificial intelligence that serves as an effective means for organizing knowledge. CBR uses big data to organize cases and solves new management decision problems by matching them with the most similar historical cases and using expert knowledge gathered from those past cases. The reasoning process of CBR closely resembles that of human decision-making [29]. When doctors encounter new problems, they often recall past experiences managing similar situations and adjust and modify that information to formulate solutions to the current problem. Furthermore, case-based reasoning is a flexible knowledge reasoning technique that can flexibly construct case libraries and provide knowledge services based on different management decision tasks and real-time temporal information, making it well suited for organizing CM case knowledge. CM case knowledge is often ambiguous and difficult to verify in the diagnosis process. Case-based reasoning can efficiently solve problems without starting from the beginning by reusing historical knowledge, significantly improving problem-solving efficiency and providing more complete and interpretable initial solutions.

Case knowledge is the core of CM knowledge and can provide useful decision-making information support for the medical process of CM. Reasoning, generation, and discovery based on medical big data, CM case knowledge, and general medical knowledge can provide fine-grained management methods for hospital-assisted decision-making and risk warning. This section mainly introduces a human–machine collaborative case knowledge generation method and a case knowledge discovery method that considers implicit feedback.

4.2.1. Human–Computer Collaborative Method for CM Case Knowledge Generation

A human–machine collaborative medical and health knowledge generation method is proposed, which is aimed at case knowledge bases and general knowledge bases, to achieve the knowledge generation method of new medical decision-making solutions. By matching, reusing, modifying, quality evaluation, and review of historical cases, the knowledge generation of new medical decision-making solutions is achieved, and human participation is used to evaluate case quality and improve case knowledge, effectively reducing medical risks [44].

Cases are represented by (x, y) vectors, where $x = (x_1, x_2, \ldots, x_n)$ is a vector of independent variables that represent the feature attributes, and $y \in Y$, where Y is a discrete variable corresponding to the class. In the case library of solved historical cases, the class values are known. Therefore, when given a new unsolved target case, it can be retrieved by finding the most similar case in the library. To generate a solution for the new case, the weighted heterogeneous value distance measure (WHVDM) is used for case matching, providing a knowledge reference for decision makers.

Specifically, the WHVDM between the new target case t and the stored case r is defined as

$$WHVDM(t,r) = \sqrt{\sum_{i=1}^{n} w_i d_i^2(t,r)}$$

where

$$d_i^2(t,r) = \begin{cases} vdm(t,r), \text{if } x_i \text{ is discrete} \\ diff^2(x_{t,i}, x_{r,i}), \text{if } x_i \text{ is continuous} \end{cases} \quad (5)$$

In Equation (5), $vdm(t,r)$ is the value difference measure (VDM) proposed by Stanfill and Waltz. The VDM between the discrete attribute x_i of the target case t and the stored case r is defined as

$$vdm_i(t,r) = \sum_{a \in Y} (Pr(y = a | x_i = x_{t,i}) - Pr(y = a | x_i = x_{r,i}))^2 \\ \cdot \sqrt{\sum_{a \in Y} Pr(y = a | x_i = x_{t,i})^2} \quad (6)$$

In Equation (5), $diff^2(x_{t,i}, x_{r,i})$ is the Euclidean distance measurement, which is part of various distance measurements used in CBR systems. Specifically, given a new target case t and a stored case r:

$$diff^2(x_{t,i}, x_{r,i}) = (x_{t,i} - x_{r,i})^2 \quad (7)$$

This method is suitable for measuring the distance between cases that contain both discrete and continuous variables, highlighting the impact of the relative importance of case attributes. This impact is reflected in the weights of the feature attribute vector, which are obtained through a genetic algorithm [45]. Compared with commonly used knowledge discovery methods such as PBF neural network, CART, logistic regression, and naïve Bayes, this method has an accuracy improvement of over 3.2% and at least a 4.5% improvement in the comprehensive F-value performance evaluation index [46].

4.2.2. A Case Knowledge Discovery Method Considering Implicit Feedback in Human–Computer Interaction

As big data technology and medical digital systems continue to mature, the interaction behavior information between doctors and systems, such as personal preferences and case evaluation scores, is recorded in databases, providing huge data support for data mining. User feedback information is mainly divided into explicit feedback data and implicit feedback data. Explicit feedback behavior is reflected mainly through case rating, collection, labeling, and other methods, whereas implicit feedback behavior refers to the personal preferences of doctors, and obtaining and using such data may involve a certain delay. Considering doctors' implicit behavior in case knowledge discovery can predict the most needed case knowledge based on the browsing preferences and behavior sequences of CM doctors, which is of great significance for personalized configuration of doctor users, improving work efficiency, as well as diagnosis and treatment levels.

The main process is as follows: The browsing sequence and rating records of the attending physician are preprocessed, and the similarity between each case is calculated. Then, the generalized matrix factorization (GMF) and multilayer perceptron (MLP) neural networks are used to extract the physician's behavior characteristics and habits from the sequence, and combined with the rating information of past cases to obtain preliminary recommendation results. Finally, after personal screening by the doctor, the final recommendation list is obtained.

Specifically, the user–case interaction matrix Y obtained from implicit feedback is defined as

$$y_{ui} = \left\langle \begin{matrix} 1 \\ 0 \end{matrix} \right\rangle \tag{8}$$

where a value of 1 means that there is human–computer interaction between user and i; a value of 0 may mean that user does not like i, or that user does not know that there is i at all.

After dimensionality reduction in the embedding layer, the input binary sparse vector is mapped to a dense vector, and then with the help of the GMF model, the inner product of the user latent vector and the case latent vector is used as the evaluation of the user's preference for i, and the first neural collaborative filtering (NCF) layer is defined:

$$\phi_1(p_u, q_i) = p_u \circ q_i \tag{9}$$

where $p_u = P^T v_u^U, q_i = Q^T v_i^I$ are latent vectors of users and cases, respectively, projected onto the output layer:

$$\hat{y}_{ui} = a_{out}\left(h^T(p_u \circ q_i)\right) \tag{10}$$

where a_{out} is the activation function of the output layer, selected by the ReLU function, and h is the weight of the output layer, which is obtained through training.

Next, with the help of the MLP model, we use the standard multilayer perceptron to learn the potential features of users and cases, and calculate the output for each layer of MLP under the NCF framework:

$$z_i = \begin{cases} a_i(W_L^T z_{lL-1} + b_L), i = 2 \cdots L \\ \varphi_i(p_u, q_i) = [\frac{p_u}{q_i}], i = 1 \end{cases} \tag{11}$$

The output of the final model is

$$y_{ui} = \sigma\left(h^T z_L\right) \tag{12}$$

The knowledge recommendation results are obtained by splicing the last layer of the hidden layers, applying the exponential mechanism for privacy preservation, and then normalizing the resulting probability values.

4.3. Dynamic Personalized Knowledge Recommendation

The CM knowledge smart service platform integrates data resources from various fields and caters to different levels of CM health service demands, including the government, medical consortiums, hospitals, and individuals. By leveraging the behavior characteristics and patterns of different patient groups, it aggregates real-time sensing information and historical health data to track health status and demands. The platform provides personalized, dynamic, visualized, and intelligent panoramic health knowledge services for different groups and scenarios under the 5P medical mode. It achieves this through the use of an individual health assessment model and demand evolution model based on real-time time-series health data.

4.3.1. Health Risk Assessment Based on Time-Series Warning Signals

Through the analysis and processing of relevant medical information data, the changing characteristics and patterns of the health status of different patients can be studied, allowing for accurate identification of disease clues and health risks. With this information, personalized health assessment and demand evolution models can be established based on real-time time-series health data. This enables the provision of a service that recommends intervention plans based on users' characteristics, ultimately meeting the individualized health needs of patients.

The specific process is as follows: When an early warning signal is given, user tag vectors are constructed based on the user's physical condition, and collaborative filtering based on singular value decomposition is performed. After obtaining a plan, the BP-DS model is used to match the plan details with the user's physical characteristic values and adjusts them accordingly. Finally, trend fitting is applied to predict the user's physical condition and decide whether to end the treatment or choose the next stage of the plan, thereby generating a complete intervention plan.

Specifically, the process involves using the Pearson correlation similarity calculation to obtain a matrix that represents the similarity between users:

$$Similarity_r(u,v) = \frac{\sum_{i \in C}(r_{u,i} - \bar{r}_u)(r_{v,i} - \bar{r}_v)}{\sqrt{\sum_{i \in C}(r_{u,i} - \bar{r}_u)^2}\sqrt{\sum_{i \in C}(r_{v,i} - \bar{r}_v)^2}} \tag{13}$$

Calculate the preference of user tag attributes and obtain the similarity of attribute preferences:

$$P_{ui} = \frac{Weight_{ui}}{Weight_u} \tag{14}$$

$$Similarity_s(u,v) = \frac{\sum_{i=1}^{k} p_{ui} p_{vi}}{\sqrt{\sum_{i=1}^{k} p_{ui}^2}\sqrt{\sum_{i=1}^{k} p_{vi}^2}} \tag{15}$$

The comprehensive similarity is determined by taking into account both the similarity between users and the similarity of attribute preferences:

$$Similarity(u,v) = w \times Similarity_r(u,v) + (1-w) \times Similarity_s(u,v) \tag{16}$$

The user rating matrix and the project tag attribute matrix T are used in calculations to obtain the predicted value for the user:

$$Pre_{u,i} = \bar{R}_u + \frac{\sum_{v \in K-neighbours} Similarity(u,v) \times (R_{v,i} - \bar{R}_v)}{\sum_{v \in K-neighbours} Similarity(u,v)} \tag{17}$$

4.3.2. A Collaborative Recommendation for Medical Research and Education Integration

The medical education and research integration model refers to the application of the industry–university–research cooperation model in higher medical education. The CM intelligent knowledge service meets the needs of doctors and hospitals for medical treatment, education, and research integration. On the one hand, it provides doctors with decision-making assistance solutions by recommending similar cases. On the other hand, it provides classic clinical teaching cases and relevant common knowledge to young and grassroots doctors who lack experience in diagnosis, treatment, research, and training, thereby offering knowledge services.

This method uses a vast amount of clinical cases and general CM medical knowledge, and applies them to the medical education and research model to proactively recommend relevant medical and health knowledge based on dynamic perception of patient health status and doctor needs. The specific process involves first extracting entities and corresponding concepts using a hierarchical segmentation processing algorithm, and constructing a mapping relationship between them to obtain the medical knowledge base in the form of an <entity, concept> binary tuple. When a new patient record appears, features are extracted from the case information using the term frequency and inverse document frequency (TF-IDF) method to build the case feature set, which is then matched with the preset clinical case library to obtain similar cases. The LSTM-CRF named entity recognition technology is then used to extract medical terms, which are associated with the general CM medical knowledge base to provide detailed medical knowledge pages. The medical knowledge recommendation content is displayed with the help of the medical knowledge graph, to achieve the purpose of knowledge learning or disease research. Figure 12 shows the flowchart for constructing the knowledge graph.

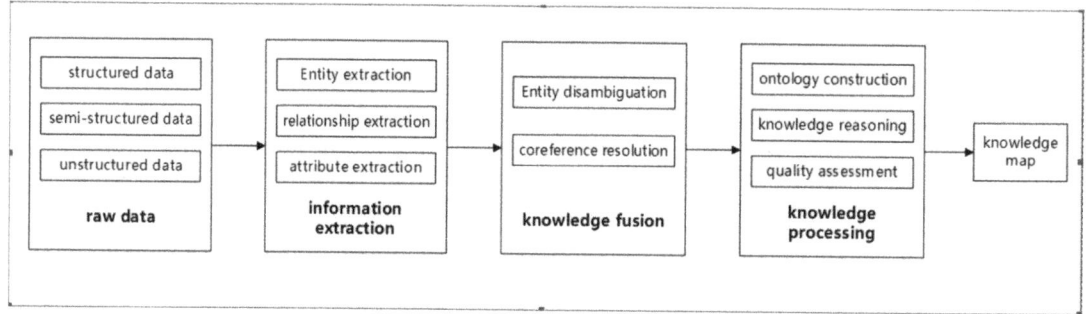

Figure 12. Flowchart for constructing the knowledge graph.

5. Innovative CM Knowledge Services Models in the Era of Digitalization

At present, the theory and practice of intelligent CM knowledge services are in their early stages. To address this, an interdisciplinary approach can be used to develop and utilize a CM knowledge service system driven by big data under the 5P medical mode. This system will be unique to China and internationally leading, and it will aid in advancing the scientific research level of CM health management in China to the forefront of the world. In addition, it will provide much-needed scientific theoretical support for CM knowledge services.

5.1. Case-Based CM Knowledge Service Model Guided by Holistic View and Dialectical

A significant part of CM knowledge is implicit, originating from the clinical practice of CM practitioners and unique to each individual, making it challenging to describe and teach using language and text. This is known as "implicit knowledge". Because of the existence of implicit CM knowledge, the transmission of CM is often achieved through observation, imitation, communication, and practical experience. The process of exploring, transmitting, and sharing implicit CM knowledge involves transforming implicit knowledge into explicit knowledge and integrating various resources, such as well-known doctors, disciplines, and information [47].

In managing implicit CM knowledge, a holistic view and dialectical approach should be used as guidance, and the diagnostic and reasoning processes of well-known doctors should be fully preserved. Effective clinical information should be obtained from a vast amount of CM knowledge to avoid knowledge deviation, omission, and distortion in the transmission of CM. This is essential for the preservation, sorting, and sharing of implicit CM knowledge. Therefore, it is urgent to construct a CM knowledge base centered on well-known CM cases, fully preserving the diagnostic and reasoning processes of well-known doctors and updating the case library automatically through self-learning and adaptive mechanisms.

5.2. Human–Machine Cooperative Medical Knowledge Recommendation Service Model

The traditional approach to knowledge reasoning relies mainly on computer models and algorithms, with limited human involvement, making it difficult to comprehend the reasoning results and limiting their practical utility [48]. Compared with the traditional clinical decision support system, the assisted diagnosis and treatment system participated by doctors has the advantages of knowledge adoption, ease of use, usefulness in improving medical quality, satisfaction of system use, interpretability of recommended schemes, perceived security, and ability enhancement have advantages [48]. To improve the acceptability of knowledge reasoning results, it is necessary to overcome the bottleneck of depending solely on machine reasoning for knowledge acquisition. Therefore, there is an urgent need to clarify the production and circulation rules for multisource and multimodal medical and health data, such as diagnosis and treatment data and health data, in the context of CM health services. Establishing a multimodal intelligent knowledge reasoning system can create a human-in-the-loop mechanism for knowledge reasoning and acquisition, providing a new paradigm for CM medical and health knowledge reasoning.

5.3. Active Knowledge Service Model for 5P Healthcare

In the era of intelligent interconnection, active health is a medical and health service model that is guided by the concept of holistic medicine and the CM principle of preventing diseases before they occur. It is supported by the new generation of information technologies such as the internet and artificial intelligence, and aims to provide proactive prevention and diagnosis services. The construction and application of a CM knowledge-based intelligent service system emphasizes personalization, precision, participation, and collaboration, with the aim of transforming disease treatment into health management. Therefore, there is an urgent need to conduct research on actively responsive and flexible disease risk judgment mechanisms, personalized treatment plan generation based on comprehensive patient health information, and health information knowledge recommendation mechanisms. Additionally, it is crucial to establish information security sharing and collaboration mechanisms among various medical institutions to provide doctors with high-quality knowledge services, and to improve medical quality and efficiency.

5.4. Panoramic and Dynamic Knowledge Service Mode Driven by Knowledge and Data

The 5P medical mode emphasizes the importance of personalized and precise medical services. To achieve dynamic health services that cater to individual needs, it is essential to overcome the limitations in relying mainly on static, local, and limited historical medical and health information for knowledge acquisition. This requires expanding the theoretical system of medical and health services and management, and establishing a new model of panoramic, dynamic, active, and personalized services based on real-time health perception and demand assessment [49]. Therefore, there is an urgent need to construct a dynamic and real-time CM case knowledge base based on CM knowledge and health and diagnostic data. This involves developing intelligent models for health risk assessment and dynamic intervention program recommendations based on time-series warning signals. By breaking down the business barriers and geographic boundaries of CM practitioners, it is possible to transform CM knowledge services from passive treatment and static services to real-time perception, dynamic evaluation, and active intervention, achieving precise and personalized health service modes.

6. Conclusions

The CM intelligent knowledge service model, driven by data and knowledge, has significantly transformed the CM knowledge service model, thereby promoting the development of the CM industry. CM is a unique medical system that requires a professional medical background and CM knowledge reserves. CM intelligent knowledge services need to provide accurate and credible content while also respecting the development process and characteristics of traditional medicine. They cannot simply be compared with modern medicine. This article reviews, summarizes, and analyzes the research status of CM intelligent knowledge service models, revealing that new information technologies such as artificial intelligence have been widely applied to CM services. However, the research on the CM knowledge service system is still incomplete. Therefore, this article proposes a CM intelligent knowledge service model under the 5P medical mode, which systematically studies the knowledge organization and data sharing mechanism, knowledge generation and discovery, knowledge reasoning, and full-cycle active personalized service mode of CM intelligent knowledge services. CM knowledge organization, knowledge generation, and knowledge service interact and depend on each other, forming a relatively complete CM knowledge management ecosystem. Through the effective integration and coordination of these three aspects, the innovation, dissemination, and utilization of CM knowledge can be realized, and the competitiveness and innovation ability of CM organizations can be improved. The CM intelligent knowledge service combines CM general knowledge, case knowledge, and reasoning knowledge to serve CM diagnosis, treatment, teaching, and research, which is of great significance for the development of CM and public welfare.

In future research, we should fully leverage the potential of the new generation of information technology to enhance and ensure the quality of CM knowledge service, focusing on the following three key aspects:

1. Facilitating the Integration of CM and Western Medicine:

Both Chinese and Western medicine boast their unique strengths and inherent weaknesses. By amalgamating the concepts and techniques of Chinese and Western medicine, we can foster cross-learning and capitalize on each system's strengths. This harmonious integration has the potential to yield more comprehensive and effective healthcare solutions [50,51].

2. Unearthing the Untapped Potential of Folk Chinese Medicine:

Folk CM holds a treasure trove of invaluable knowledge, encompassing secret recipes and unique techniques that have been passed down through generations. It is of paramount importance to actively preserve and retrieve this traditional wisdom. Thoroughly researching and documenting folk remedies and practices allows us to incorporate these hidden gems into mainstream CM knowledge, thus enriching the array of healthcare practices and treatments.

3. Enhancing CM Diagnosis and Medication through Standardized and Precise Prediction:

To bolster the accuracy of CM diagnosis and medication, adopting standardized and precise prediction methods plays a pivotal role. Harnessing cutting-edge technologies and data analysis tools enables the optimization of diagnostic processes [52], leading to more accurate treatment plans and improved patient outcomes.

Author Contributions: Conceptualization and methodology, X.W.; formal analysis and data curation, X.Y.; writing—original draft preparation, Y.X.; writing—review and editing, D.G. All authors have read and agreed to the published version of the manuscript.

Funding: This research was funded by the National Social Science Foundation of China under grant number 21BTQ102.

Institutional Review Board Statement: Not applicable.

Informed Consent Statement: Not applicable.

Data Availability Statement: Data are available in a publicly accessible repository. The data presented in this study are openly available in https://www.webofscience.com/wos/alldb/basic-search accessed on 10 February 2023.

Conflicts of Interest: The authors declare no conflict of interest.

References

1. EB152/20 Well-Being and Health Promotion. Available online: https://apps.who.int/gb/e/e_eb152.html (accessed on 18 May 2023).
2. Heng, J.; Yang, Z.; Xu, Z. The Current Situation and Countermeasures of the New Forms of Traditional Chinese Medicine Health Services. *Chin. Hosp.* **2022**, *26*, 18–21. [CrossRef]
3. Outline of the "Healthy China 2030" Initiative. Available online: http://www.gov.cn/zhengce/2016-10/25/content_5124174.htm (accessed on 18 May 2023).
4. Cheung, F. TCM: Made in China. *Nature* **2011**, *480*, S82–S83. [CrossRef] [PubMed]
5. Wang, K.; Chen, Q.; Shao, Y.; Yin, S.; Liu, C.; Liu, Y.; Wang, R.; Wang, T.; Qiu, Y.; Yu, H. Anticancer Activities of CM and Their Active Components against Tumor Metastasis. *Biomed. Pharmacother.* **2021**, *133*, 111044. [CrossRef]
6. Banik, K.; Ranaware, A.M.; Deshpande, V.; Nalawade, S.P.; Padmavathi, G.; Bordoloi, D.; Sailo, B.L.; Shanmugam, M.K.; Fan, L.; Arfuso, F.; et al. Honokiol for Cancer Therapeutics: A Traditional Medicine that Can Modulate Multiple Oncogenic Targets. *Pharmacol. Res.* **2019**, *144*, 192–209. [CrossRef]
7. Weng, W.; Goel, A. Curcumin and Colorectal Cancer: An Update and Current Perspective on this Natural Medicine. *Semin. Cancer Biol.* **2022**, *80*, 73–86. [CrossRef]
8. Martel, J.; Ojcius, D.M.; Chang, C.; Lin, C.; Lu, C.; Ko, Y.; Tseng, S.; Lai, H.; Young, J.D. Anti-Obesogenic and Antidiabetic Effects of Plants and Mushrooms. *Nat. Rev. Endocrinol.* **2017**, *13*, 149–160. [CrossRef] [PubMed]
9. Feng, X.; Sureda, A.; Jafari, S.; Memariani, Z.; Tewari, D.; Annunziata, G.; Barrea, L.; Hassan, S.T.S.; Smejkal, K.; Malanik, M.; et al. Berberine in Cardiovascular and Metabolic Diseases: From Mechanisms to Therapeutics. *Theranostics* **2019**, *9*, 1923–1951. [CrossRef]
10. Hao, X.; Pu, Z.; Cao, G.; You, D.; Zhou, Y.; Deng, C.; Shi, M.; Nile, S.H.; Wang, Y.; Zhou, W.; et al. Tanshinone and Salvianolic Acid Biosynthesis Are Regulated by Smmyb98 in Salvia Miltiorrhiza Hairy Roots. *J. Adv. Res.* **2020**, *23*, 1–12. [CrossRef] [PubMed]
11. Chen, Y.; Liu, Q.; An, P.; Jia, M.; Luan, X.; Tang, J.; Zhang, H. Ginsenoside Rd: A Promising Natural Neuroprotective Agent. *Phytomedicine* **2022**, *95*, 153883. [CrossRef] [PubMed]
12. Liu, Y.; Gao, J.; Peng, M.; Meng, H.; Ma, H.; Cai, P.; Xu, Y.; Zhao, Q.; Si, G. A Review on Central Nervous System Effects of Gastrodin. *Front. Pharmacol.* **2018**, *9*, 24. [CrossRef]
13. Yan, F.; Li, F.; Liu, J.; Ye, S.; Zhang, Y.; Jia, J.; Li, H.; Chen, D.; Mo, X. The Formulae and Biologically Active Ingredients of Chinese Herbal Medicines for the Treatment of Atopic Dermatitis. *Biomed. Pharmacother.* **2020**, *127*, 110142. [CrossRef]
14. Mu, Z.; Guo, J.; Zhang, D.; Xu, Y.; Zhou, M.; Guo, Y.; Hou, Y.; Gao, X.; Han, X.; Geng, L. Therapeutic Effects of Shikonin on Skin Diseases: A Review. *Am. J. Chin. Med.* **2021**, *49*, 1871–1895. [CrossRef]
15. O'Reilly, E.; Sevigny, M.; Sabarre, K.; Phillips, K.P. Perspectives of Complementary and Alternative Medicine (Cam) Practitioners in the Support and Treatment of Infertility. *BMC Complement. Altern. Med.* **2014**, *14*, 394. [CrossRef]
16. Yang, G.; Tan, Z.; Zhou, L.; Yang, M.; Peng, L.; Liu, J.; Cai, J.; Yang, R.; Han, J.; Huang, Y.; et al. Effects of Angiotensin Ii Receptor Blockers and Ace (Angiotensin-Converting Enzyme) Inhibitors on Virus Infection, Inflammatory Status, and Clinical Outcomes in Patients with COVID-19 and Hypertension a Single-Center Retrospective Study. *Hypertension* **2020**, *76*, 51–58. [CrossRef] [PubMed]
17. Wan, S.; Xiang, Y.; Fang, W.; Zheng, Y.; Li, B.; Hu, Y.; Lang, C.; Huang, D.; Sun, Q.; Xiong, Y.; et al. Clinical Features and Treatment of COVID-19 Patients in Northeast Chongqing. *J. Med. Virol.* **2020**, *92*, 797–806. [CrossRef]
18. Lung, J.; Lin, Y.; Yang, Y.; Chou, Y.; Shu, L.; Cheng, Y.; Liu, H.T.; Wu, C. The Potential Chemical Structure of Anti-Sars-Cov-2 Rna-Dependent Rna Polymerase. *J. Med. Virol.* **2020**, *92*, 693–697. [CrossRef]
19. Statistical Bulletin on the Development of China's Health and Medical Care in 2021. Available online: http://www.gov.cn/xinwen/2022-07/12/content_5700670.htm (accessed on 18 May 2023).
20. Enhancement of Chinese Citizens' Health Literacy in Traditional Chinese Medicine During the 13th Five-Year Plan Period. Available online: https://zhongyi.gmw.cn/2021-10/26/content_35262425.htm (accessed on 18 May 2023).
21. Liu, X.; Zhang, B.; Susarla, A.; Padman, R. Go to Youtube and Call Me in the Morning: Use of Social Media for Chronic Conditions. *MIS Q.* **2020**, *44*, 257–283. [CrossRef]
22. Bobroske, K.; Freeman, M.; Huan, L.; Cattrell, A.; Scholtes, S. Curbing the Opioid Epidemic at Its Root: The Effect of Provider Discordance After Opioid Initiation. *Manag. Sci.* **2022**, *68*, 2003–2015. [CrossRef]

23. Hajjar, A.; Alagoz, O. Personalized Disease Screening Decisions Considering a Chronic Condition. *Manag. Sci.* **2022**. [CrossRef]
24. Basajja, M.; Suchanek, M.; Taye, G.T.; Amare, S.Y.; Nambobi, M.; Folorunso, S.; Plug, R.; Oladipo, F.; van Reisen, M. Proof of Concept and Horizons on Deployment of Fair Data Points in the COVID-19 Pandemic. *Data Intell.* **2022**, *4*, 917–937. [CrossRef]
25. Du, H.; Le, Z.; Wang, H.; Chen, Y.; Yu, J. Cokg-Qa: Multi-Hop Question Answering over COVID-19 Knowledge Graphs. *Data Intell.* **2022**, *4*, 471–492. [CrossRef]
26. Ghardallou, M.; Wirtz, M.; Folorunso, S.; Touati, Z.; Ogundepo, E.; Smits, K.; Mtiraoui, A.; van Reisen, M. Expanding Non-Patient COVID-19 Data: Towards the Fairification of Migrants' Data in Tunisia, Libya and Niger. *Data Intell.* **2022**, *4*, 955–970. [CrossRef]
27. Jia, T.; Yang, Y.; Lu, X.; Zhu, Q.; Yang, K.; Zhou, X. Link Prediction Based on Tensor Decomposition for the Knowledge Graph of COVID-19 Antiviral Drug. *Data Intell.* **2022**, *4*, 134–148. [CrossRef]
28. Wang, X.; Liu, J.; Wu, C.; Liu, J.; Li, Q.; Chen, Y.; Wang, X.; Chen, X.; Pang, X.; Chang, B.; et al. Artificial Intelligence in Tongue Diagnosis: Using Deep Convolutional Neural Network for Recognizing Unhealthy Tongue with Tooth-Mark. *Comp. Struct. Biotechnol. J.* **2020**, *18*, 973–980. [CrossRef]
29. Chu, X.; Sun, B.; Huang, Q.; Peng, S.; Zhou, Y.; Zhang, Y. Quantitative Knowledge Presentation Models of Traditional Chinese Medicine (CM): A Review. *Artif. Intell. Med.* **2020**, *103*, 101810. [CrossRef]
30. Li, C.; Zhang, D.; Chen, S. Research about Tongue Image of Traditional Chinese Medicine(CM) Based on Artificial Intelligence Technology. In Proceedings of the IEEE 5th Information Technology and Mechatronics Engineering Conference (ITOEC), Chongqing, China, 12–14 June 2020.
31. Yang, S.; Ding, S.; Gu, D.; Li, X.; Liu, Z. Data-Driven Knowledge Discovery And Knowledge Service Methods in Medical and Health Big Data. *J. Manag. World* **2022**, *38*, 219–229. [CrossRef]
32. Li, L.; Wang, Z.; Wang, J.; Zheng, Y.; Li, Y.; Wang, Q. Enlightenment About Using TCM Constitutions for Individualized Medicine and Construction of Chinese-Style Precision Medicine: Research Progress with TCM Constitutions. *Sci. China-Life Sci.* **2021**, *64*, 2092–2099. [CrossRef]
33. Mingers, J.; Leydesdorff, L. A Review of Theory and Practice in Scientometrics. *Eur. J. Oper. Res.* **2015**, *246*, 1–19. [CrossRef]
34. Chen, C. Citespace II: Detecting and Visualizing Emerging Trends and Transient Patterns in Scientific Literature. *J. Am. Soc. Inf. Sci. Technol.* **2006**, *57*, 359–377. [CrossRef]
35. Donthu, N.; Kumar, S.; Mukherjee, D.; Pandey, N.; Lim, W.M. How to Conduct a Bibliometric Analysis: An Overview and Guidelines. *J. Bus. Res.* **2021**, *133*, 285–296. [CrossRef]
36. Ren, X.; Guo, Y.; Wang, H.; Gao, X.; Chen, W.; Wang, T. The Intelligent Experience Inheritance System for Traditional Chinese Medicine. *J. Evid.-Based Med.* **2023**, *16*, 91–100. [CrossRef] [PubMed]
37. Ma, S.; Liu, J.; Li, W.; Liu, Y.; Hui, X.; Qu, P.; Jiang, Z.; Li, J.; Wang, J. Machine Learning in Tcm with Natural Products and Molecules: Current Status and Future Perspectives. *Chin. Med.* **2023**, *18*, 43. [CrossRef]
38. Niu, Q.; Li, H.; Tong, L.; Liu, S.; Zong, W.; Zhang, S.; Tian, S.; Wang, J.; Liu, J.; Li, B.; et al. Tcmfp: A Novel Herbal Formula Prediction Method Based on Network Target's Score Integrated with Semi-Supervised Learning Genetic Algorithms. *Brief. Bioinform.* **2023**, *24*, bbad102. [CrossRef]
39. Liu, Z.; Luo, C.; Fu, D.; Gui, J.; Zheng, Z.; Qi, L.; Guo, H. A Novel Transfer Learning Model for Traditional Herbal Medicine Prescription Generation from Unstructured Resources and Knowledge. *Artif. Intell. Med.* **2022**, *124*, 102232. [CrossRef] [PubMed]
40. Yang, Z.; Zhang, Q.; Yu, L.; Zhu, J.; Cao, Y.; Gao, X. The Signaling Pathways and Targets of Traditional Chinese Medicine and Natural Medicine in Triple-Negative Breast Cancer. *J. Ethnopharmacol.* **2021**, *264*, 113249. [CrossRef] [PubMed]
41. Wang, Y.; Shi, X.; Li, L.; Efferth, T.; Shang, D. The Impact of Artificial Intelligence on Traditional Chinese Medicine. *Am. J. Chin. Med.* **2021**, *49*, 1297–1314. [CrossRef] [PubMed]
42. Zhou, X.; Peng, Y.; Liu, B. Text Mining for Traditional Chinese Medical Knowledge Discovery: A Survey. *J. Biomed. Inform.* **2010**, *43*, 650–660. [CrossRef]
43. Zhang, Q.; Zhou, J.; Zhang, B. Computational Traditional Chinese Medicine Diagnosis: A Literature Survey. *Comput. Biol. Med.* **2021**, *133*, 104358. [CrossRef]
44. Gu, D.; Liang, C.; Zhao, H. A Case-Based Reasoning System Based on Weighted Heterogeneous Value Distance Metric for Breast Cancer Diagnosis. *Artif. Intell. Med.* **2017**, *77*, 31–47. [CrossRef]
45. Yang, S.; Ding, S.; Gu, D.; Li, X.; Ouyang, B.; Qi, J. Healthcare Internet of Things: Transformation and Innovative Development of Medical and Health Models in the New Era. *J. Manag. Sci. China* **2021**, *24*, 1–11. [CrossRef]
46. Gu, D.; Su, K.; Zhao, H. A Case-Based Ensemble Learning System for Explainable Breast Cancer Recurrence Prediction. *Artif. Intell. Med.* **2020**, *107*, 101858. [CrossRef] [PubMed]
47. Ma, J.; Li, H.; Hu, M.; Sun, H. Research on Mining Medical Implicit Knowledge based on CART Algorithm: A Case Study on Traditional Chinese Medicine Case Records. *Inf. Sci.* **2021**, *39*, 84–91. [CrossRef]
48. Xie, Y.; Gu, D.; Wang, X.; Yang, X.; Zhao, W.; Khakimova, A.K.; Liu, H. A Smart Healthcare Knowledge Service Framework for Hierarchical Medical Treatment System. *Healthcare* **2021**, *10*, 32. [CrossRef]
49. Guo, X.; Wang, H.; Xu, M. Research on the Impact of Healthcare Information Sharing on Patient Transfer Quantity and Service Quality Level. *Chin. J. Manag. Sci.* **2021**, *29*, 226–236. [CrossRef]
50. Liu, B.; Hu, J.; Xie, Y.; Weng, W.; Wang, R.; Zhang, Y.; Li, X.; Zhang, K.; Ren, A.; Li, J.; et al. Effects of Integrative Chinese and Western Medicine on Arterial Oxygen Saturation in Patients with Severe Acute Respiratory Syndrome. *Chin. J. Integr. Med.* **2004**, *10*, 117–122. [CrossRef]

51. Tu, B.; Johnston, M.; Hui, K. Elderly Patient Refractory to Multiple Pain Medications Successfully Treated with Integrative East-West Medicine. *Int. J. Gen. Med.* **2008**, *1*, 3–6. [PubMed]
52. Gu, D.; Li, M.; Yang, X.; Gu, Y.; Zhao, Y.; Liang, C.; Liu, H. An Analysis of Cognitive Change in Online Mental Health Communities: A Textual Data Analysis Based on Post Replies of Support Seekers. *Inf. Process. Manage.* **2023**, *60*, 103192. [CrossRef]

Disclaimer/Publisher's Note: The statements, opinions and data contained in all publications are solely those of the individual author(s) and contributor(s) and not of MDPI and/or the editor(s). MDPI and/or the editor(s) disclaim responsibility for any injury to people or property resulting from any ideas, methods, instructions or products referred to in the content.

MDPI
St. Alban-Anlage 66
4052 Basel
Switzerland
www.mdpi.com

Healthcare Editorial Office
E-mail: healthcare@mdpi.com
www.mdpi.com/journal/healthcare

Disclaimer/Publisher's Note: The statements, opinions and data contained in all publications are solely those of the individual author(s) and contributor(s) and not of MDPI and/or the editor(s). MDPI and/or the editor(s) disclaim responsibility for any injury to people or property resulting from any ideas, methods, instructions or products referred to in the content.

www.ingramcontent.com/pod-product-compliance
Lightning Source LLC
LaVergne TN
LVHW070726100526
838202LV00013B/1184